Topics in Environmental Physiology and Medicine
edited by Karl E. Schaefer

High Altitude Physiology and Medicine

Edited by
Walter Brendel
Roman A. Zink

With 159 Figures

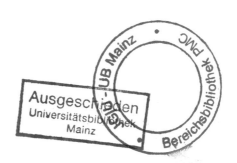

Springer-Verlag
New York   Heidelberg   Berlin

Walter Brendel, M.D.
Institute for Surgical Research of the LM-University
Klinkum Grosshadern
D-8000 Munich 70
Federal Republic of Germany

Roman A. Zink, M.D.
Documentation Center for High
  Altitude Medicine
and Urological Clinic of the
  LM-University
Klinkum Grosshadern
D-8000 Munich 70
Federal Republic of Germany

Library of Congress Cataloging in Publication Data
Main entry under title:
High altitude physiology and medicine.
  (Topics in environmental physiology and medicine)
  Bibliography: p.
  Includes index.
  1. Altitude, Influence of. 2. Oxygen in the body.
3. Acclimatization. I. Brendel, Walter, 1922-   .
II. Zink, Roman A. [DNLM: 1. Acclimatization.
2. Adaptation, Physiological. 3. Altitude. WD 710
H638]
QP82.2.A4H53   612'.0144   80-28376

© 1982 by Springer-Verlag New York, Inc.
All rights reserved. No part of this book may be translated or reproduced in any form without written permission from Springer-Verlag, 175 Fifth Avenue, New York, New York 10010, U.S.A.
The use of general descriptive names, trade names, trademarks, etc. in this publication, even if the former are not especially identified, is not to be taken as a sign that such names, as understood by the Trade Marks and Merchandise Marks Act, may accordingly be used freely by anyone.

Printed in the United States of America.

9 8 7 6 5 4 3 2 1

ISBN 0-387-**90482-4**   Springer-Verlag   New York   Heidelberg   Berlin
ISBN 3-540-**90482-4**   Springer-Verlag   Berlin   Heidelberg   New York

Contents

Preface   xi

Contributors   xv

Part I: Physiology of Adaptation

Oxygen Uptake in the Lungs

1. Sleep Hypoxemia at Altitude   3
   John R. Sutton, Gary W. Gray, Murray D. McFadden, Charles S. Houston, and A.C. Peter Powles

2. O_2 Breathing at Altitude: Effects on Maximal Performance   9
   Paolo Cerretelli

3. Oxygen Uptake at High Altitude: Limiting Role of Diffusion in Lungs   16
   Johannes Piiper

4. Respiratory and Cardiocirculatory Responses of Acclimatization of High Altitude Natives (La Paz, 3500 m) to Tropical Lowland (Santa Cruz, 420 m)   21
   M. Paz Zamora, J. Coudert, J. Ergueta Collao, E. Vargas, and N. Gutierrez

5. Chemoreflex Ventilatory Responses at Sea Level in Subjects with Past History of Good Acclimatization and Severe Acute Mountain Sickness   28
   Shu-Tsu Hu, Shao-Yung Huang, Shou-Cheng Chu, and Cheng-Fung Pa

6. Dysoxia (Abnormal Cell O_2 Metabolism) and High Altitude Exposure   33
   Eugene D. Robin

Oxygen Affinity and Oxygen Unloading

7. Minimal P_{O_2} in Working and Resting Tissues 45
   D.W. Lübbers

8. Effects of High Altitude (Low Arterial P_{O_2}) and of Displacements of the Oxygen Dissociation Curve of Blood on Peripheral O_2 Extraction and P_{O_2} 54
   Jochen Duhm

9. Influence of the Position of the Oxygen Dissociation Curve on the Oxygen Supply to Tissues 66
   F. Kreuzer and Z. Turek

10. Carbon Dioxide and Oxygen Dissociation Curves During and After a Stay at Moderate Altitude 73
    D. Böning, F. Trost, K.-M. Braumann, H. Bender, and K. Bitter

Hypoxia and Anaerobic Metabolism

11. Ventilatory, Circulatory, and Metabolic Mechanisms During Muscular Exercise at High Altitude (La Paz, 3500 m) 81
    M. Paz Zamora, J. Coudert, J. Arnaud, E. Vargas, J. Ergueta Collao, N. Gutierrez, H. Spielvogel, G. Antezana, and J. Durand

12. The Effects of Hypoxia on Maximal Anaerobic Alactic Power in Man 88
    P.E. di Prampero, P. Mognoni, and A. Veicsteinas

13. Anaerobic Metabolism at High Altitude: The Lactacid Mechanism 94
    P. Cerretelli, A. Veicsteinas, and C. Marconi

14. Oxygen Deficit and Debt in Submaximal Exercise at Sea Level and High Altitude 103
    J. Raynaud and J. Durand

Flow Distribution and Oxygen Transport

15. Blood Rheology in Hemoconcentration 109
    H. Schmid-Schönbein

16. Oxygen Transport Capacity 117
    K. Messmer

17. Skeletal Muscle Perfusion, Exercise Capacity, and the Optimal Hematocrit 123
    P. Gaehtgens and F. Kreutz

18. Cardiac Output and Regional Blood Flows in Altitude Residents 129
    J. Durand, P. Varene, and C. Jacquemin

19. The Pulmonary Circulation of High Altitude Natives 142
    G. Antezana, L. Barragan, J. Coudert, L. Coudkowicz, J.

Durand, A. Lockhart, J. Mensch-Dechene, M. Paz Zamora, H. Spielvogel, E. Vargas, and M. Zelter

20. Comparison Between Newcomer Rats and First Generation of Rats Born at High Altitude, Particularly Concerning the Oxygen Supply to the Heart 150
    F. Kreuzer and Z. Turek

21. Circulatory Flow of Oxygen Returning to the Lung During Submaximal Exercise in Altitude Residents 157
    J. Durand and J. Mensch-Dechene

22. Effect of the α-Adrenergic Blocking Agent Phentolamine (Regitine) on Acute Hypoxic Pulmonary Hypertension in Awake Dogs 159
    Shu-Tsu Hu, Hsueh-Han Ning, Chao-Nien Chou, and Hua-Yu Huang

Hormonal, Hematologic, and Electrolyte Changes

23. Hormonal Responses to Exercise at Altitude in Sea Level and Mountain Man 165
    John R. Sutton and Fausto Garmendia

24. Time Course of Plasma Growth Hormone During Exercise in Man at Altitude 172
    J. Raynaud, L. Drouet, J. Coudert, and J. Durand

25. Transcapillary Escape Rate of Albumin After Exposure to 4300 m 176
    G. Coates, G.W. Gray, C. Nahmias, A.C. Powles, and J.R. Sutton

26. Platelet Survival and Sequestration in the Lung at Altitude 179
    G. Coates, G.W. Gray, C. Nahmias, A.C. Powles, and J.R. Sutton

27. Electrolyte Changes in the Blood and Urine of High Altitude Climbers 183
    C. Rupp, R.A. Zink, and W. Brendel

28. The Influence of Trekking on Some Hematologic Parameters and Urine Production 187
    R.A. Zink, H.P. Lobenhoffer, B. Heimhuber, C. Rupp, and R. Schneider

Part II: Disturbances Due to High Altitude and Therapy of High Altitude Complaints

Cerebral and Ophthalmologic Changes

29. High Altitude Complaints, Diseases, and Accidents in Himalayan High Altitude Expeditions (1946-1978) 193
    H.R. Weingart, R.A. Zink, and W. Brendel

30. Cerebral Edema: The Influence of Hypoxia and Impaired Microcirculation 199
A. Baethmann

31. Physiologic Adaptation to Altitude and Hyperexis 209
J. Durand

32. Eye Problems at High Altitudes 212
F. Brandt and O.K. Malla

33. Cotton-Wool Spots: A New Addition to High Altitude Retinopathy 215
Peter Hackett and Drummond Rennie

High Altitude Pulmonary Edema

34. High Altitude Pulmonary Edema: Analysis of 166 Cases 219
H.P. Lobenhoffer, R.A. Zink, and W. Brendel

35. Hemodynamic Study of High Altitude Pulmonary Edema (12,200 ft) 232
G. Antezana, G. Leguía, A. Morales Guzman, J. Coudert, and H. Spielvogel

36. Pathogenesis of High Altitude Pulmonary Edema (HAPE) 242
R. Viswanathan

37. Subclinical Pulmonary Edema with Hypobaric Hypoxia 248
G. Coates, G. Gray, A. Mansell, C. Nahmias, A. Powles, J. Sutton, and C. Webber

38. Mechanism of Pulmonary Edema Following Uneven Pulmonary Artery Obstruction and Its Relationship to High Altitude Lung Injury 255
Norman C. Staub

39. Vasopressin in Acute Mountain Sickness and High Altitude Pulmonary Edema 261
P.H. Hackett, Mary L. Forsling, J. Milledge, and D. Rennie

40. Hypoxic Pulmonary Vasoconstriction and Ambient Temperature 263
J. Durand, J. Coudert, J.D. Guieu, and J. Mensch-Dechene

41. Pathophysiology of Acute Mountain Sickness and High Altitude Pulmonary Edema: An Hypothesis 266
J.R. Sutton and N. Lassen

42. Use of Furosemide in Prevention of HAPE 268
S.K. Kwatra and R. Viswanathan

Chronic Mountain Sickness and Performance

43. Chronic Mountain Sickness: A Pulmonary Vascular Disease? 271
Julio C. Cruz and Sixto Recavarren

44. Predicting Mountaineering Performance at Great Altitudes 278
    Hsueh-Han Ning, Shao-Yung Huang, Mei-Chuen Gung, Chung-Yuan Shi, and Shu-Tsu Hu

45. Effect of Ambient Temperature, Age, Sex, and Drugs on Survival Rate of Rats 284
    R. Viswanathan

High Altitude Expeditions

46. Hemodilution: Practical Experiences in High Altitude Expeditions 291
    R.A. Zink, W. Schaffert, K. Messmer, and W. Brendel

47. How to Stay Healthy While Climbing Mount Everest 298
    Oswald Oelz

48. Proposals for International Standardization in the Research and Documentation of High Altitude Medicine 301
    R.A. Zink, W. Schaffert, and H.P. Lobenhoffer

49. Equipment Requirements for High-Altitude Studies (Personal Experiences) 307
    W. Schaffert and R.A. Zink

    Index 310

Preface

High altitude physiology and medicine has again become important. The exceptional achievements of mountaineers who have climbed nearly all peaks over 8,000 m without breathing equipment raise the question of maximal adaptation capacity of man to low oxygen pressures. More importantly, the increase in tourism in the Andes and the Himalayas brings over 10,000 people to sites at altitudes above 4,000 and 5,000 m each year. At such heights several kinds of high altitude diseases are likely to occur, and these complications require detailed medical investigations.

Medical authorities need to inform both mountaineers and tourists as to how great a physical burden can be taken in the mountain environment without risk to health. Physicians need to know what kind of prophylaxis is to be employed at high altitudes to prevent the development of diseases and what therapeutic measures should be used once high altitude diseases have occurred. Moreover, the physical condition of the indigenous population living at higher altitudes such as the Andes and the Himalayas, who are exposed continuously to the stress of high altitude, requires our attention. We have become familiar with symptoms characteristic of chronic high-altitude disease: under special conditions this population has a tendency to develop pulmonary hypertension, which is associated with pulmonary edema, pulmonary congestion, and right heart failure.

This book will provide the latest up-to-date information on the recent developments in physiology and medicine of high altitudes for the medical researcher and the physician who treat patients suffering from high altitude sickness or give advice on preventive measures to their clients. The book is also intended to provide the newest teaching material on high altitude physiology and medicine to medical schools.

This volume contains contributions from many countries: Bolivia, China, France partly with investigators from Bolivia, Holland, India, Switzerland, USA, and West Germany. The publication is organized in two parts: (A) physiology of adaptation to high altitude and (B) high altitude sickness and therapy. In the first part the majority of articles deal with basic problems of oxygen transport and respiratory and circulatory control in rest and exercise at high altitude. New aspects of sleep at high altitude and hormonal responses and electrolyte changes at high altitude are included.

The second part is concerned with altitude sickness and therapy and contains the latest contributions towards understanding and managing the two most important pathological disturbances at high altitude: cerebral edema and eye problems, and pulmonary edema. Moreover, acute and chronic mountain sickness is treated extensively. Preventive medicine measures at high altitude are presented in two reports. An overview of accidents and diseases at high altitude (Himalaya and Andes) is given, which makes the reader aware of the new dimension which modern tourism to the Himalayas and Andes has brought to high altitude medicine. A special section contains a proposal for international standardization in high altitude research and documentation of high altitude medicine.

Medical scientists have carried out in the past laborious and extensive investigations of high altitude physiology and medicine and continue to do so at present with more sophisticated methods and broader scope, opening up new fields of knowledge.

In recent years there has been a shift in the focus and emphasis on high altitude research. Previously the main areas of interest in high altitude physiology were external respiration, in particular ventilation and gas exchange of the lungs and an exploration of the mechanisms responsible for the increased ventilation observed at high altitudes under condition of reduced partial pressure of oxygen. Mechanisms sensitizing the respiratory center were found to depend on altitude and length of sojourn at high altitudes. As a result of increased ventilation a respiratory alkalosis develops. Its influence on the O_2-dissociation curve, the Bohr effect, and cerebral blood flow has been thoroughly investigated. The increase in hematopoiesis found under conditions of oxygen deficiency at high altitudes, which results in an increased number of red cells in the circulating blood and therefore in a larger oxygen-carrying capacity, has long been considered the basic mechanism of adaptation to low oxygen pressure.

Hematopoietin was "discovered" as a hormonal factor controlling erythropoiesis, and the influence of the kidneys on the formation of hematopoietin was established. However, the significance of these findings about adaptive mechanisms to high altitude oxygen deficiency related to respiration, gas exchange, and erythropoiesis declined with the advent of equipment providing oxygen at high altitudes and in space.

In the past comparatively little attention has been given to peripheral respiration, i.e., oxygen transport in capillaries and oxygen diffusion into tissues and mitochondria within cells. This lack of knowledge was largely due to the unavailability of specific methods required for the investigation of peripheral oxygen transport processes. In recent years new methods have been developed and dramatic advances in the knowledge of microcirculation at high altitude have been made. It is now recognized that high altitude disorders and diseases such as pulmonary edema and brain edema are related to disturbances of blood flow in capillaries and impairment of oxygen transport to mitochondria. The better understanding of microcirculation has made it possible to use more effective measures for prophylaxis and therapy of high altitude disorders.

Pioneers in high altitude physiology and medicine came from many different countries. Paul Bert (1878) was the first to become interested in the effect of reduced atmospheric pressure; Mosso (1897) and Cohnheim (1903) published books on altitude physiology. Before and after World War I significant advances in high altitude physiology were made by Barcroft (1914), Haldane and Priestly (1935), Henderson (1938), Zuntz (1906), and Loewy (1932), as well as by Dill (1938), and Keys (1938), and Fleisch (1944), von Muralt (1948), and Verzar (1945).

Before World War II, in Germany, an active group of young physiologists, as researchers in aviation medicine or as mountaineers, became very interested in high altitude physiology. This group included Balke, Benzinger, Gauer, Hartmann, Hepp, Kramer, Luft, Noell, Opitz, Schneider, and Strughold. Their work has been published in *German Aviation Medicine, World War II*. Fundamental investigations were conducted, especially on respiration during acute and chronic hypoxia, hypoxia and anoxia of the brain, survival time of the brain and hypoxia tolerance. After World War II most of these young investigators went on a temporary or permanent basis to the United States to work on space physiology.

Significant progress in establishing the limits of adaptation to high altitude were made during high altitude expeditions carried out by Pugh at Mount Cho Oyu and Mount Everest (1957, 1964, 1968) and by Houston and Riley (1947) and West (1962a,b). In the 1950s, research on problems of high altitude adaptation in Germany was limited to few investigators (Brendel, 1956). Later a group formed in Munich, associated with the Institute for Surgical Research, studied problems of oxygen supply to tissues at high altitudes. Investigations were carried out during expeditions to the Kantschen Szönga by Zink et al. (1978) and at the Lhotse by Schaffert and Zink (1979). These investigations were based on our newly acquired understanding of the physiology and pathophysiology of the microcirculation and on the method of isovolemic hemodilution developed in Munich (Messmer, 1971, 1975) to combat disturbances of the microcirculation.

The encouraging experience with this method during expeditions to the Kantschen Szönga and Lhotse was the basic reason for organizing this symposium in Murnau, West Germany, and for inviting contributions to this volume by experts on capillary oxygen transport to tissue, especially to skeletal muscle and brain. We hope that the combined knowledge of these experts, specialists in respiratory physiology and our colleagues with practical experience in high altitude medicine of mountaineers and high altitude residents, will result in a better understanding of the pathophysiological processes of high altitude-induced diseases and complications.

The International Symposium on High Altitude Physiology and Medicine was organized by the Documentation Center for High Altitude Medicine (a section of the Association for Comparative Alpine Research, Munich) and the Institute for Surgical Research, LM-University of Munich. It was supported by the Volkswagen Foundation (I135 610).

References

1. Barcroft, J. (1925): The Respiratory Function of the Blood. In: Lessons from High Altitude. Cambridge, Massachusetts: Cambridge University Press.
2. Bert, P. (1943): Barometric Pressure (English translation of Paris edition, 1877.) Columbus, Ohio: College Book Company.
3. Brendel, W. (1956): Anpassung von

Atmung, Hämoglobin, Körpertemperatur, und Kreislauf bei langfristigem Aufenthalt in grossen Höhen (Himalaya). Arch. Ges. Physiol. *263*:227.
4. Cohnheim, O. (1903): Physiologie des Alpinismus. Wiesbaden, West Germany: Bergmann.
5. Dill, P.B. (1938): Life, Heat, and Altitude. Cambridge, Massachusetts: Harvard University Press.
6. Fleisch, A. and von Muralt, A. (1949): Klimaphysiologische Untersuchungen in der Schweiz, 1944–1948. Basel, Switzerland: Benno Schwabe.
7. Haldane, J.S. and Priestley, J.G. (1935): Respiration. Oxford, Great Britain: Clarendon Press.
8. Henderson, Y. (1938): Adventures in Respiration. London, Great Britain: Balliere.
9. Houston, C.G. and Riley, R.L. (1947): Respiratory and circulatory changes during acclimatization to high altitude. Am. J. Physiol. *149*:565.
10. Keys, A. (1938): Die Wirkung des Höhenklimas. In: Ergebnisse der inneren Medizin. Berlin, West Germany: Springer.
11. Loewy, A. (1932): Physiologie des Höhenklimas. Berlin, West Germany: Springer.
12. Messmer, K. (1975): Hemodilution. Surg. Clin. North Am. *55*:659. *Also* Sunder-Plassmann, L., Klövekorn, W.P., Holper, K., Hase, U., and Messmer, K. (1971): The physiological significance of acutely induced hemodilution. In: 6th European Conference on Microcirculation. Basel, Switzerland: Karger, pp. 23–28.
13. Mosso, A. (1898): Fisiologia dell 'Uomo sulle Alpi, 1897. German translation: Der Mensch auf den Hochalpen. Veit: Leipzig.
14. Pugh, C.G. (1957): Resting ventilation and alveolar air on Mt. Everest. J. Physiol. *135*:590.
15. Pugh, C.G. (1968): muscular exercise on Mount Everest. In: Joke, Exercise and Altitude. Basel, Switzerland: Karger.
16. Pugh, C.G. (1964): Man above 5000 meters—mountain exploration. In: Handbook of Physiology. Washington, D.C.: American Physiological Society, Chap. 55.
17. Schaffert, W. and Zink, R.A. (1979): How should we manage high altitude illness? In: Proceedings of the Hypoxia Symposium. Arch. Inst. North America, Calgary.
18. Verzar, F. (1945): Höhenklima-Forschung des Baseler Physiologischen Instituts. Basel, Switzerland: Benno Schwage.
19. West, J.B. (1962): Diffusing capacity of the lung for carbon monoxide at high altitude. J. Appl. Physiol. *17*:421.
20. West, J.B., et al. (1962): Arterial oxygen saturation during exercise at high altitude. J. Appl. Physiol. *17*:617.
21. Zink, R.A., Schaffer, W., Brendel, W., Messmer, K., Schmiedt, E., and Bernett, P. (1978): Hemodilution in high altitude mountain climbing. Proced. Amer. Soc. Anesthesiologists.
22. Zuntz, N., Loewy, A., Müller, D., and Caspari, W. (1906): Höhenklima und Bergwanderungen. Berlin, West Germany: Deutsches Verlagshaus.

Contributors

G. Antezana *Chapters 11, 19, 35*
Instituto Boliviano de Biología de Altura
Facultad de Medicina
Universidad Mayor de San Andres
La Paz, Bolivia

J. Arnaud *Chapter 11*
Instituto Boliviano de Biología de Altura
Facultad de Medicina
Universidad Mayor de San Andres
La Paz, Bolivia

A. Baethman *Chapter 30*
Institute for Surgical Research
Ludwig-Maximilians Universität München
D-8000 Munich 70, Federal Republic of
 Germany

L. Barragan *Chapter 19*
Instituto Boliviano de Biología de Altura
Facultad de Medicina
Universidad Mayor de San Andres
La Paz, Bolivia

K. Bitter *Chapter 10*
Abteilung Sportmedizin und Arbeitsphysiologie
Medizinische Hochschule Hannover
D-3000 Hannover 61, Federal Republic of
 Germany

H. Bender *Chapter 10*
Abteilung Sportmedizin und Arbeitsphysiologie
Medizinische Hochschule Hannover
D-3000 Hannover 61, Federal Republic of
 Germany

D. Böning *Chapter 10*
Abteilung Sportmedizin und Arbeitsphysiologie
Medizinische Hochschule Hannover
D-3000 Hannover 61, Federal Republic of
 Germany

F. Brandt *Chapter 32*
Eye Clinic
Ludwig-Maximilians Universität München
D-8000 Munich 2, Federal Republic of
 Germany

K.-M. Braumann *Chapter 10*
Abteilung Sportmedizin und Arbeitsphysiologie
Medizinische Hochschule Hannover
D-3000 Hannover 61, Federal Republic of
 Germany

W. Brendel *Chapters 27, 29, 34, 46*
Institute for Surgical Research
 of the LM-University
Klinkum Grosshadern
D-8000 Munich 70
Federal Republic of Germany

Contributors

P. Cerretelli *Chapters 2, 13*
Department of Physiology
University of Geneva
CH-1211 Geneva, Switzerland

C.-H. Chou *Chapter 22*
Shanghai Institute of Physiology
Academia Sinica
Shanghai, People's Republic of China

S.-C. Chu *Chapter 5*
Shanghai Institute of Physiology
Academia Sinica
Shanghai, People's Republic of China

G. Coates *Chapters 25, 26, 37*
Department of Medicine
McMaster University
Hamilton, Ontario L8S 4K1, Canada
and
Defense Institute of Environmental
   Medicine
Downsview, Ontario M3J 1P3, Canada

J. Coudert *Chapters 4, 11, 19, 24, 35, 40*
Instituto Boliviano de Biología de Altura
Facultad de Medicina
Universidad Mayor de San Andres
La Paz, Bolivia

L. Coudkowicz *Chapter 19*
Instituto Boliviano de Biología de Altura
Facultad de Medicina
Universidad Mayor de San Andres
La Paz, Bolivia

J.C. Cruz *Chapter 43*
Cardiovascular Pulmonary Research
   Laboratory
University of Colorado Medical Center
Denver, Colorado 80208, U.S.A.

P.E. di Prampero *Chapter 12*
Department of Physiology
University of Geneva
CH-1211 Geneva, Switzerland

J. Drouet *Chapter 24*
Département de Physiologie Humaine
Faculté de Médecine Paris XI
Paris, France

J. Duhm *Chapter 8*
Physiologisches Institut
   der Universität München
D-8000 Munich 2, Federal Republic of
   Germany

J. Durand *Chapters 11, 14, 18, 19, 21, 24,
   31, 40*
Département de Physiologie Humaine
Faculté de Médecine Paris XI
Paris, France

J. Ergueta Collao *Chapters 4, 11*
Instituto Boliviano de Biología de Altura
Facultad de Medicina
Universidad Mayor de San Andres
La Paz, Bolivia

M.L. Forsling *Chapter 39*
The Middlesex Hospital School
London WIP 6DB, United Kingdom

P. Gaehtgens *Chapter 17*
Institute for Normal and Pathological
   Physiology
University of Cologne
Cologne, Federal Republic of Germany

F. Garmendia *Chapter 23*
Institute of Andean Biology
University of San Marcos
Lima, Peru

G.W. Gray *Chapters 1, 25, 26, 37*
Department of Medicine
McMaster University
Hamilton, Ontario L8S 4K1, Canada
and
Defense Institute of Environmental
   Medicine
Downsview, Ontario M3J 1P3, Canada

J.D. Guieu *Chapter 40*
Instituto Boliviano de Biología de Altura
Facultad de Medicina
Universidad Mayor de San Andres
La Paz, Bolivia

M.-C. Gung *Chapter 44*
Shanghai Instutute of Physiology
Academia Sinica
Shanghai, People's Republic of China

N. Gutierrez *Chapters 4, 11*
Instituto Boliviano de Biología de Altura
Facultad de Medicina
Universidad Mayor de San Andres
La Paz, Bolivia

A. Morales Guzman *Chapter 35*
Instituto Boliviano de Biología de Altura
Facultad de Medicina
Universidad Mayor de San Andres
La Paz, Bolivia

P.H. Hackett *Chapters 33, 39*
Cardiovascular Pulmonary Research Laboratory
University of Colorado Medical Center
Denver, Colorado 80208, U.S.A.

B. Heimhuber *Chapter 28*
Documentation Center for High Altitude
D-8000 Munich 19, Federal Republic of Germany

C.S. Houston *Chapter 1*
Department of Medicine
McMaster University
Hamilton, Ontario L8S 4K1,
Canada

S.-T. Hu *Chapters 5, 22, 44*
Shanghai Institute of Physiology
Academia Sinica
Shanghai, People's Republic of China

H.-Y. Huang *Chapters 5, 22, 44*
Department of Respiration and Circulation
Shanghai Institute of Physiology
Academia Sinica
Shanghai, People's Republic of China

C. Jacquemin *Chapter 18*
Instituto Boliviano de Biología de Altura
Facultad de Medicina
Universidad Mayor de San Andres
La Paz, Bolivia

F. Kreutz *Chapter 17*
Institute for Normal and Pathological Physiology
University of Cologne
Cologne, Federal Republic of Germany

F. Kreuzer *Chapters 9, 20*
Department of Physiology
University of Nijmegen
Nijmegen, Holland

S.K. Kwatra *Chapter 42*
Voillabhbhai Patel Chest Institute
Dehli 110007, India

N. Lassen *Chapter 41*
Department of Clinical Physiology
Bispebjerg Hospital
Copenhagen, Denmark

G. Leguia *Chapter 35*
Instituto Boliviano de Biología de Altura
Facultad de Medicina
Universidad Mayor de San Andres
La Paz, Bolivia

H.P. Lobenhoffer, *Chapters 28, 34, 48*
Documentation Center
   for High Altitude
D-8000 Munich 19,
Federal Republic of Germany

A. Lockhart *Chapter 19*
Instituto Boliviano de Biología de Altura
Facultad de Medicina
Universidad Mayor de San Andres
La Paz, Bolivia

D.W. Lübbers *Chapter 7*
Max-Planck-Institut für Arbeitsphysiologie
D-3400 Göttingen, Federal Republic of Germany

O.K. Malla *Chapter 32*
Nepal Eye Hospital
Katmandu, Nepal

A. Mansell *Chapter 37*
Departments of Radiology and Medicine
McMaster University
Hamilton, Ontario L8S 4K1,
Canada

C. Marconi *Chapter 13*
Department of Physiology
University of Geneva
CH-1211 Geneva, Switzerland

J. Mensch-Dechene *Chapters 19, 21, 40*
Instituto Boliviano de Biología de Altura
Facultad de Medicina
Universidad Mayor de San Andres
La Paz, Bolivia

M.D. McFadden *Chapter 1*
Department of Medicine
McMaster University
Hamilton, Ontario L8S 4K1,
Canada

K. Messmer *Chapters 16, 46*
Institute for Surgical Research
Ludwig-Maximilians Universität München
D-8000 Munich 70,
Federal Republic of Germany

J. Milledge *Chapter 39*
Northwick Park Hospital
Harrow, Middlesex,
England

A. Mognoni *Chapter 12*
Department of Physiology
University of Geneva
CH-1211 Geneva, Switzerland

C. Nahmias *Chapters 25, 26, 37*
Department of Medicine
McMaster University
Hamilton, Ontario L8S 4K1,
Canada
and

Defense Institute of Environmental
  Medicine
Downsview, Ontario M3J 1P3
Canada

H.-H. Ning *Chapters 22, 44*
Shanghai Institute of Physiology
Academia Sinica
Shanghai, China

O. Oelz *Chapter 47*
Department of Medicine
University Hospital
Zurich, Switzerland

C.-F. Pa *Chapter 5*
Shanghai Instutute of Physiology
Academia Sinica
Shanghai
People's Republic of China

M. Paz Zamora *Chapters 4, 11, 19*
Institute Boliviano de Biología de Altura
Facultad de Medicina
Universidad Mayor de San Andres
La Paz, Bolivia

J. Piiper *Chapter 3*
Abteilung Physiologie
Max-Planck-Institut für Experimentelle
  Medizin
D-3000 Göttingen
Federal Republic of Germany

A.C. Powles *Chapters 1, 25, 26, 37*
Department of Medicine
McMaster University
Hamilton Ontario L8S 4K1
Canada
and
Defense Institute of Environmental
  Medicine
Downsview, Ontario M3J 1P3
Canada

J. Raynaud *Chapters 14, 24*
Département de Physiologie Humaine
Faculté de Médecine Paris XI
Paris, France

S. Recavarren *Chapter 43*
Department of Pathology
Universidad Peruana Cayetano Heredia
Lima, Peru

D. Rennie *Chapters 33, 39*
Rush Medical Center
Chicago, Illinois, U.S.A.

E.D. Robin *Chapter 6*
Stanford University
School of Medicine
Stanford, California 95305, U.S.A.

C. Rupp *Chapters 27, 28*
Documentation Center
  for High Altitude Medicine
D-8000 Munich 19,
Federal Republic of Germany

W. Schaffert *Chapters 46, 48, 49*
Documentation Center for High Altitude
  Medicine
D-8000 Munich 19
Federal Republic of Germany

H. Schmid-Schönbein *Chapter 15*
Institute of Physiology
Rhein-Westfälische Technische Universität
D-5100 Aachen,
Federal Republic of Germany

R. Schneider *Chapter 28*
Institute of Physiology
Rhein-Westfälische Technische Universität
D-5100 Aachen,
Federal Republic of Germany

C.-Y. Shi *Chapter 44*
Shanghai Institute of Physiology
Academia Sinica
Shanghai, People's Republic of China

H. Spielvogel *Chapters 11, 19, 35*
Instituto Boliviano de Biología de Altura
Facultad de Medicina
Universidad Mayor de San Andres
La Paz, Bolivia

N.C. Staub *Chapter 38*
Cardiovascular Research Institute
  and Department of Physiology
University of California
San Francisco, California, U.S.A.

J.R. Sutton *Chapters 1, 23, 25, 26, 37, 41*
Department of Medicine
McMaster University
Hamilton, Ontario L8S 4K1,
Canada
and
Defense Institute of Environmental
  Medicine
Downsview, Ontario M3J 1P3
Canada

F. Trost *Chapter 10*
Abteilung Sportmedizin und Arbeitsphysiologie
Medizinische Hochschule Hannover
D-3000 Hannover 61
Federal Republic of Germany

Z. Turek *Chapters 9, 20*
Department of Physiology
University of Nijmegen
Nijmegen, Holland

E. Vargas *Chapters 4, 11, 19*
Instituto Boliviano de Biología de Altura
Facultad de Medicina
Universidad Mayor de San Andres
La Paz, Bolivia

P. Varene *Chapter 18*
Departement de Physiologie Humaine
Faculté de Médecine Paris XI
Paris, France

A. Veicsteinas *Chapters 12, 13*
Department of Physiology
University of Geneva
CH-1211 Geneva, Switzerland

R. Viswanathan *Chapters 36, 42, 45*
Voillabhbhai Patel Chest Institute
Dehli 110007, India

C. Webber *Chapter 37*
Departments of Radiology and
 Medicine
McMaster University
Hamilton, Ontario L8S 4K1, Canada

H.R. Weingart *Chapter 29*
Documentation Center for High Altitude
 Medicine
D-8000 Munich 19
Federal Republic of Germany

M. Zelter *Chapter 19*
Instituto Boliviano de Biología de Altura
 Facultad de Medicina
Universidad Mayor de San Andres
La Paz, Bolivia

R.A. Zink *Chapters 27, 28, 29, 34, 46, 48, 49*
Documentation Center for High Altitude
 Medicine
D-8000 Munich 19
and Urological Clinic of the LM-University
Klinkum Grosshadern
D-8000 Munich 70
Federal Republic of Germany

Part I
Physiology of Adaptation

I. Oxygen Uptake in the Lungs

1
Sleep Hypoxemia at Altitude

JOHN R. SUTTON, GARY W. GRAY, MURRAY D. MCFADDEN, CHARLES S. HOUSTON, AND A.C. PETER POWLES

Disturbance of sleep is common at high altitude and is frequently due to cold and other physical discomforts. However, it may also be a manifestation of acute mountain sickness (12,20,23). The tragic story of Dr. Jacottet, who died in the Vallot hut on Mont Blanc, illustrates this point. In a letter to his brother in Vienna, Sept. 1, 1891, he said, "I was unable to sleep and passed so bad a night that I would not wish it on my worst enemy." The next night he died of mountain sickness and an autopsy showed that he also probably suffered high altitude pulmonary edema.

Climbers have frequently noticed irregular breathing in their sleeping colleagues at high altitude. Angelo Mosso first noted the occurrence of periodic breathing of the Cheyne-Stokes type during sleep at high altitude and reported his findings to the Reale Accademia dei Lincei on January 4, 1885. However, the reasons for such periodic breathing and many of the central nervous system events that occur during sleep are still poorly understood.

Periodic breathing also occurs at sea level in apparently normal persons, but is much more common in patients with cerebrovascular disease in whom cerebral oxygen delivery is impaired. Irregular breathing also occurs in patients with abnormalities of pulmonary gas exchange; these include patients with intermittent upper airway obstructions who develop episodic hypoxia and hypercapnia and those with chronic obstructive pulmonary disease with hypoxia and hypercapnia which worsen with sleep.

At altitude, periodic breathing is often seen in subjects who do not breathe periodically at sea level and has been observed at altitudes as low as 8000 ft (2400 m) (Houston, unpublished observation).

What causes periodic breathing? Is there any difference in the mechanisms operating at sea level and higher altitudes?

Phillipson suggests that periodic breathing at sea level is a "hypercapnic"-related phenomenon, whereas at altitude, the mechanism is primarily hypoxic (13). Sleep at sea level results in decreased ventilation; arterial $P{CO_2}$ rises and arterial $P{O_2}$ falls with little change in arterial oxygen saturation. It is the small increase in $P{CO_2}$ which results in little or no periodic breathing at sea level in normals. Similar decreases in ventilation during sleep at altitude will result in a significant fall in $Sa{O_2}$ and may cause hypoxic arousal which will result in increased ventilation and improvement in oxygenation.

Sleep then resumes, ventilation again decreases, and the cycle will repeat, giving rise to periodic breathing. Thus, periodic breathing may occur at altitude in subjects who do not breathe periodically at sea level.

Sleep itself is not a homogeneous state, but rather it is organized temporally into a cyclic pattern of sequential stages with each cycle lasting approximately 90 min in the adult. The staging of sleep is generally defined by a combination of behavioral, electroencephalographic, electromyographic, and electrooculographic criteria. On this basis, two distinct types of sleep can be identified: the quiet, synchronized, or non-rapid eye movement sleep and the active, paradoxical, or rapid eye movement sleep. Within the non-REM sleep, four subdivisions or four stages representing progressively deeper stages of sleep are described. The deepest are stages 3 and 4, referred to as slow wave sleep, because the EEG is dominated by high voltage waves of low frequency. Rapid eye movement sleep is a distinct state which, in many ways, resembles wakefulness, but is associated with frequent bursts of rapid eye movements.

In 1974, on Mount Logan, altitude 5360 m, one of us observed marked sleep disturbances in a subject with acute mountain sickness (21). We performed an EEG during sleep and noticed numerous arousals and an almost complete absence of stages 3 and 4 sleep. These findings prompted us to perform further and more detailed studies during sleep at high altitude. These investigations included studies of the effect of acetazolamide on sleep. Acetazolamide, a carbonic anhydrase inhibitor, markedly slows the hydration of carbon dioxide (11) and has been used for many years to prevent acute mountain sickness (1,4,6,7). Sleep disturbance is a very common feature of acute mountain sickness and we have noted a subjective improvement in the quality of sleep following administration of acetazolamide. Thus, a formal study of the effects of acetazolamide on sleep seemed warranted.

Subjects and Methods

These studies were conducted over two summers; 13 subjects were studied in 1976 and 16 subjects were studied in 1977. The protocol was essentially the same for both years, with only minor differences.

Twenty-nine fit young men and women, aged 20-36 years, were studied at sea level and at an altitude of 5360 m. Ventilatory responses to hypoxia (16), hypercapnia (15), and exercise (17) were measured at sea level prior to ascent. At altitude, half the subjects were studied after they had been at high altitude for 3-8 days and half were studied after more than 30 days. All climbed from a staging camp at 3290 m in 8 to 13 days. Although none were experiencing marked symptoms of acute mountain sickness, sleep disturbances were common, especially in those subjects who had been at this altitude the shortest time. Some subjects also experienced morning headache.

The subjects slept in pairs in a heated tent and an observer was present and awake throughout the night. Arterial oxygen saturation was measured continuously with a fiberoptic ear oximeter (Hewlett-Packard) (25). The electroencephalogram (EEG), electrooculogram (EOG), and submental electromyogram (EMG) were recorded and combined with behavioral criteria to determine sleep stage (18). Breathing frequency was recorded by a pair of magnetometers attached to the anterior and posterior thorax. This procedure was only performed in 1977. The highest, lowest, and average arterial oxygen saturations were noted every 5 min and the values meaned for each subject during the night. The lowest oxygen saturation during the night was also noted. These results were referred to as average, mean high, mean low, and lowest.

Table 1-1. Average Arterial Oxygen Saturation, Awake and Asleep, in Acclimatized and Partially Acclimatized Subjects.

Subject	Awake Sa_{O_2}	Sleep Sa_{O_2}(%)	p
Acclimatized	75.2 ± 6.9	64.3 ± 9.7	$p < 0.01$
Partially acclimatized	75.4 ± 6.4 $p = $ N.S.	64.4 ± 8.6 $p = $ N.S.	$p < 0.01$

In a second study, the effects of acetazolamide on sleep hypoxemia were examined. Two studies were conducted 6 to 8 days apart and the order randomized so that five subjects were studied first while taking acetazolamide (250 mg every 8 h for 5 doses) and four subjects were studied first while taking no drugs. The data were analyzed by a person who was unaware which subjects were taking the acetazolamide.

Results

Study 1

All subjects were markedly hypoxemic while awake at altitude, with a hypocapnic alkalosis. Average blood gases were as follows: Po_2, 41.6 ± 0.01 mmHg; Pco_2, 21.2 ± 0.9 mmHg; pH, 7.47 ± 0.01 (mean ± 1 SEM). The arterial oxygen saturations while awake and asleep were similar in the two groups of subjects (Table 1-1), there was a significant decrease in saturation of approximately 11% during sleep, and breathing was periodic up to 90% of the time, an example of which is given in Fig. 1-1. There was no relationship between the degree of sleep hypoxemia and sleep stage (Fig. 1-2). The ventilatory responses to hypercapnia, hypoxia, and exercise, measured at sea level before ascent, were within the normal range reported from this laboratory and have been reported previously (14). The responses were not related to the severity of sleep hypoxemia, EEG changes, the severity of the acute

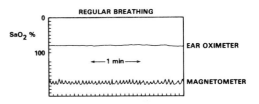
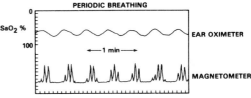

Fig. 1-1. Examples of regular and periodic breathing and the simultaneous arterial oxygen saturation. In the lower example, periodic breathing at 2 cycles/min is associated with oscillation in arterial oxygen saturation; hyperventilation follows the decrease and apnea the increase in arterial oxygen saturation.

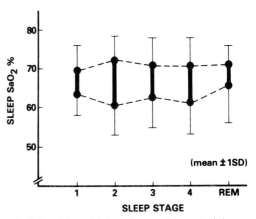

Fig. 1-2. Mean high and mean low arterial oxygen saturation during different sleep stages.

mountain sickness (AMS), or the response to acetazolamide.

Study 2

Acetazolamide reduced the irregular breathing and increased the mean high, the mean low, average, and lowest arterial oxygen saturations (Figs. 1-3 and 1-4). Profound hypoxemia to less than 50% arterial oxygen saturation was prevented. All subjects stated that they slept better while taking acetazolamide and had fewer morning headaches; they also experienced paresthesias.

Discussion

The results indicate that all subjects at 5360 m are hypoxemic with a hypocapnic alkalosis, findings similar to those previously reported from this altitude (5,6,22). However, previous studies of ventilatory responsiveness at sea level cannot predict the individual's susceptibility to profound hypoxemia during sleep at altitude. At altitude, there is a decrease in the duration of time spent in the deeper stages of sleep (stages 3 and 4) and an increased number of arousals (19). However, the present study

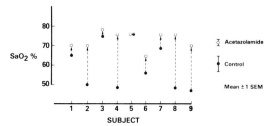

Fig. 1-4. Effect of acetazolamide on the lowest arterial oxygen saturation recorded during the night.

demonstrates that the degree of hypoxemia is similar in all sleep stages (Fig. 1-2).

In the second study, the observations of increased arterial oxygen saturation during sleep in subjects taking acetazolamide may well explain one of the mechanisms whereby acetazolamide, long used in the prevention and treatment of AMS, produces its beneficial effect. The most obvious way that acetazolamide could diminish sleep hypoxemia is by increasing ventilation. Acetazolamide impedes carbon dioxide transport, producing an increased cerebrospinal fluid and cerebral carbon dioxide tension and an increased hydrogen ion concentration, which would result in a sustained increase in alveolar ventilation. There are other possible mechanisms for the improvement in oxygen tension. Acetazolamide might reduce periodic breathing and apnea or it may have improved sleep oxygenation by altering sleep pattern. We have limited information on the latter, but our preliminary findings and those of Weil et al. (26) suggest that acetazolamide reduces sleep hypoxemia similarly in all sleep stages. Furthermore, in this study, we demonstrated that there was no relationship between arterial oxygen saturation and sleep stage.

Carbonic anhydrase, the enzyme inhibited by acetazolamide, is widely distributed in the body, and therefore the use of acetazolamide may have other effects than those mentioned above. For instance, acetazolamide reduces cerebrospinal fluid production (9) and may thus minimize the intracranial circulatory effects of cerebral

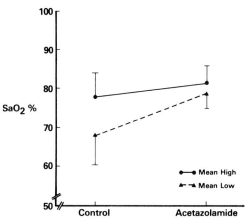

Fig. 1-3. The effect of acetazolamide on mean high and mean low sleep arterial oxygen saturation.

edema, which has been postulated as the underlying mechanism of AMS (8,24).

These considerations may well be important, as hypoxemia, per se, cannot be equated with AMS, and even after many weeks at altitude some subjects will exhibit profound sleep hypoxemia but have no symptoms of AMS (14). Thus, although improvement in arterial oxygenation during sleep may be important to minimize the problems of AMS in the newcomer to high altitude, other processes are also important in the mechanism of acclimatization.

Most investigators have administered acetazolamide before ascent to altitude in order to prevent AMS (1,4,6,7). The present study suggests that acetazolamide will also be an effective prophylaxis against AMS even when given after a person reaches high altitude. Our findings also suggest that acetazolamide should be considered in the treatment of persons with established AMS as it will improve arterial oxygenation, especially during sleep. Although a dose of 250 mg of acetazolamide every 8 h was used in the present study, more recent observations suggest that as little as 250 mg acetazolamide taken 1 h before sleep may be adequate to improve the quality of sleep and reduce the incidence and severity of morning headaches at altitude (Sutton, 1979, unpublished observation). Other drugs might also improve sleep at altitude. Medroxyprogesterone reduces periodic breathing and improves sleep oxygenation (26); however, it appears to increase the sensation of breathing (Weil, personal communication), making its use undesirable to mountaineers already approaching a ventilatory limit to exercise. Theophylline derivatives may also be useful during sleep at altitude, as they are well known to abolish periodic breathing of the Cheyne-Stokes type (3,10). However, these drugs have the potential to increase perfusion to underventilated areas of lung and could theoretically worsen the hypoxemia in certain individuals, especially those with preclinical pulmonary edema (2). Until definitive studies are performed with other drugs, acetazolamide would appear to be the drug of choice to prevent AMS and improve sleep hypoxemia.

Acknowledgment

This study was supported by N.I.H. grant HL 14102-05, under the auspices of the Arctic Institute of North America, with logistic support from the Canadian Armed Forces.

We are indebted to Hewlett-Packard Ltd. for the loan of the Hewlett-Packard 47201A ear oximeter which made these studies possible; to M. Basalygo, F. Clarke, S. Coons, E. Head, E. Inman, J. Kane, A. Menkis, M. Robertson, P. Rondi, and B. Weatherstone for technical help; to Dr. C.W. Dunnet and Mr. W. Taylor for statistical assistance; and to Drs. A.C. Bryan, E.J.M. Campbell, N.L. Jones, E.A. Phillipson, and C.E. Sullivan for their criticism of the manuscript.

References

1. Cain, S.M. and Dunn, J.E., II: Low doses of acetazolamide to aid accommodation of men to altitude. J. Appl. Physiol. 21:1195, 1966.
2. Coates, G., Gray, G., Mansell, A., Nahmias, C., Powles, A.C.P., Sutton, J.R., and Webber, C.: Changes in lung volume, lung density, and distribution of ventilation during hypobaric decompression. J. Appl. Physiol. Respirat. Environ. Exercise Physiol. 46:752, 1979.
3. Dowell, A.R., Heyman, A., Sieker, H.O., and Tripathy, K.: Effect of aminophylline on respiratory center sensitivity in Cheyne-Stokes respiration and in pulmonary emphysema. N. Engl. J. Med. 273:1447, 1965.
4. Forwand, S.A., Landowne, M., and Follansbee, J.N.: Effect of acetazolamide on acute mountain sickness. N. Engl. J. Med. 279:839, 1968.
5. Frayser, R., Rennie, I.D., Gray, G., and Houston, C.S.: Hormonal and electrolyte response to exposure to 17,500 feet. J. Appl. Physiol. 38:636, 1975.

6. Gray, G.W., Bryan, A.C., Frayser, R., Rennie, I.D., and Houston, C.S.: Prevention of acute mountain sickness. Aerospace Med. 42:81, 1971.
7. Hackett, P.H., Rennie, D., and Levine, H.D.: The incidence, importance, and prophylaxis of acute mountain sickness. Lancet 2:1149, 1976.
8. Houston, C.S. and Dickinson, J.: Cerebral forms of high altitude illness. Lancet 2:758, 1975.
9. Kiste, S.J.: Carbonic anhydrase inhibition: effect of acetazolamide on cerebrospinal fluid flow. J. Pharmacol. Exp. Ther. 117:402, 1956.
10. Marais, O.A.S. and McMichael, J.: Theophylline-ethylenediamine in Cheyne-Stokes respiration. Lancet 2:437, 1937.
11. Maren, T.H.: Carbonic anhydrase. Physiol. Rev. 47:595, 1967.
12. Nupse and Die: Editorial. Lancet 2:1177, 1976.
13. Phillipson, E.A.: Respiratory adaptation in sleep. Ann. Rev. Physiol. 40:133, 1978.
14. Powles, A.C.P., Sutton, J.R., Gray, G.W., Mansell, A.L., McFadden, M., and Houston, C.S.: Sleep hypoxemia at altitude: its relationship to acute mountain sickness and ventilatory responsiveness to hypoxia and hypercapnia. In Folinsbee, L.J., Wagner, J.A., Borgia, J.D., Drinkwater, B.L., Gliner, J.A., and Bedi, J.F. (eds.): Environmental Stress. Individual Human Adaptation. New York, Academic Press, 1978, pp. 373–381.
15. Read, D.J.C.: A clinical method for assessing the ventilatory response to carbon dioxide. Aust. Ann. Med. 16:20, 1967.
16. Rebuck, A.S. and Campbell, E.J.M.: A clinical method for assessing ventilatory response to hypoxia. Am. Rev. Resp. Dis. 109:345, 1974.
17. Rebuck, A.S., Jones, N.L., and Campbell, E.J.M.: Ventilatory response to exercise and to CO_2 rebreathing in normal subjects. Clin. Sci. 43:861, 1972.
18. Rechtschaffen, A. and Kales, A. (eds.): A Manual of Standardized Terminology, Techniques and Scoring System in Sleep Stages of Human Subjects. Washington, D.C., Public Health Service, U.S. Government Printing Office, 1968.
19. Reite, M., Jackson, D., Cahoon, R.L., and Weil, J.V.: Sleep physiology at high altitude. Electroencephalogr. Clin. Neurophysiol. 38:463, 1975.
20. Sutton, J.R.: Acute mountain sickness: An historical review with some experiences from the Peruvian Andes. Med. J. Aust. 2:243, 1971.
21. Sutton, J.R., Bryan, A.C., Gray, G.W., Horton, E.S., Rebuck, A.S., Woodley, W., Rennie, I.D., and Houston, C.S.: Pulmonary gas exchange in acute mountain sickness. Aviat. Space Environ. Med. 47:1032, 1976.
22. Sutton, J.R., Gray, G., McFadden, M., Bryan, A.C., Horton, E.S., and Houston, C.S.: Nitrogen washout studies in acute mountain sickness. Aviat. Space Environ. Med. 48:108, 1977.
23. Sutton, J.R., Houston, C.S., Mansell, A.L., McFadden, M.D., Hackett, P.M., Rigg, J.R.A., and Powles, A.C.P.: Effect of acetazolamide on hypoxemia during sleep at high altitude. N. Engl. J. Med. 301:1329, 1979.
24. Sutton, J.R. and Lassen, N.: Pathophysiology of acute mountain sickness and high altitude pulmonary oedema: an hypothesis. Bull. Eur. Physiopathol. Respir. 15:1045, 1979.
25. Sutton, J.R., Powles, A.C.P., Gray, G.W., Kane, J., Mansell, A., McFadden, M., Robertson, M., Rondi, P., and Houston, C.S.: Arterial hypoxemia during maximum exercise at altitude. Clin. Res. 25:673A, 1977.
26. Weil, J.V., Kryger, M.H., and Scoggin, C.H.: Sleep and breathing at high altitude. In Guilleminault, C. and Dement, W.C. (eds.): Sleep Apnea Syndromes. New York, Liss, 1978, pp. 119–123.

2
O_2 Breathing at Altitude: Effects on Maximal Performance

P. CERRETELLI

Maximal oxygen consumption ($\dot{V}_{O_2max}$) undergoes a progressive reduction in both acute and chronic hypoxia. This is shown in Fig. 2-1, which summarizes data from various authors. Such decrease, for a pressure drop of 0.5 atm (corresponding to an altitude of about 5500 m) ranges between 30 and 45%, independent of the degree of acclimatization and of the ethnic characteristics of the subjects. Common factors known to change $\dot{V}_{O_2max}$ in opposite directions in hypoxia are (1) the decreased arterial O_2 saturation (% HbO_2) due to decreased inspired oxygen pressure ($P_{I_{O_2}}$) and perhaps to an impairment of the diffusion properties of the lung; and (2) the increased blood hemoglobin (Hb) concentration. This, after prolonged exposure to high altitude, may attain 1.4 times the sea level control value (Fig. 2-2).

Additional factors involved in the control of $\dot{V}_{O_2max}$ to an extent varying with the degree of acclimatization and/or hypoxia are (1) a decrease of maximal cardiac output ($\dot{Q}_{max}$) likely to occur after prolonged exposure to hypoxia as a consequence of increased blood viscosity due to higher hematocrit; (2) a decreased maximal O_2 flow through the muscle capillaries due to impairments in the microcirculation; and (3) possibly, changes of the respiratory potential of the working tissues due to a reduction of muscle mass.

A shift from air to oxygen breathing leads to full saturation of arterial blood and

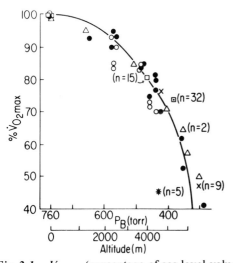

Fig. 2-1. $\dot{V}_{O_2max}$ (percentage of sea level value) as a function of P_B and of altitude. *Open symbols*, acute hypoxia; *filled symbols*, chronic hypoxia; *crosses*, altitude natives. Redrawn from ref. 1, with the addition of data from ref. 7 (*), ref. 9 (×), and ref. 3 (□, in a decompression chamber; △, breathing hypoxic mixtures; ⊡ on 32 lowlanders acclimatized to 5350 m).

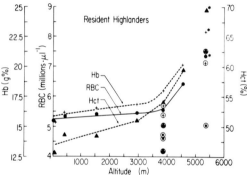

Fig. 2-2. RBC, Hb, Hct values as a function of altitude in resident highlanders (*filled circles and triangles, crosses*; from *Biology Data Book*, 1974). Circled dots, crosses, and *triangles*: measurements of RBC, Hb, and Hct, respectively, on Sherpas (refs. 4 and 15, and unpublished observations). Symbols with asterisk refer to measurements on acclimatized lowlanders carried out at 5350 m after most subjects were exposed to altitudes up to 7500 m. (From ref. 5.)

should increase the O_2 transport to the tissues with a substantial improvement in performance. It has long been known, however, particularly to mountaineers, that oxygen breathing during altitude climbing is not as beneficial as expected on physiologic grounds. This rather mysterious finding was tentatively explained by Barcroft (2) in the 1920s as representing a permanent left-shift of the oxyhemoglobin dissociation curve. Apart from the fact that an increased hemoglobin oxygen affinity seems to improve rather than impair the O_2 transport in acclimatized subjects exposed to moderate altitudes (12), the blood O_2 affinity of acclimatized lowlanders appears indeed slightly decreased. In fact, the average P_{50} value was found to be 2 to 4 torr higher than that at sea level (16).

The aim of this presentation is to describe the mechanisms and degree by which (1) a sudden shift of P_{IO_2} from 80 to 390 torr, (2) a rapid increase of barometric pressure (P_B) from 390 to 540 torr, and (3) a 4-week sojourn at sea level following altitude exposure affect the maximal aerobic power of acclimatized lowlanders.

The Effect of Sudden Hyperoxia on $\dot{V}_{O_2max}$

The results of blood measurements at rest together with the most significant exercise parameters recorded for 10 subjects in three environmental situations, i.e., breathing air at sea level and at 5350 m as well as breathing O_2 at 5350 m, appear in Table 2-1 (4). The effects of chronic hypoxia (6–8 weeks exposure at altitudes up to 6500 m) on blood composition are comparable to those reported in the literature for similar conditions; in particular, Hb concentration shows a 37% increase. $\dot{V}_{O_2max}$ determined at 5350 m by a closed circuit system dropped to 70% of the sea level value, resuming only 92% of the control when breathing oxygen in spite of only a slight reduction of maximal cardiac output ($\dot{Q}_{max}$) (Fig. 2-3 *left* and *center*). In fact, $\dot{Q}$, determined in two of the subjects cited in Table 2-1 by the N_2-CO_2 rebreathing method (6) during an exercise requiring 90% of $\dot{V}_{O_2max}$ at 5350 m, was 19 and 20.7 liter/min, respectively, i.e., 91 and 87% of the sea level

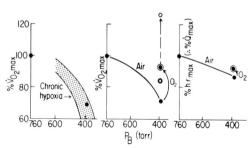

Fig. 2-3. $\dot{V}_{O_2max}$ and h.r.$_{max}$ (percentage of sea level averages) as a function of P_B. **Left** The dotted area indicates the range of $\dot{V}_{O_2max}$ as a function of altitude (Fig. 2-1). The solid dot designates the average value found by the author in the group under study. **Center** The effect of breathing oxygen is indicated by the arrow (*circled dots*). The asterisk indicates the estimated $\dot{V}_{O_2max}$ in hypoxia calculated on actual $\dot{Q}_{max}$ and maximal potential ($a - \bar{v}$) O_2 difference. The latter is based on measured Ca_{O_2} and on the $C\bar{v}_{O_2}$ value found at sea level during maximal exercise. The open circle indicates the expected $\dot{V}_{O_2max}$ when breathing O_2 on the assumptions made in the text. **Right** Effect of breathing oxygen on h.r.$_{max}$. (From ref. 5.)

Table 2-1. Age, Body Weight, RBC Count, Hemoglobin Concentration, Hematocrit, Arterial O_2 Saturation at Rest and at Exercise, Workload, Maximum Heart Rate, Ventilation Frequency, Ventilation (BTPS), and Maximum O_2 Uptake when Breathing Air at Sea Level and at 5350 m as well as O_2 at 5350 m.

Subjects (n = 10)	Age (years)	Weight (kg)	RBC (millions/μl)	Hb (g%)	Hct (%)	% HbO_2 (rest)	W_{max} (kpm/min)	VE_{max} (l/min)	f_{max} (br/min)	$h.r._{max}$ (beats/min)	% HbO_2 (maximum exercise)	$\dot{V}_{O_2max}$ (direct method) (l/min)	$\dot{V}_{O_2max}$ (direct method) (ml/kg/min)
Sea level	26.1 ±4.7 (SD)	72.7 ±10.6	4.73 ±0.23	15.0 ±0.9	44.7 ±1.3	98.2 ±2.1	1500	97.4 ±13.6	36.5 ±5.7	185 ±9.4	97.8 ±2.8	3.21 ±0.27	45.3 ±8.6
5350 m		67.4 ±10.3	6.57 ±0.68	20.6 ±1.4	63.8 ±4.6	82.0[a] ±4.2	1200	145.1 ±21.5	47.0 ±10.1	161 ±15.4	77.4[b] ±4.0	2.26 ±0.24	34.4 ±6.7
5350 m in O_2						98.3 ±2.0	1440	141.7 ±43.1	50.2 ±10.1	169 ±14.5	98.0 ±1.5	2.94 ±0.41	43.1 ±9.2

[a] n = 26 subjects.
[b] n = 15 subjects.
From reference 4.

controls. The corresponding heart rate (h.r.) values were 148 and 162 beats/min, respectively, i.e., 93 and 91% of the values found at sea level. $\dot{Q}_{max}$, estimated by extrapolation of the $\dot{Q}/\dot{V}_{O_2}$ relationship to $\dot{V}_{O_2max}$, was found to be 10% less than in control conditions. Cardiac output was not measured during oxygen breathing. It appears conceivable, however, that $\dot{Q}_{max}$ may be, if anything, somewhat higher than in hypoxia. This is also compatible with the higher h.r.$_{max}$ levels attained (169 vs 161 beats/min) (Fig. 2-3 *right*). In conclusion, considering the 37% increase of Hb concentration and assuming a 10% drop of $\dot{Q}_{max}$, the maximal aerobic power of an acclimatized subject breathing O_2 should be 25 to 30% higher than at sea level. On the contrary, as previously pointed out, upon O_2 breathing $\dot{V}_{O_2max}$ attains only 92% of the sea level control.

The Effects of a Rapid Increase of P_B from 390 to 540 torr

The results of measurements of $\dot{V}_{O_2max}$, Hb concentration, and h.r.$_{max}$ carried out on 13 subjects 6 to 24 h after descent by helicopter from 5350 m to 2850 m appear in Table 2-2 (4). Average $\dot{V}_{O_2max}$, assessed by an indirect method based on the extrapolation of the h.r./$\dot{V}_{O_2}$ relationship, rose significantly ($p < 0.001$) from 2.36 to 3.03 liter/min, i.e., to 97% of the sea level control. h.r.$_{max}$ rose 8 beats/min to 95% of the control. Again, the improvement of $\dot{V}_{O_2max}$ at higher P_B was less than could be expected from the increase of Hb concentration and the moderate decrease of maximal cardiac output.

The Relationship Between $\dot{V}_{O_2max}$ and Hb Concentration Before and After Altitude Exposure

The results of the measurements of $\dot{V}_{O_2max}$ and related variables carried out on 13 subjects at sea level before altitude exposure, after a 12- to 16-week sojourn at altitude, and again 25 to 28 days after return to sea level, are summarized in Table 2-3 (4). The average 5% increase of $\dot{V}_{O_2max}$ (liter/min) found in connection to the 11.6% increase in Hb concentration when comparing return to departure values is not statistically significant ($p > 0.3$). Thus, increased Hb concentration does not necessarily raise $\dot{V}_{O_2max}$.

Discussion

The blood O_2 partial pressure values and the systemic pressure levels in the various experimental conditions may be of some relevance in the interpretation of the results

Table 2-2. Age, Body Weight, RBC Count, Hemoglobin Concentration, Maximal Heart Rate, and Maximum O_2 Uptake Breathing Air at Sea Level, at 5350 m, and at 2850 m.

Subjects ($n = 13$)	Age (years)	Weight (kg)	RBC (millions/μl)	Hb (g%)	h.r.$_{max}$ (beats/min)	$\dot{V}_{O_2max}$ (direct method)	
						(l/min)	(ml/kg/min)
Sea level	29.1 ±5.9 (SD)	71.5 ±9.7	4.70 ±0.22	15.0 ±0.9	187 ±13	3.13 ±0.29	44.3 ±5.6
5350 m					160 ±9	2.36 ±0.29	35.7 ±6.2
		66.9 ±8.7	6.65 ±0.7	21.6 ±2.4			
2850 m					168 ±15	3.03 ±0.25	45.8 ±5.3

From reference 4.

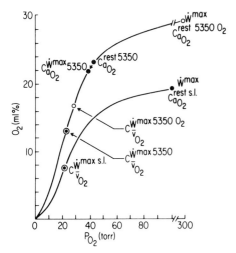

Fig. 2-4. "Physiologic" O_2 dissociation curves for normal (Hb = 15 g%) and acclimatized lowlanders (Hb = 21 g%). Arterial (*a*) and mixed venous ($\bar{v}$) points are indicated for resting (*rest*) and maximal working ($\dot{W}_{max}$), at sea level (s.l.) or at 5350 m (5350), breathing ambient air or O_2 (O_2). The points characterizing Ca_{O_2} at sea level (at rest and at maximal exercise) are superimposed; the same is done for the points indicating resting and exercise Ca_{O_2} at 5350 m while breathing O_2. (From ref. 5.)

appearing in Tables 2-1 and 2-2. In Fig. 2-4 "physiologic" (at actual P_{CO_2} and 2,3-DPG concentration levels) oxyhemoglobin dissociation curves are drawn for Hb concentrations of 15 and 21 g%, respectively. Measured arterial (*a*) and mixed venous ($\bar{v}$) points (with the exception of the venous point at exercise during O_2 breathing, which was calculated) are indicated at rest and during maximal work (W) for both air and O_2 breathing conditions. The $P\bar{v}_{O_2}$ values during maximal exercise at altitude (both in air and O_2 breathing) appear relatively high when compared to sea level controls and to conditions of acute hypoxia. The products between measured or estimated maximal cardiac output ($\dot{Q}_{max}$) and maximal "potential" arteriovenous O_2 differences, i.e., calculated on actual Pa_{O_2} and $P\bar{v}_{O_2}$ values measured during maximal exercise at sea level, yield $\dot{V}_{O_2max}$ figures much higher than those actually found at altitude both when breathing air and oxygen. In these conditions, $\dot{V}_{O_2max}$ should attain levels of 85 and 125%, respectively, of the sea level controls (Fig. 2-3 *center*, *asterisk* and *open circle*), as compared to the actual values of 70 and 92%, respectively.

The systemic pressure does not increase during heavy exercise in acclimatized lowlanders more than it does at sea level. This finding, considering the relatively high cardiac output and blood viscosity, necessarily implies reduced peripheral resistance to flow. This could be prompted by a dilatation of the metarterioles coupled with a contraction of the precapillary sphincters and/or by the opening of an adequate number of non-nutritional circuits. Such changes would cause a reduction of the O_2 flow to the muscles and would also justify (1) the relatively low $\dot{V}_{O_2max}$ values observed in acclimatized lowlanders both when breathing ambient air or O_2 in spite of an adequate O_2 transport to the periphery, and (2) the high $P\bar{v}_{O_2}$ values found during maximal exercise in both conditions.

The reduction of $\dot{V}_{O_2max}$ found in acclimatized lowlanders when breathing air as well as O_2 could alternatively be attributed to a failure of the "power plant," i.e., to a deterioration of the respiratory function of the mitochondria induced by chronic hypoxia and/or to a decrease of their absolute number caused by a reduction of muscle mass. Experiments carried out in acclimatized rats (11), however, indicate that the respiratory function of the mitochondria is not affected by chronic hypoxia. In addition, a decrease of muscle mass of the same order of magnitude as observed in man following prolonged exposure to altitude does not seem to affect the $\dot{V}_{O_2max}$ of dogs to any measurable extent (personal observation).

The conclusion that may be drawn from the first two experiments is that the limit to $\dot{V}_{O_2max}$ in acclimatized lowlanders is mostly peripheral and may probably be attributed to a lower "effective" perfusion of the muscle. Whether this change is the consequence of a primitive increase of the central vasomotor tone due to hypoxia or of more complex adaptive mechanisms aimed at decreasing the load on the heart is still a matter for investigation.

As shown in Table 2-3, an 11.6% in-

Table 2-3. Age, Body Weight, RBC Count, Hemoglobin Concentration, Maximum Heart Rate, and Maximum O_2 Consumption at Sea Level Before Altitude Exposure, at 5350 m, and at Sea Level 4 Weeks After Return.

Subjects (n = 13)	Age (years)	Weight (kg)	RBC (millions/µl)	Hb (g%)	h.r.$_{max}$ (beats/min)	$\dot{V}_{O_2max}$ (indirect method) (l/min)	(ml/kg/min)
Sea level (departure)	25.3 ±6.1 (SD)	72.8 ±7.9	4.48 ±0.30	14.6 ±0.7	191 ±9	3.23 ±0.31	44.3 ±4.5
5350 m		67.8 ±7.5	6.52 ±0.48	23.4 ±2.6	162 ±16	2.35 ±0.44	34.3 ±3.7
Sea level (return)		67.6 ±7.5	5.01 ±0.66	16.3 ±2.1	186 ±9	3.39 ±0.46	50.7 ±6.8

From reference 4.

crease in Hb concentration a month after return to sea level is not paralleled by a significant increase of $\dot{V}_{O_2max}$. Training conditions of the subjects were the same before and after altitude exposure. The 7% reduction in body weight probably reflects more a loss of body fat than a decrease of muscle mass and should therefore not influence $\dot{V}_{O_2max}$. The latter, moreover, does not seem to depend, within broad limits, on muscle mass.

On the other hand, the effects of blood infusion on maximal aerobic power appear rather controversial. Ekblom et al. (8) found a 9% increase in $\dot{V}_{O_2max}$ following a 13% increase of Hb concentration by reinfusion of homologous blood. In contrast, Williams et al. (17) did not find differential effects of whole blood (500 ml), RBC (275 ml), or plasma (225 ml) infusion on endurance capacity or resting, submaximal, and maximal heart rate. In the present experimental conditions the increased Hb concentration is a consequence of altitude exposure, thus involving different mechanisms of adjustment of blood composition as well as systemic and peripheral circulation. The failure of higher Hb to increase the maximal aerobic performance of the subject by no means implies that in ordinary normoxic conditions the factors limiting aerobic work are peripheral. It is well known that $\dot{V}_{O_2max}$ increases significantly when increasing O_2 partial pressure in inspired air (10,13,14) which indicates that the oxidative potential of the muscles exceeds the capacity for O_2 transport by the circulation. Rather, a change of red cell concentration in circulating blood could be counterbalanced at the muscle level by a reduction of "effective" blood flow. This drop, in conditions of very high hematocrit, could be more than compensatory, thus causing a reduction of the O_2 available at the muscle level.

Conclusions

The failure of sudden hyperoxia to raise $\dot{V}_{O_2max}$ of acclimatized lowlanders to sea level or even higher values in the absence of a drastic reduction of maximal cardiac output could be explained by a reduction of "effective" blood flow to the working muscles. A similar limitation, even though mediated by different mechanisms, could also explain the controversial effects of blood infusion on the maximal aerobic power of subjects at sea level. By contrast, the recently described beneficial effects of hemodilution on the maximal performance of acclimatized lowlanders could originate from an increase of "effective" blood flow to working muscles that would more than compensate for the decreased hematocrit.

References

1. Åstrand, P.O. and Rodahl, K.: Textbook of Work Physiology. New York, McGraw-Hill, 1970, p. 573.
2. Barcroft, I.: Features in the Architecture of Physiological Function. New York, Hafner, 1972, p. 222.
3. Cerretelli, P.: Metabolismo ossidativo e anaerobico nel soggetto acclimatato all'altitudine. Minerva Aerosp. 67:11, 1976.
4. Cerretelli, P.: Limiting factors to oxygen transport on Mount Everest. J. Appl. Physiol. 40:658, 1976.
5. Cerretelli, P.: Gas exchange at high altitude. In West, J.B. (ed.): Pulmonary Gas Exchange. New York. Academic Press, 1980.
6. Cerretelli, P., Cruz, J.C., Farhi, L.E., and Rahn, H.: Determination of mixed venous O_2 and CO_2 tensions and cardiac output by a rebreathing method. Respir. Physiol. 1:258, 1966.
7. Cerretelli, P. and Margaria, R.: Maximum oxygen consumption at altitude. Int. Z. Angew. Physiol. 18:460, 1961.
8. Ekblom, B., Goldbarg, A.N., and Gullbring, B.: Response to exercise after blood loss and reinfusion. J. Appl. Physiol. 33:175, 1972.
9. Elsner, R.W., Bolstad, A., and Forno, C.: Maximum oxygen consumption of Peruvian Indians native to high altitude. In Weihe, W.H. (ed.): The Physiological Effects of High Altitude. Oxford, Pergamon, 1964, pp. 217-223.
10. Fagraeus, L., Karlsson, J., Linnarson, D., and Saltin, B.: Oxygen uptake during maximal work at lowered and raised ambient air pressures. Acta Physiol. Scand. 87:411, 1973.
11. Gold, A.J., Johnson, T.F., and Costello, L.C.: Effects of altitude stress on mitochondrial function. Am. J. Physiol. 224:946, 1973.
12. Hebbel, R.P., Eaton, J.W., Kroenenberg, R.S., Zanjani, E.D., Moore, L.G., and Berger, E.M.: Human llamas. Adaptation to altitude in subjects with high hemoglobin oxygen affinity. J. Clin. Invest. 62:593, 1978.
13. Margaria, R., Camporesi, E., Aghemo, P., and Sassi, G.: The effect of O_2 breathing on maximal aerobic power. Pflügers Arch. 336:225, 1972.
14. Margaria, R., Cerretelli, P., Marchi, S., and Rossi, L.: Maximum exercise in oxygen. Int. Z. Angew. Physiol. 18:465, 1961.
15. Morpurgo, G., Battaglia, B., Carter, N.D., Modiano, G., and Passi, S.: The Bohr effect and the red cell 2,3-DPG and Hb content in Sherpas and Europeans at low and high altitude. Experientia 28:1280, 1972.
16. Samaja, M., Veicsteinas, A., and Cerretelli, P.: Oxygen affinity of blood in altitude Sherpas. J. Appl. Physiol.: Respirat. Environ. Exercise Physiol. 47:337, 1979.
17. Williams, M.H., Godwin, A.R., Perkins, R., and Bocrie, J.: Effect of blood reinjection upon endurance capacity and heart rate. Med. Sci. Sports 5:181, 1973.

3

Oxygen Uptake at High Altitude: Limiting Role of Diffusion in Lungs

JOHANNES PIIPER

It is generally agreed that in normoxia O_2 uptake in normal lungs at rest is not limited by alveolar-capillary diffusion. During severe exercise the cardiac output, and not alveolar-capillary diffusion, is believed to limit the maximum O_2 uptake. On the other hand, it is also well established that diffusion limitation in alveolar O_2 transfer becomes more important in hypoxia, and is expected to be particularly important when the O_2 uptake is simultaneously increased as in physical exercise. Indeed, West et al. [8] measured a marked drop in arterial O_2 saturation, despite a simultaneous rise in alveolar P_{O_2}, when subjects exercised at 5800 m. This must have meant a considerable increase of the alveolar-arterial P_{O_2} difference which the authors regarded as most probably due to diffusion limitation [7]. Similar results have been reported by Saltin et al. [5] for subjects exercising at 4300 m.

In this paper the quantitative role of diffusion limitation in alveolar-capillary O_2 transfer in conditions prevailing in high altitude hypoxia will be examined. For the sake of simplicity, the subscript "O_2" is generally omitted from symbols in the text and in equations.

Lung Model and Theory

The simple lung model adopted for analysis of alveolar-capillary gas transfer is depicted in Fig. 3-1A. On the basis of this model the following relationship is obtained for the end-capillary O_2 pressure, $P_{c'}$ (reached at $x = x_0$):

$$\frac{P_{c'} - P_{\bar{v}}}{P_A - P_{\bar{v}}} = 1 - e^{-D/(\dot{Q}\beta)} \quad [3\text{-}1]$$

This fundamental relationship expresses the completeness of alveolar-capillary equilibration as a function of the parameter $D/(\dot{Q}\beta)$. This parameter contains the variables (1) pulmonary diffusing capacity for O_2, D; (2) pulmonary blood flow = cardiac output, $\dot{Q}$; and (3) slope of the blood O_2 dissociation curve, termed "capacitance coefficient" of blood for O_2, β [3]. According to Eq. 3-1 the O_2 partial pressure difference, $(P_{c'} - P_{\bar{v}})$, is diminished, i.e., the equilibration efficiency is reduced, by a decrease in D, by an increase in $\dot{Q}$, or by an increase in β.

It appears from the calculations depicted in Fig. 3-1B that with $D/(\dot{Q}\beta) > 3$, the equilibration is more than 95% complete,

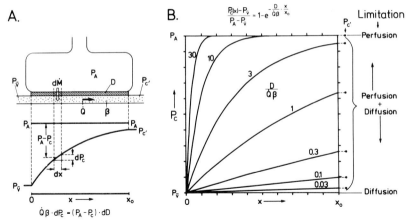

Fig. 3-1. Model analysis of gas transfer between alveolar gas and pulmonary capillary blood. The model (**A**) consists of an alveolar space with constant O_2 partial pressure, P_A, and a pulmonary capillary, perfused at the rate $\dot{Q}$ and separated from the alveolar space by a barrier ("alveolar membrane") which has the O_2 conductance or O_2 diffusing capacity, D. Mixed venous blood, with O_2 partial pressure $P\bar{v}$, enters the capillary; O_2 uptake ($\dot{M}$) increases its O_2 pressure ($P_{c'}$) which reaches the end-capillary value $P_{c'}$ upon leaving the capillary. Considering the O_2 dissociation curve (close to) linear in the region between $P_{\bar{v}}$ and $P_{c'}$, the O_2 concentration increment within the element dx is equal to $\beta \cdot dP_c$ (β = slope of blood O_2 dissociation curve). The differential equation at the bottom of **A**, which follows from mass balance, yields upon integration the equation on top of **B**. The partial pressure profiles following therefrom are plotted in **B** for several values of the parameter $D/(\dot{Q}\beta)$.

and the O_2 uptake (for given D, $\dot{Q}$, β, P_A, and $P_{\bar{v}}$) is not limited by diffusion, but only by perfusion. Conversely, with $D/(\dot{Q}\beta) < 0.1$ the diffusion limitation is so predominant that an increase of perfusion to infinity would not be able to increase the O_2 uptake by more than 5%. Between these limiting cases of pure perfusion limitation and pure diffusion limitation there is a range of $D/(\dot{Q}\beta)$, from 0.1 to 3, where both diffusion and perfusion exert a limiting effect on O_2 uptake. This is the range of physiologic interest.

It follows from Fig. 3-1 and Eq. 3-1 that the O_2 uptake, $\dot{M} = \dot{Q} \cdot \beta \cdot (P_{c'} - P_{\bar{v}})$, may be expressed by the following relationship:

$$\dot{M} = \dot{Q}\beta \cdot (1 - e^{-D/(\dot{Q}\beta)}) \cdot (P_A - P_{\bar{v}}) \quad [3\text{-}2]$$

By introducing the (alveolar–mixed venous) O_2 conductance, G,

$$G = \frac{\dot{M}}{P_A - P_{\bar{v}}} \quad [3\text{-}3]$$

the relationship of Eq. 3-2 is transformed to

$$G = \dot{Q}\beta \cdot (1 - e^{-D/(\dot{Q}\beta)}) \quad [3\text{-}4]$$

One may define an index of diffusion limitation, L_{diff}, as the fractional decrease of the O_2 conductance from the non–diffusion-limited case ($D = \infty$; $G = \dot{Q}\beta$) to the actual value, G. One obtains from Eq. 3-4 the following:

$$L_{\text{diff}} \equiv \frac{\dot{Q}\beta - G}{\dot{Q}\beta} = e^{-D/(\dot{Q}\beta)} \quad [3\text{-}5]$$

Similarly, a corresponding expression for perfusion limitation, L_{perf}, may be derived, considering that for the condition $\dot{Q}\beta = \infty$, Eq. 3-4 yields $G = D$:

$$L_{\text{perf}} \equiv \frac{D - G}{D} = 1 - \frac{1 - e^{-D/(\dot{Q}\beta)}}{\frac{D}{\dot{Q}\beta}} \quad [3\text{-}6]$$

Furthermore, the limiting effects of diffusion and perfusion can be expressed as the increment in O_2 conductance G

produced by an increment in D or $\dot{Q}\beta$, respectively:

$$\frac{\delta G}{\delta D} = e^{-D/(\dot{Q}\beta)} \qquad [3\text{-}7]$$

$$\frac{\delta G}{\delta(\dot{Q}\beta)} = 1 - e^{-D/(\dot{Q}\beta)} \cdot \left(1 + \frac{D}{\dot{Q}\beta}\right) \qquad [3\text{-}8]$$

The determinant role of the parameter $D/(\dot{Q}\beta)$ is apparent from Eqs. 3-5–3-8.

Application to Experimental Data

The measurements performed on members of the Italian 1973 Mount Everest Expedition in their base camp (5350 m, $P_B = 390$ torr), reported by Cerretelli (1), are used for calculations. The basic quantities characterizing O_2 transport and the variables $\dot{Q}$ and β required for our analysis are presented in Table 3-1.

For O_2 diffusing capacity, D, the following values recently determined in our laboratory (2) are used:

rest, $D_{O_2} = 48$ ml · min^{-1} · torr^{-1}
exercise (O_2 uptake, 2 l/min), $D_{O_2} = 64$ ml · min^{-1} · torr^{-1}

From these values the following $D/(\dot{Q}\beta)$ values are obtained:

rest, $D/(\dot{Q}\beta) = 1.9$
maximum exercise, $D/(\dot{Q}\beta) = 0.43$

The perfusion and diffusion limitation parameters calculated from the index $D/(\dot{Q}\beta)$ according to Eqs. 3-5–3-8 are presented in Table 3-2. The main features are the following:

1) Both at rest and during exercise there is considerable diffusion limitation in alveolar-capillary O_2 transfer.
2) In resting conditions, the limiting role of perfusion is quantitatively more important than that of diffusion. Thus, a certain relative increase of the cardiac output, $\dot{Q}$, leads to an increase in O_2 transport conductance, G, which is $0.56/0.14 = 4$ times greater than that achieved by an equal relative increase of the O_2 diffusing capacity, D.
3) During heavy exercise, however, diffusion limitation has definitely become predominant. $L_{\text{diff}} = 0.65$ means that due to diffusion limitation the O_2 conductance is reduced to 35% of the value which would be predicted for same $\dot{Q}\beta$, but for $D = \infty$. Moreover, a certain relative increase of D would produce an increase of the O_2 conductance, G,

Table 3-2. *Analysis of Diffusion and Perfusion Limitation in Alveolar-Capillary O_2 Transfer at High Altitude (5350 m).*

	Rest	Maximum exercise
$D/(\dot{Q}\beta)$	1.9	0.43
L_{diff}	0.15	0.65
L_{perf}	0.55	0.19
$\delta G/\delta(D)$	0.15	0.65
$\delta G/\delta(\dot{Q}\beta)$	0.56	0.07

$\dot{Q}$ and β data from ref. 1; D data from ref. 2. All quantities are dimensionless.

Table 3-1. *Measured and Estimated Data on O_2 Uptake and Transport at High Altitude (5350 m).*

Quantity	Units	Rest	Maximum exercise
O_2 uptake	l O_2 · min^{-1}	0.3[a]	2.26
$(C_a - C_{\bar{v}})_{O_2}$	vol. %	5[a]	8.9
$\dot{Q}$	l · min^{-1}	6	25
O_2 capacity	vol. %	28	28
P_{aO_2}	torr	43	39
$P_{\bar{v}O_2}$	torr	31	24
β_{O_2}	ml O_2 · l^{-1} · torr^{-1}	4.2	5.9

[a] Assumed.
From reference 1.

which is $0.65/0.07 \approx 9$ times higher than that effected by the same relative increase of the cardiac output, $\dot{Q}$.

Critical Remarks

Validity of Model

The model used in this study may be oversimplified for several reasons, including the following.

1) *Nature of D.* Part of the resistance to O_2 uptake is probably located not in the tissue layers separating blood from gas, but in the blood itself. Furthermore, the finite reaction kinetics of O_2 with hemoglobin contribute to the resistance to O_2 uptake. It is doubtful if the sum of all these processes can be adequately described in terms of an overall conductance traditionally termed "pulmonary O_2 diffusing capacity."

2) *Functional Inhomogeneity.* The O_2 transfer efficiency of lungs is known to be reduced by the presence of unequal distribution of ventilation, blood flow, and diffusing capacity. Although the effects of such "functional inhomogeneities" are expected to be reduced in hypoxia, their neglect may lead to serious errors. Furthermore, the effects of stratification, pulsatile blood flow, and respiratory cyclic variations should be considered.

3) *Constancy of β.* Even in the P_{O_2} range of 10 to 35 torr the O_2 dissociation curve of blood is not exactly linear, i.e., β is not constant. At higher P_{O_2}, recourse has to be taken to the Bohr integration technique, which may be considered equivalent to calculating the correct "effective β" for given conditions.

Validity of Values

The estimation of the index $D/(\dot{Q}\beta)$ is subject to criticism for a number of reasons, as follows. Therefore, the values should be regarded as rough estimates only.

Pulmonary Diffusing Capacity for O_2, D_{O_2}

Measurements of the pulmonary diffusing capacity for O_2 (D_{O_2}) and CO (D_{CO}) in man at rest and during exercise have yielded very variable results. We chose our own results on D_{O_2}, measured in healthy young males, because the rebreathing method used appears to have a number of advantages and, therefore, to yield more reliable values than other methods (4). Moreover, during the rebreathing the arterial P_{O_2} was in the hypoxic range of 25 to 45 torr which is close to the arterial P_{O_2} expected at high altitude.

According to the results of West (6) there seems to be no specific effect of high altitude adaptation on D_{CO}.

Particularly for resting conditions the overall literature averages of D_{CO} and D_{O_2} are much lower, about one-half of our values. If these average values are used, $D/\dot{Q}\beta$ would drop from 1.9 to 1, meaning about equal extent of diffusion and perfusion limitation in O_2 transfer.

According to the results obtained by Meyer and Piiper (2), D_{O_2} reaches a plateau at moderate exercise, being about equal at O_2 uptake levels of 1 and 2 liters/min. This appears to justify the use of D_{O_2}, measured at medium heavy exercise, for exercise with maximum O_2 uptake. The following calculation, however, suggests that this may be erroneous.

For the level of maximum O_2 uptake, the mean alveolar-pulmonary capillary P_{O_2} difference ($\dot{M}_{O_2}/D_{O_2}$) is calculated to equal 35 torr. Using this value and the mean capillary P_{O_2} estimated from Pa_{O_2} and $P\bar{v}_{O_2}$ of Table 3-1, 32 torr, an alveolar P_{O_2} of 67 torr is obtained, which value is only 5 torr below the P_{O_2} of water vapor–saturated inspired gas (72 torr). Therefore, either the blood P_{O_2} values estimated by Cerretelli (1) are too high or the D_{O_2} used for maximum exercise is too low. It may be relevant to mention that Saltin et al. (5) required a D_{O_2} of 80 ml · torr^{-1} · min^{-1} to account for O_2 transfer in heavy exercise at 4300 m of altitude.

Capacitance Coefficient, β_{O_2}

The slope of the blood O_2 dissociation curve or capacitance coefficient, β_{O_2}, is proportional to the hematocrit and depends strongly on the P_{O_2} range. The considerable difference between the β values for rest and exercise (Table 3-1) is entirely due to the change in the P_{O_2} range. For an accurate estimate of β_{O_2} all the factors acutely or chronically influencing the O_2 dissociation curve (pH, P_{CO_2}, temperature, 2,3-diphosphoglycerate) must be considered. The values of Table 3-1 are crude estimates only.

Cardiac Output, $\dot{Q}$

The $\dot{Q}$ values of Cerretelli (1) are based mainly on heart rate measurements. In other experimental studies (cited in ref. 1) different values have been reported.

Conclusion

Although the $D/(\dot{Q}\beta)$ values used in this study have a broad margin of uncertainty, qualitatively (and semiquantitatively) they may be taken to reveal the important role of diffusion limitation in alveolar-pulmonary capillary transfer of O_2 at high altitude, particularly during heavy exercise.

Summary

Calculations on models for alveolar-capillary gas equilibration in lungs show that O_2 uptake in hypoxia encountered at high altitude is limited both by perfusion and diffusion. Whereas in resting conditions perfusion limitation appears to be more important, in strenuous physical exercise diffusion limitation becomes predominant.

References

1. Cerretelli, P.: Limiting factors to oxygen transport on Mount Everest. J. Appl. Physiol. 40:658, 1976.
2. Meyer, M. and Piiper, J.: Pulmonary diffusing capacity for carbon monoxide and oxygen in man during heavy exercise. Pflügers Arch. 373:R37, 1978.
3. Piiper, J., Dejours, P., Haab, P., and Rahn, H.: Concepts and basic quantities in gas exchange physiology. Respir. Physiol. 13:292, 1971.
4. Piiper, J., Meyer, M., and Scheid, P.: Pulmonary diffusing capacity for O_2 and CO at rest and during exercise. Advantages of rebreathing techniques using stable isotopes. Bull. Eur. Physiopathol. Respir. 15:145, 1979.
5. Saltin, B., Grover, R.F., Blomqvist, C.G., Hartley, H., and Johnson, R.L., Jr.: Maximal oxygen uptake and cardiac output after 2 weeks at 4300 m. J. Appl. Physiol. 25:400, 1968.
6. West, J.B.: Diffusing capacity of the lung for carbon monoxide at high altitude. J. Appl. Physiol. 17:421, 1962.
7. West, J.B.: Gas diffusion in the lung at altitude. In Margaria, R. (ed.): Exercise at Altitude. Amsterdam, Excerpta Medica, 1967, pp. 75–83.
8. West, J.B., Lahiri, S., Gill, M.B., Milledge, J.S., Pugh, L.G.C.E., and Ward, M.P.: Arterial oxygen saturation during exercise at high altitude. J. Appl. Physiol. 17:617, 1962.

4
Respiratory and Cardiocirculatory Responses of Acclimatization of High Altitude Natives (La Paz, 3500 m) to Tropical Lowland (Santa Cruz, 420 m)

M. Paz Zamora, J. Coudert, J. Ergueta Collao, E. Vargas, and N. Gutierrez

Socioeconomic development of the countries of the Andean region necessitates massive migration of population groups through colonization programs, which mobilize high altitude natives as well as lowlanders. Health problems of these natives need to be studied and solved since they must work and remain economically active. It is the objective of this study to contribute to the better understanding of the acclimatization pattern of highlanders when transferred to lowland.

Materials and Methods

Seventeen native Aymaras, sedentary males born and living at the Andean high plateau (3800 m), of an average age of 22 years, were first studied at La Paz (3500 m) and then in Santa Cruz (420 m) during the first 18 days of acclimatization to low altitude (group HL). A control group of 10 sedentary males, born and living at low altitude, with an average age of 19 years, were studied under the same conditions, i.e., 420 m (group LL). Several parameters were measured in each group by means of the same techniques.

Results and Discussion

The temperature conditions and relative humidity are different at low and high altitude (dryer and colder weather in La Paz, hotter and more humid conditions in Santa Cruz). The modifications observed during the acclimatization to low altitude and the existing differences between the HL and LL groups can be described as follows.

Lung Volumes

Contrary to the studies performed at high altitude during the first week of acclimatization, in which a transitory decrease in vital capacity (VC) (22) was noted, the static pulmonary volumes do not suffer any important change during acclimatization to low altitude. But the state of pulmonary expansion, together with an increase of residual functional capacity (CRF) and residual volume (VR) described at high altitude in the natives of the high plateau (9), is maintained during the first week at low altitude. Observations made on rats have given evidence of anatomic modifications of pulmonary tissue, such as an increase of the al-

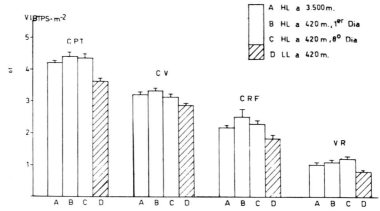

Fig. 4-1. Static pulmonary volumes obtained at high altitude (3500 m) and during the first week of acclimatization to low altitude (420 m) of a group of 17 males, natives of the Andes high plateau (3800 m) (group HL). CPT, total pulmonary capacity. CV, vital capacity. CRF, residual functional capacity. VR, residual volume.

veolar interchange areas and pulmonary capillaries when neonates are placed for a sufficient period in a hypoxic environment (2). In view of the above facts, the increase in pulmonary volumes of high altitude natives might be considered as a sign of adaptation to the chronic hypoxia of the environment. The decrease in the dynamic spirometric volumes—forced expiratory volume (VEMS) and maximum midexpiratory flow rate (VMM)—during acclimatization to lowland is due to the changes in the physical properties of air (increase of molecular density) (Fig. 4-1).

Ventilation and Gas in Arterial Blood

Ventilation during rest diminishes progressively in the first 2 weeks of acclimatization to low altitude. The initial drop in $\dot{V}E$ minute volume might be explained as a consequence of the suppression of hypoxic drive on the peripheral chemoreceptors. The latter have been studied in natives of high altitudes and are reputed to be hyposensitive to P_{O_2} variations (12,14,16,20). To explain the decrease in ventilation during acclimatization to low altitude, which is progressive in time and is parallel to a progressive increase in Pa_{CO_2} which stimulates ventilation, other mechanisms have to be invoked. This question is comparable to the one that arises during acclimatization to high altitude; i.e., the progressive increase in ventilation does not immediately reach maximum value (21). At high altitude changes in the hydrogen ion concentration of cerebrospinal fluid could also explain the progressive ventilatory response of what could be called the "hyperoxia of low altitudes." At first the increase in hydrogen ion concentration, together with respiratory acidosis (Fig. 4-2), will limit the effect of "hyperoxia" on ventilation. During subsequent days, active transport of hydrogen ions from the cerebrospinal fluid with an increased bicarbonate concentration would permit the decrease of induced ventilation due to the higher O_2 tension in the air (Fig. 4-3).

Alveolar-Arterial Oxygen Gradient

The decreased difference of alveolar-arterial O_2 tensions already described at high altitudes in the natives of such regions (10) is also maintained at low altitude. It was established that the pulmonary diffusion capacity is higher in natives of high altitude

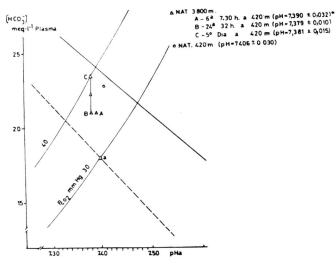

Fig. 4-2. HCO$_3$/pHa relationship (dashed line) and the arterial normal point established for natives of La Paz studied at La Paz (from ref. 12). Arterial blood data obtained during the first 5 acclimatization days at low altitude in a group of 17 natives of the Andean high plateau (3800 m) correspond, in relation to the point of origin (La Paz, 3500 m), to an incompletely compensated state of gaseous acidosis.

(5,19). The pulmonary morphologic modifications that show, as a result, an increase of the interchange areas between the alveoli and pulmonary capillaries as well as an increase of pulmonary arterial pressures are also maintained at low altitude during more than a year (23). This could ensure a better perfusion of the pulmonary apex. All these factors contribute, during the acclimatization of natives of high altitude to low altitude, toward attaining a small alveolar-arterial oxygen gradient. It should be noted that these observations are controversial. Other authors have stated (11,18), on the contrary, that an increase in the alveolar-arterial oxygen gradient is evidence of ventilation perfusion mismatching as well as an increase in arteriovenous shunts in high altitude natives. Furthermore, recent studies on the distribution of pulmonary blood flow

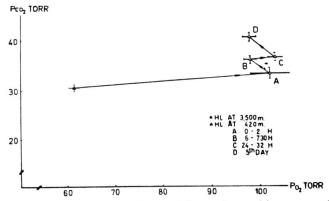

Fig. 4-3. Evolution of relationship between Pa_{O_2} and Pa_{CO_2} observed in a group of natives from 3800 m (group HL) studied at 3500 m and during the first 5 days of acclimatization to 420 m.

do not show any improvement in perfusion of the pulmonary apex at high altitude (Table 4-1) (3).

Heart Rate

The bradycardia observed during acclimatization at low altitude can not only be explained by the action of the higher P_{O_2}. In effect, inhalation of pure O_2 at high altitude produces a moderate diminution in heart rate, smaller than that observed in high altitude natives descending to sea level (23). In addition the bradycardia appears belated. It was not observed in this study on the first and second days at low altitude (Table 4-2). This pattern of acclimatization can be compared to acclimatization to high altitude with the accompanying tachycardia. An increase in activity of the sympathetic nervous system with an increase in elimination of urinary catecholamines (4) were invoked. During acclimatization to low altitude an inverse process may occur. In fact, a diminution of sympathetic activity is observed, as evident by the weak urinary elimination of vanillylmandelic acid on the 18th and 19th day at low altitude.

Maximal Aerobic Capacity ($\dot{V}_{O_2max}$)

To be sure that the improvement of V_{O_2max} observed during the acclimatization to low altitude was not related to the training, new measurements were taken in La Paz in 17 persons of the HL group 1 month after their return from Santa Cruz. The differences in the results obtained before the descent (39 ± 4 ml/kg) and 1 month after return (40 ± 5 ml/kg) are not significant ($p < 0.15$). Among the factors contributing to the improved physical capacity of high altitude natives during acclimatization to low altitude is the increase in stroke volume (SV) and hence the increase in maximum cardiac output. It has been reported that during exercise at high altitudes a diminution of the maximum cardiac output as well as SV has been observed in persons coming from sea level, as well as in high altitude natives (1,6,8,17). The diminution of the urinary elimination of 17,21-hydroxy-20-ketosteroids, observed on the sixth day at low altitude, corresponds to the inverse phenomenon already reported by other authors who have noted an increase of suprarenal cortical function (7,13,15) during the acclimatization period (Table 4-2).

Evaluation of the results obtained during this study shows that acclimatization of high altitude natives to low altitudes causes approximately the reverse response, point by point, of all the phenomena observed during acclimatization of sea level natives to high altitudes. After the 18th day at low altitude, the high plateau native shows responses quite close to those of the lowlander. The definitive persistence of certain

Table 4-1. *Partial Pressures of O_2 and CO_2 and Differences in Alveolar-Capillary O_2, Measured in Natives Born and Residing at La Paz (3500 m) (Group HL) and Then During Acclimatization at Santa Cruz (420 m).*

	Group HL					Group LL
	3500 m	420 m (6-7.5 h)	420 m (23-32 h)	420 m (Day 5)	p	420 m
n	17	12	9	11		10
Pa_{CO_2} (torr)	30.5 0.7	35.8 0.7	36.2 0.8	40.3 0.7	0.005	37.7 0.4
Pa_{O_2} (torr)	61.5 1.1	98.9 1.2	103.9 1.4	98.8 1.4	NS	95.0 1.7
n	10		p	10	p	10
PA_{O_2} (torr)	63.1 0.9		0.005	100.7 0.8	0.01	104.3 1.0
$PA_{O_2} \cdot Pa_{O_2}$	3.67 0.81		NS	1.08 1.54	0.0025	9.53 1.77

Table 4-2. Ventilation, Respiratory Rate, Tidal Volume, Heart Rate, and Oxygen Consumption Measured at La Paz (3500 m) and at Santa Cruz (420 m) During Acclimatization.

	Group HL						Group LL
	3500 m	420 m (Day 1-2)	420 m (Day 5-6)	420 m (Day 10-11)	420 m (Day 16-17)	p	420 m
V_E 1, BTPS · min^{-1}/m^2	6.05 0.24	5.72 0.15	5.12 0.10	5.06 0.13	4.91 0.16	NS	4.87 0.14
f · min^{-1}	18.6 0.7	17.9 0.6	17.7 0.7	17.2 0.4	16.9 0.5	NS	16.6
V_T 1, BTPS	0.528 0.02	0.519 0.02	0.472 0.02	0.476 0.16	0.496 0.021	NS	0.514 0.024
$\dot{V}_{O_2}$ ml/min	254 8	264 11	250 10	250 7	247 8	0.05	269 9
HR · min^{-1}	66 1.7	66 2.9	60 1.6	60 1.1	60 0.8	0.025	66 1.8
HR$_{max}$ · min^{-1}	185 2.8	— 3.3	170	194 1.9	197 1.0	NS	200 2.8
V_{O_2max} ml/min/kg STPD	39.4 1.0	44.4 0.9	43.4 1.3	45.1 1.0	44.5 1.1	NS	42.4 1.3

morphologic and biologic characteristics genetically induced by life at high altitude for many generations can be clearly identified.

References

1. Alexander, J.K., Hartley, L.H., Modelski, M., and Grover, R.F.: Reduction of stroke volume during exercise in man following ascent to 3100 m. altitude. J. Appl. Physiol. 23:849-858, 1967.
2. Burri, P.H. and Weibel, E.R.: Morphometric evaluation of changes in lung structure due to high altitude. In Porter, R. and Knight, J. (eds.): High Altitude Physiology; Cardiac and Respiratory Aspects. CIBA Foundation Symposium. Edinburgh and London, CIBA, 1971.
3. Coudert, J., Paz Zamora, M., Barragan, L., Briancon, L., Spielvogel, H., and Cudkowicz, L.: The regional distribution of pulmonary blood flow in normal high altitude dwellers. Respiration 32:189, 1975.
4. Cunningham, W.L., Becker, E.J., and Kreuzer, F.: Catecholamines in plasma and urine at high altitude. J. Appl. Physiol. 20:607, 1965.
5. DeGraff, A.C., Jr., Grover, R.F., Johnson, R.L., Jr., Hammond, J.W., Jr., and Miller, J.M.: Diffusing capacity of the lung in Caucasians native to 3100 m. J. Appl. Physiol. 29:71, 1970.
6. Grover, R.F., Reeves, J.T., Grover, E.B., and Leathers, J.E.: Muscular exercise in young men native to 3100 m altitude. J. Appl. Physiol. 22:555, 1967.
7. Halhuber, M.J. and Gabl, F.: In Weihe, W.H. (ed.): Physiological Effects of High Altitude. New York, Pergamon Press, 1964, p. 131.
8. Hartley, L.H., Alexander, J.K., Modelski, H., and Grover, R.F.: Subnormal cardiac output at rest and during exercise in residents at 3100 m altitude. J. Appl. Physiol. 23:849, 1967.
9. Hurtado, A.: Enfermedad de Los Andes. Algunas observaciones sobre el volúmen del Tórax, la capacidad vital y el metabolismo básico, en la altura. An. Fac. Med. Lima 14:166, 1928.
10. Hurtado, A.: Animals in high altitudes: Residents man. In: Handbook of Physiology, Sect. 4. Adaptation to Environment. Washington, D.C., American Physiological Society, 1964, pp. 843-860.
11. Kreuzer, F., Tenney, S.M., Mithoefer, J.C., and Remmers, J.: Alveolar arterial oxygen gradient in Andean natives at high altitude. J. Appl. Physiol. 419:13, 1964.
12. LeFrancois, R., Gautier, H., and Pasquis, P.: Ventilatory oxygen drive in acute and chronic hypoxia. Respir. Physiol. 4:217, 1968.
13. MacKinnon, P.C.B., Monk-Jones, M.L., and Fotherby, K.A.: Study of various indices of adrenocortical activity during 23 days at high altitude. J. Endocrinol. 26:555, 1963.
14. Milledge, J.S. and Lahiri, S.: Respiratory control in lowlanders and Sherpa highlanders at altitude. Respir. Physiol. 2:310, 1967.
15. Moncloa, F., Donayre, J., and Guerra-Garcia, R.: Endocrine studies at high altitude. II. Adrenal-cortical function in sea level natives exposed to high altitudes (4300 m for 2 weeks). J. Clin. Endocrinol. Metab. 25:1640, 1965.
16. Paz Zamora, M., Coudert, J., and LeFrancois, R.: Sensibilidad respiratoria al oxígeno con relación al tiempo de estadía en la altura (La Paz, 3500 m). Rev. Inst. Boliviano Biol. Altura 4(2):5, 1972.
17. Pugh, L.G.C.E.: Cardiac output in muscular exercise at 5800 m (19,000 ft). J. Appl. Physiol. 19:441, 1964.
18. Reeves, J.T., Halpin, J., Chon, J.E., and Daoud, F.: Increased alveolar-arterial oxygen difference during simulated high-altitude exposure. J. Appl. Physiol. 27:658, 1969.
19. Remmers, J.E. and Mithoefer, J.C.: The carbon monoxide diffusing capacity in permanent residents at high altitudes. Respir. Physiol. 6:233, 1969.
20. Severinghaus, J.W., Bainton, C.R., and Carcelen, A.: Respiratory insensitivity to hypoxia in chronically hypoxic man. Respir. Physiol. 1:308, 1966.
21. Severinghaus, J.W., Mitchel, R.A., Richardson, B.W., and Singer, M.M.: Respiratory control at high altitude suggesting active transport regulation of CSF pH. J. Appl. Physiol. 18:1155, 1963.

22. Tenney, S.M., Rahn, H., Stroud, R.C., and Mithoefer, J.C.: Adaptation to high altitude: Changes in lung volumes during the seven days at Mt. Evans, Colorado. J. Appl. Physiol. 5:607, 1953.

23. Zijlstra, W.C. and Van Kampen, E.J.: Clin. Chim. Acta 5:719, 1960.

5

Chemoreflexive Ventilatory Response at Sea Level in Subjects With Past History of Good Acclimatization and Severe Acute Mountain Sickness

SHU-TSU HU, SHAO-YUNG HUANG, SHOU-CHENG CHU, AND CHENG-FUNG PA

It is well established that an initial hyperpneic response is of great importance in the early acclimatization to high altitude, in which the peripheral chemoreceptors play a prominent role. It would be interesting to determine whether susceptibility to severe acute mountain sickness (AMS) is related to some impairment of peripheral chemoreception. In order to throw light on this problem, subjects with past histories of severe AMS (AMS-susceptible subjects) were sought and assessed at sea level for their peripheral chemoreceptor sensitivity, with a parallel study on subjects who had been shown to acclimatize normally at altitude (well-acclimatized subjects).

The observations were made at sea level (Shanghai, altitude 4 m). The AMS-susceptible group consisted of four lowlander males who suffered from severe AMS each time they ascended to 3680 m or higher in the past several years. In the other group were six lowlander males who were shown to acclimatize well on exposure to 4000 m or higher altitudes. All the subjects had returned from their last high altitude trip for at least 20 months before they were invited to perform the test, and they were in good health during the observation period (Table 5-1). They are all nonathletes and received no physical training either before or after their high altitude sojourns.

Methods

The 3-breath method follows in principle Dejour's original method (2,3), which selectively stimulates the peripheral chemoreceptors. The test gases used were N_2 (for hypoxia); 12.5% CO_2 plus 50% O_2 with the remainder, N_2 (for hypercapnia); and 12.5% CO_2 in N_2 (for hypoxic hypercapnia). Ventilation was recorded with a spirometer, and end tidal Po_2 and P_{CO_2} were monitored with a paramagnetic O_2 analyzer and an infrared CO_2 analyzer, respectively. The subject first breathed room air through a bag-in-box circuit. After his breathing became even, the 3-way stopcock was suddenly turned to connect the breathing valve with a reservoir of the test gas without the subject being aware of the manipulation. He was then allowed to inhale three tidal volumes of the test gas. The transient ventilatory response was measured and calculated from the sum of the fifth, sixth (which usually appears as the greatest tidal

Table 5-1. Past History of Acclimatization and AMS.

Subjects	Age	Altitude of ascents	Number of ascents	Number of years before test	Evident AMS symptoms	Activities at altitude
C.F.P.	37	4850 m	1 (with 4 reascents)	2 years	—	Active scientific work (ASW)
S.C.C.	37	4000–5900 m	5	13, 12, 8, 6, and 3 years	—	ASW
S.Y.H.	45	4000–5000 m	4	12, 11, 10 and 8 years	—	ASW
C.M.L.	43	4000–5000 m	2	11 and 10 years	—	ASW
T.S.C.	42	4000–5000 m	3	10, 9, and 8 years	—	ASW
C.C.C.	42	5900 m	1	12 years	—	ASW
C.N.C.	38	3680 m	2	12 and 11 years	Severe AMS; hospitalized during 2nd ascent	Incapacitated
		4300 m	2	8 and 4 years	Severe AMS; hospitalized during both ascents	Incapacitated
K.T.Y.	35	4500 m	2	7.5 and 8 years	Severe AMS; sent down	Incapacitated
H.C.Y.	25	4000 m	1	20 months	Severe AMS; sent down	Unconscious for 2 days
K.T.L.	40	5000 m	2	14 and 13 years	Severe AMS; pulmonary edema during 2nd ascent	Incapacitated

response to the test gas), and seventh tidal volumes divided by the time elapsed during these three breaths. The calculated results were expressed as minute ventilation volumes and compared with the pretest values to get the relative percentage change of minute ventilation ($\Delta V\%$).

All the subjects received the 3-breath test. An additional single vital capacity breath test (4), which elicits more distinct chemoreceptive ventilatory response with the test gas, was performed on several subjects.

Results

Table 5-2 summarizes the data from the two groups undertaking the 3-breath test. It can be seen that in all of the well-acclimatized subjects ventilation increased markedly after inhalation of hypoxic, hypercapnic, or hypoxic hypercapnic test gases. The average ΔV were 54, 55, and 159%, respectively. The results showed distinct responsiveness of the peripheral chemoreception to the test gases. In contrast, the AMS-susceptible group showed remarkably blunted responses to all of the stimuli. There was only feeble ventilatory response to hypoxia. The increase in ventilation during hypercapnia was distinctly small, and the superimposed hypoxia failed to greatly augment the hypercapnic response as seen in the case of the well-acclimatized subjects.

Parts of both groups of subjects were again invited 1.5 years later to repeat the test under the same conditions, except that five instead of three breaths of N_2 inhalation were used in order to intensify the

Table 5-2. 3-Breath Test.

Subjects	N_2 (hypoxia)		12.5% CO_2, 51% O_2, 36.5% N_2 (hypercapnia)		12.5% CO_2 in N_2 (hypoxic hypercapnia)		
	$P_{ET}O_2$ (mmHg)	ΔV (%)	$P_{ET}CO_2$ (mmHg)	ΔV (%)	$P_{ET}O_2$ (mmHg)	$P_{ET}CO_2$ (mmHg)	ΔV (%)
Well acclimatized							
C.F.P.	44	33	63	58	40	64	279
S.C.C.	44	60	64	61	49	59	91
S.Y.H.	46	94	60	59	40	61	249
C.M.L.	49	66	65	39	49	60	117
T.S.C.	29	48	70	62	39	71	102
C.C.C.	30	21	67	53	46	69	115
Average	41	54	65	55	44	64	159
Severe AMS							
C.N.C.	62	−6	54	23	63	59	23
K.T.Y.	61	2	66	19	64	66	45
H.C.Y.	29	15	69	20	29	72	55
K.T.L.	29	19	65	11	29	65	77
Average	45	8	64	18	46	66	50
Repetition tests[a]							
Well acclimatized							
C.F.P.	34	96	63	119	45	60	185
S.C.C.	31	44	58	31	40	51	109
S.Y.H.	44	378	60	84	59	57	288
C.C.C.	29	95	63	72	29	61	123
Severe AMS							
C.N.C.	29	−13	51	−1	41	52	20
K.T.Y.	51	5	56	3	69	64	25

[a] Five breaths instead of three breaths of N_2 inhalation were used during the repetition tests.
$P_{ET}O_2$ = partial pressure of O_2 in end-tidal expiratory gas.
$P_{ET}CO_2$ = partial pressure of CO_2 in end-tidal expiratory gas.

Table 5-3. *Single Vital Capacity Breath Test.*

Subjects	N_2 (hypoxia)		12.5% CO_2, 51% O_2, 36.5% N_2 (hypercapnia)		12.5% CO_2 in N_2 (hypoxic hypercapnia)		
	$P_{ET}O_2$ (mmHg)	ΔV (%)	$P_{ET}CO_2$ (mmHg)	ΔV (%)	$P_{ET}O_2$ (mmHg)	$P_{ET}CO_2$ (mmHg)	ΔV (%)
Well acclimatized							
C.F.P.	29	41	81	96	29	86	508
S.C.C.	29	87	82	60	29	81	498
S.Y.H.	44	194	81	342	49	81	a
Severe AMS							
C.N.C.	29	23	65	24	39	69	17
K.T.Y.	42	8	85	68	60	78	90
Repetition tests							
C.N.C.	44	−2	67	18	29	65	38
K.T.Y.	29	48	69	64	34	69	90

[a] Violent ventilatory response intolerable.

hypoxic stimulus within the alveoli. It can be seen from the lower part of Table 5-2 that the situation was very similar to that of the previous test. Noteworthy is the even more striking contrast between the distinct hyperventilation in the well-acclimatized and the practical absence of response in the AMS-susceptible subjects, when five instead of three breaths of the hypoxic test gas were inhaled.

Table 5-3 shows the results with the single vital capacity breath test which was applied to a part of the two groups. There was a much stronger response in all three well-acclimatized subjects studied. In one of them the hyperpneic response to the hypoxic hypercapnic test gas was so intolerably violent that the test had to be stopped right away. On the other hand, in the two AMS-susceptible subjects studied the ventilatory responses were very much weaker, especially in the case of the hypoxic hypercapnic gas inhalation. This low responsiveness was confirmed with repetition of the test on the same subjects 1.5 years later (Table 5-3).

Discussion

The observation of the consistent feeble responses in all four AMS-susceptible subjects as assessed by more than one method does not seem to be fortuitous. This preliminary study brings to attention the possibility of extra-low responsiveness of the chemoreflexive mechanisms to hypoxia and hypercapnia to be a factor in susceptibility to AMS. Of course the study needs further amplification before any definite conclusion can be drawn. It would be tempting to see if detection of extra-low chemoreflexive ventilatory sensitivity could be helpful in predicting the occurrence of severe AMS in an individual.

It has been reported (1,5) that athletes tend to show low ventilatory responsiveness, attributable either to familial influences or to an athletic training effect. The present study was concerned mainly with the "innate" chemoreflexive attributes of the subjects, and the training factor was purposely excluded from this study by restricting the selected subjects to nonathletes who received no physical training for their ascents or in daily life.

The contrasting difference in the magnitude of ventilatory response between the two groups of subjects assessed by the 3- or 5-breath and single vital capacity breath methods appears to be based chiefly upon differences of peripheral chemoreceptor sensitivities, although the methods may possibly be contaminated by some central stimulation or secondary events. Unfortunately, no test has been done in this study to evaluate the ventilatory response of central origin, which nevertheless cannot be ruled out.

References

1. Byrne-Quinn, E., Weil, J.V., Sodal, I.E., Filley, G.F., and Grover, R.F.: Ventilatory control in the athlete. J. Appl. Physiol. 30:91, 1971.
2. Dejours, P.: Chemoreflexes in breathing. Physiol. Rev. 42:335, 1962.
3. Dejours, P.: Intérêt méthodologique de l'étude d'un organisme vivant à la phase initiale de rupture d'un équilibre physiologique. C. R. Acad. Sci. (Paris) 245:1946, 1957.
4. Gabel, R.A., Kronenberg, R.S., and Severinghaus, T.W.: Vital capacity breaths of 5% or 15% CO_2 in N_2 or O_2 to test carotid chemosensitivity. Respir. Physiol. 17:195, 1973.
5. Miyamura, M., Yamashina, T., and Honda, Y.: Ventilatory responses to CO_2 rebreathing at rest and during exercise in untrained subjects and athletes. Jpn. J. Physiol. 26:245, 1976.

6

Dysoxia (Abnormal Cell O_2 Metabolism) and High Altitude Exposure

EUGENE D. ROBIN

The primary abnormality in high altitude exposure is the development of O_2 depletion. Oxygen depletion is a common abnormality found in a large variety of biologic systems under many different circumstances. High altitude exposure thus provides a specific example of a more general and extremely important group of disorders.

An analysis of the basic nature of the abnormalities associated with O_2 depletion indicates that there is a single central abnormality. Oxygen depletion leads to alterations of cellular O_2 utilization which are fundamentally biochemical in nature. Despite the central position of altered O_2 utilization in O_2 depletion states, until recently there has been no term which specifically denotes abnormal cellular O_2 metabolism. The term "dysoxia," defined as abnormal cellular O_2 metabolism, provides a single word for this concept (24a,26).

Given such a definition, this discussion will focus on five aspects of abnormal cellular O_2 metabolism: (1) the causes of dysoxia; (2) a summary of the biochemical pathways involving O_2 utilization; (3) stages of O_2 depletion; (4) substrate (glucose) supply and O_2 depletion; and (5) nature of the mechanisms which are evoked to cope with O_2 depletion.

Causes of Dysoxia

Table 6-1 summarizes the causes of dysoxia. Altered cellular utilization of O_2 may be grouped into three general types. One general type is *hypoxic dysoxia*, defined as those conditions characterized by abnormal cellular utilization because of cellular O_2 depletion. Oxygen depletion may develop as a result of decreases in arterial O_2 tension and/or saturation; as a result of decreases in red cell mass; or as a result of abnormalities of blood, systemic capillary, interstitial, plasma membrane, or intracellular O_2 transport. Exposure to high altitude is but one of many forms of hypoxic dysoxia operating to reduce arterial P_{O_2} and saturation.

The second general category of abnormal cell O_2 metabolism is *normoxic dysoxia*. This includes those disorders in which the supply of O_2 is entirely normal, but O_2 utilization is altered because the subcellular sites at which O_2 is used have functional and/or structural abnormalities. For example, over 20 separate mi-

Table 6-1. Causes of Dysoxia.

Hypoxic dysoxia

1. Hypoxemia
   Reduced arterial P_{O_2} or S_{O_2}
   Reduced red cell mass
2. Abnormal blood O_2 transport
   Reduced cardiac output
   Maldistribution of cardiac output
   Ischemia (reduced regional blood flow)
   Systemic arteriovenous communications
   Increased affinity of hemoglobin for O_2
3. Abnormalities of systemic capillaries
4. Reduced interstitial O_2 transport
5. Abnormal plasma membranes
6. Abnormal intracellular O_2 transport

Normoxic dysoxia

1. Intrinsic mitochondrial disease
   Increased mitochondrial O_2 consumption
   Decreased mitochondrial O_2 consumption
2. Disorders of endoplasmic reticulum, Golgi body, nucleus, liposomes, etc. (chronic granulomatous disease)

Hyperoxic dysoxia

1. Free radical injury
2. H_2O_2 injury
3. cAMP overload

---

tochondrial diseases have already been described in which there are abnormalities of mitochondrial O_2 utilization because of intrinsic mitochondrial disease (25). Chronic granulomatous disease is an example of extramitochondrial normoxic dysoxia. This disorder is characterized by recurrent bacterial infections in children. The granulocytes from affected individuals fail to show the usual burst of O_2 consumption during phagocytosis. Molecular O_2 is not converted into H_2O_2 at a normal rate. This leads to deficient release of H_2O_2 which is bactericidal. As a result, there is impaired bacterial killing (6).

The third general category is *hyperoxic dysoxia*, defined as abnormalities of cellular O_2 utilization caused by activities of O_2 which are higher than normal. *Hyperoxic dysoxia* is often described as O_2 toxicity. The exact mechanisms by which increased activities of O_2 produce cell injury are not clear. However, attention has focused on direct injury to cells produced by free radicals of O_2 (O_2^-, OH, singlet O) (9) and hydrogen peroxide (H_2O_2), or hyperoxic augmentation of inflammatory or metabolic free radical or H_2O_2 injury (22). Recent studies suggest that abnormal cell function may result from chronic increases in cAMP evoked by high O_2 activities (cAMP overload), leading to inappropriate phosphorylation of regulatory and structural proteins (11).

Biochemical Pathways Involving O_2 Utilization

Cellular O_2 utilization can be divided into two components. Approximately 80% of total O_2 utilization occurs within the mitochondrion. Molecular O_2 combines with substrate to provide free energy, which in turn is used for the active transport of various ions. As elegantly shown by Mitchell (21), protons are pumped from the inside to the outside of the mitochondrial membrane. The free energy provided by substrate oxidation is used to maintain a H^+ gradient across the mitochondrial membrane. As H^+ diffuses back into the mitochondrion down a favorable electrochemical gradient, the free energy which is made available is released and generates ATP as follows:

$$\text{Energy} + \text{ADP} + P_i \rightleftarrows \text{ATP}$$

The ATP so generated is used as the major source of energy for most biologic processes.

Several aspects of these processes should be emphasized. The free energy liberated by oxidation of substrate is not directly used for the generation of ATP. Thus, theoretically, any process which produces a H^+ gradient across the mitochondrial membrane would provide continued generation of ATP. This could occur, for example, even in the absence of molecular O_2. Since phosphorylation of ATP depends on the generation of a H^+ gradient, agents which dissipate the gradient will result in oxygen utilization which is not linked to ATP production. For example, the agent dinitrophenol is an ionophore which increases the permeability of the mitochondrial membrane to H^+, thus tending to abolish the H^+ gradient. This process leads to a large increase in O_2 consumption (and heat) without appropriate ATP generation (21).

Free energy directly obtained from oxidation of substrate is also used in the active transport of other ions, for example, Ca^{2++} (17). Given an increase in the amount of nonproton ion transport, less energy is available for H^+ pumping and thus less energy is available for ATP production. Disorders involving such a mechanism have been described. For example, in Luft's syndrome, there is abnormally high permeability of the mitochondrial membrane to Ca^{2++}. Ca^{2++} leaks into the inner mitochondrial compartment and is actively pumped out. This results in excess O_2 consumption not linked to ATP generation and excess heat production (8).

An important question concerns the degree of O_2 depletion which evokes abnormalities in mitochondrial O_2 utilization. Classical approaches based on studies of isolated mitochondria have suggested that depletion must be very severe. This conclusion is based on the finding that the affinity of the terminal enzyme complex in the mitochondrial electron transport chain (cytochrome aa_3) for O_2 is very high (low K_m, i.e., concentration of O_2 at which the velocity of reaction is half maximal (2)). Thus, it has been accepted that ATP generation does not become impaired until P_{O_2} is, say, less than 1 torr. This view has recently been challenged by studies in intact tissues and cells which suggest that in intact cells, the K_m of cytochrome aa_3 is considerably higher for O_2 than previously determined (15). If so, the implication would be that more moderate degrees of O_2 depletion could result in deficient ATP generation.

With respect to the other general biochemical functions of molecular O_2, approximately 20% of utilization takes place extramitochondrially. Extramitochondrial utilization of O_2 takes place in the nucleus, endoplasmic reticulum, Golgi body, peroxisome, etc. Oxygen functions in these pathways as an obligatory substrate in a series of biosynthetic, biodegradative, detoxification, and miscellaneous reactions (3). For a number of these reactions, the K_m appears to be quite high. Thus, even moderate O_2 depletion would be anticipated to produce abnormalities. For example, the enzyme, tyrosine hydroxylase, is the rate-limiting step in the biosynthesis of dihydroxyphenylalanine (DOPA) and thus controls the rate of synthesis of the catecholamines. Molecular O_2 is required for the reaction. The K_m of O_2 for the reaction is about 13 torr (in a relatively physiologic preparation), as shown in isolated synaptosomes by Davis (7).

It would be anticipated that catecholamine synthesis would be abnormal with even moderate reductions in ambient P_{O_2}, e.g., those found at high altitude. Anaerobiosis has also been shown to abolish the generation of superoxide ion and H_2O_2 from O_2 by granulocytes and macrophages during phagocytosis (5,22). It is possible that similar but milder abnormalities might occur in subjects at high altitude and impair antibacterial defense mechanisms.

Impairment of extramitochondrial O_2-requiring reactions may well account for the many so-called "nonspecific" manifestations found in most subjects with mild O_2 depletion. We can no longer accept the dictum that "hypoxia wrecks the machinery" (10). There are mild and easily reversible forms of hypoxia. The impact of O_2 depletion on O_2-requiring nonmitochondrial reactions is obviously an important frontier area for future studies. Alterations of these reactions are probably involved in the pathogenesis of various forms of high altitude disease. It may be speculated that the biochemical basis of abnormal function with high altitude exposure and adaptations to high altitude usually do not involve bioenergetics. Rather, there are alterations of neuroendocrine, neurohumoral, endocrine, and paracrine factors as well as other O_2-dependent reactions which are affected by moderate O_2 depletion.

Stages of O_2 Depletion

A systematic approach to O_2 depletion can be facilitated by classifying the general stages that occur during progressive O_2 depletion. Such a classification might include the following stages:

1) Basal or control stage: The biologic unit (cell, organ, intact organism) has an adequate O_2 supply for a sufficiently long period so that biochemical/physiologic functions are normal.
2) Aerobic stage: O_2 depletion has developed but the biologic unit maintains normal cellular O_2 utilization by utilizing O_2 stores. This stage may be considered as representing hypoxia without dysoxia.
3) Aerobic-anaerobic transition stage: O_2 stores become sufficiently depleted so that biochemical/physiologic mechanisms are evoked which are anaerobic in nature. This stage is the most common one associated with high altitude exposure.
4) Anaerobic stage: O_2 is totally depleted and the biologic unit now survives entirely on anaerobic mechanisms. This stage does not invariably occur and is only found with very profound O_2 depletion.
5) Anaerobic-aerobic transition stage: O_2 supply is returned toward normal. The biologic unit now functions aerobically but also repairs abnormalities incurred in stages 3 and 4.
6) Basal or control stage: All abnormalities evoked by O_2 depletion are entirely reversed and the unit now returns to stage 1.

This analysis can be applied to a large number of different forms of O_2 depletion. It applies, for example, to an ischemic brain, to normal mammalian skeletal nuscle during severe exercise, to mammals under diving conditions, to vertebrates such as the freshwater turtle during prolonged diving, and in the present context, to high altitude exposure.

The duration of the various stages differs in each circumstance and the mechanisms involved in each stage are often different, but the general nature of the outline is constant. There are a series of adaptations which operate in a given stage which buffer the effects of O_2 depletion. The nature of these adaptations can only be understood in the context of a given stage. Finally, stage 5 may require a long period when stages 2-4 are relatively brief. As a result, biologic systems often require prolonged periods of recovery before a basal state is regained. During stage 5, the time required for a given adaptation to be reversed may impose new biologic handicaps on the system (postadaptive cost) (29).

Substrate (Glucose) Supply

A critical feature of O_2 depletion, especially in stages 3-5, involves the problem of provision of adequate substrate (glucose). This is particularly important with respect to bioenergetics.

A major advantage of oxidative phosphorylation is that relatively small amounts of glucose are required to produce a given amount of ATP. During oxidative phosphorylation, the oxidation of 1 mol of glucose generates 36 mol of ATP. During glycolysis, the oxidation of 1 mol of glucose provides only 2 mol of ATP. It is scarcely surprising that during the recovery period (stage 5), there are important mechanisms which subserve the conservation of glucose rather than directly involving O_2 metabolism (27). In the context of high altitude exposure, it would be anticipated that nutritional status, adequacy of glucose supply, and the ability to convert other substrates into glucose would play important roles in determining the pattern of response.

Mechanisms for Coping with O_2 Depletion

Mechanisms evoked by O_2 depletion can be classified as follows:

I) Mechanisms which increase O_2 transport/supply
   a) Obvious
      1) Increased ventilation
      2) Increased red cell mass
      3) Altered hemoglobin affinity
      4) Increased cardiac output
      5) Increased capillary bed
   b) Not obvious
      1) O_2 conservation for key cells
      2) Substrate conservation for key cells
II) Altered cell metabolism: mechanisms involving bioenergetics and ATP supply
   a) Increased oxidative phosphorylative capacity at a given level of O_2 availability
      1) Extramitochondrial acidosis
      2) Increased mitochondrial enzymes (exercising, mammalian skeletal muscle)
   b) Increased glycolysis or glycolytic capacity
      1) Acute (increased enzyme activity produced by low molecular weight regulators: Pasteur effect).
      2) Chronic (increased content of rate-limiting enzymes—most eukaryotic cells): increased biosynthesis; decreased biodegradation
      3) Enzymes with altered kinetics (diving turtle): altered genomic expression
   c) Non-O_2–nonglycolytic ATP generation (phosphoenolpyruvate pathway in bivalves)
   d) Decreased energy requirements
      1) Decreased temperature
      2) Diving turtle
      3) Hibernation
      4) Use of barbiturates
      5) Fetal brain
   e) Glucose conservation (glucagon elevation in postdiving period in aquatic mammals)
III) Altered cell metabolism: mechanisms involving nonbioenergetic functions of O_2

The physiologic mechanisms (listed under I) are generally well known and are extensively discussed by others at this meeting. No further description is required.

Alterations in Cell Metabolism

Increased oxidative phosphorylative capacity at a given level of O_2 availability can occur under several circumstances. As previously noted, mitochondrial ATP generation depends on a proton gradient between the extra- and intramitochondrial phase. At a given level of ADP and P_i, any process which increases $(H^+)e/(H^+)m^1$ should provide ATP. Theoretically, this would occur even in the total absence of O_2. For example, brisk intracellular lactic acidosis could provide such a gradient and generate ATP.

Another example is provided by chronic exercise in mammalian skeletal muscle. It has been shown that chronic exercise leads to increased mitochondrial protein and mass and a specific increase in the content of cytochrome aa_3, out of proportion to increases in mitochondrial protein in the exercised muscle. The increase in cytochrome aa_3 content is paralleled by an increase in mitochondrial O_2 consumption.

---

[1]$(H^+)e$, extramitochondrial proton concentration; $(H^+)m$, mitochondrial proton concentration.

Glycolytic enzyme activities are not increased and glycolytic capacity is presumably not increased (13). These changes should result in an increased rate of maximal O_2 uptake to lactate production. Similar findings have been reported in respiratory muscles during chronic hyperventilation (16). Perhaps the increased $\dot{V}_{O_2max}$/lactate ratios reported by Cerretelli (4) in adapted vs nonadapted subjects exposed to high altitude during exercise reflect similar mechanisms.

Increases in the rate of glycolysis or glycolytic capacity can occur as a result of several mechanisms. The best known is the Pasteur effect, an acute increase in the rate of glycolysis associated with O_2 depletion. Although the mechanism is not entirely clear, the Pasteur effect is probably mediated by increased phosphofructokinase (PFK) activity (the rate-limiting step in glycolysis) which results from changes in ADP, P_i, etc. (23). It should be emphasized that the increase in the rate of glycolysis does not depend on an increase in enzyme content. Indeed, the upper limit of augmented glycolysis may be set by the cellular content of phosphofructokinase (28).

Our laboratory has shown in a number of different cellular systems that chronic hypoxic dysoxia is associated with increased activities of phosphofructokinase and pyruvate kinase (32). It appears that the increased activity is related to an increase in enzyme content, which must be related to either increased biosynthesis or decreased biodegradation of the two enzymes. There is a parallel increase in glycolytic capacity of the cells. Thus, the increased enzyme content is functionally important.

Hypoxic dysoxia may result in changes in the form of glycolytic enzymes (isozymes). A given isozyme may provide a more favorable kinetic pattern for providing energy through glycolysis. In mammals, pyruvate kinase generally exists in the form of four isozymes, each with specific kinetic properties. One of these, pyruvate kinase M_1, shows Michaelis-Menton kinetics with phosphoenolpyruvate and is not inhibited by phenylalanine. Such an enzyme is well adapted to tissues with a high requirement for generating ATP through glycolysis (18,34).

Another example is provided by studies in the freshwater turtle. This animal can dive for days with P_{O_2}s equal to zero and survive (14,30). Storey and Hochachka (33) showed that turtle heart PFK was insensitive to inhibition by ATP, whose role in inhibitory control was taken over by creatine phosphate. Such an enzyme supplies a highly sensitive control mechanism for glycolysis during the aerobic-anaerobic transition (33). Changes in isozyme pattern on a species level would be genomically controlled. Intraspecies changes presumably operate by means of altered enzyme biosynthesis or biodegradation.

Pathways for the anaerobic generation of ATP other than that of classical glycolysis have been reported in a number of invertebrates. In these pathways, there is the simultaneous use of both glucose and amino acids. Lactate is either a minor product or may not be produced at all. For example, in the oyster and other bivalves (which periodically develop dysoxia), the process of glycolysis proceeds only to phosphoenolpyruvate. Pyruvate kinase is lacking and alanine, propionate, succinate, and an unknown compound formed from pyruvate and alanine rather than lactate accumulate. Approximately 8 mol of ATP are formed from each mole of oxidized glucose (12).

One of the most intriguing and important mechanisms for dealing with hypoxic dysoxia is a decrease in energy requirements. Unfortunately, this mechanism can only be studied indirectly. A decrease in energy requirements is usually deduced indirectly from decreases in energy supply. However, this deduction is only valid during a prolonged steady state. Despite the practical problems involved in measuring precise energy requirements, there are a number of examples which presumably reflect decreases in energy requirements. Decreases in ambient temperature do reduce energy requirements but obviously reduce energy supply as well. There is indirect but convincing evidence that prolonged diving in the turtle is associated with decreased energy requirements (14,28). Hibernation is associated with decreased energy supply and this reduction is almost certainly not solely dependent on the reduction in body temperature (19). Fetuses and neonates of a number of mammalian species show enhanced resistance of the brain to hypoxic dysoxia produced by N_2 inhalation. During maturation, increases in rate-limiting enzymes of both oxidative phosphorylation and glycolysis occur which parallel decreased resistance to anoxia. The enhanced resistance of the fetal brain to total anoxia must then reflect low energy requirements during anaerobiosis (23).

Recently attention has been focused on the use of barbiturates to maintain cerebral integrity following ischemic brain insults. Although barbiturates may produce a number of alterations which explain enhanced survival (20,31), reduction in energy requirements is one possible mechanism.

The importance of adequate substrate in meeting the problems of hypoxic dysoxia has already been emphasized. In dogs made acutely hypoxic by breathing low-O_2 mixtures, there is an increase in circulating glucagon with associated hyperglycemia, which appears to be mediated by alpha-adrenergic transmitters (1). This model, incidentally, is very close to that of acute high altitude exposure.

In the harbor seal, following a 6-min dive, there is a sharp increase in plasma glucose levels which is mediated by an increase in circulating glucagon. However, plasma insulin levels do not change. This has the effect of conserving glucose for the brain and prevents a decreased brain glucose supply related to glucose uptake by the glucose-depleted peripheral tissues. Preloading the animal with glucose prevents the rise in glucagon during the post-diving period (27).

Studies of the metabolic adaptations evoked by hypoxic dysoxia are relatively new. We know little about this area in subjects at high altitude. Essentially nothing is known about the mechanisms by which biologic units cope with abnormalities of nonbioenergetic functions of O_2 during O_2 depletion.

As previously emphasized, of the many unanswered questions, this specific area may be expected to provide the greatest insight with respect to intimate details of the changes during high altitude exposure, adaptations evoked by these changes, and effects of hypoxic dysoxia in disease.

We may summarize as follows:

1) Oxygen depletion is harmful to biologic systems because it produces abnormalities of cellular O_2 utilization (dysoxia). Oxygen depletion is fundamentally a biochemical disorder.
2) Oxygen is used within the cell for both energy transduction and for the biosynthesis, biodegradation, and detoxification of important substances.
3) There is an orderly sequence of changes produced in biologic systems by progressive O_2 depletion.
4) Substrate availability is compromised by O_2 depletion.
5) There is a wide variety of adaptive mechanisms which operate both acutely and chronically to blunt the impact of O_2 depletion.
6) The major biochemical abnormalities found in high altitude exposure and the biochemical basis for high altitude adaptations probably do not involve bioenergetics. Rather, these probably involve O_2-dependent biosynthesis and biodegradation of neuroendocrine, neurohumoral, endocrine, and paracrine agents as well as other O_2-dependent reactions.
7) The biochemistry of O_2 depletion is a frontier area for future work in high altitude physiology and pathophysiology.

Acknowledgment

This work was supported in part by a grant (HL 23701) from the Lung Division, National Heart, Lung and Blood Division of the National Institutes of Health.

References

1. Baum, D., Porte, D., Jr., and Ensinck, J.: Hyperglucagonemia and the α-adrenergic receptor in acute hypoxia. J. Appl. Physiol. (in press).
2. Bienfait, H.F., Jacobs, J.M.C., and Slater, E.C.: Mitochondrial oxygen affinity as a function of redox and phosphate potentials. Biochim. Biophys. Acta 376:446, 1975.
3. Bloch, K.: Oxygen and biosynthetic patterns. Fed. Proc. 21:1058, 1962.
4. Cerretelli, P.: O_2 breathing at altitude: effects on maximal performance. Hypoxia Symposium, Banf, Canada, 1979.
5. Curnette, J.T. and Barbior, B.M.: Effects of anaerobiosis and inhibitors on O_2^- production by human granulocytes. Blood 45:851, 1975.
6. Curnette, J.T., Whitten, D.M., and Barbior, B.M.: Defective superoxide production by granulocytes from patients with chronic granulomatous disease. N. Engl. J. Med. 290:593, 1974.
7. Davis, J.N.: Synaptosomal tyrosine hydroxylation: affinity for oxygen. J. Neurochem. 28:1043, 1977.
8. Dimauro, S., Bonilla, E., Lee, C.P., Schotland, D.L., Scarpa, A., Conn, H.L., and Chance, B.: Luft's disease: Further biochemical and ultrastructural studies of skeletal muscle in the second case. J. Neurol. Sci. 27:217, 1976.
9. Fridovich, I.: Superoxide dismutases. Ann. Rev. Biochem. 44:147, 1975.
10. Haldane, J.S.: Respiration. New Haven, Yale University Press, 1927.
11. Hance, A.J., Theodore, J., Robin, E.D., and Raffin, T.: Cyclic AMP overload and cell O_2 injury. (in press).
12. Hochachka, P.W. and Sumero, G.N.: Strategies of Biochemical Adaptation. Philadelphia, Saunders, 1976.
13. Holloszy, J.O.: Biochemical adaptations in muscle. J. Biol. Chem. 242:2278, 1967.
14. Jackson, D.C.: Metabolic depression and O_2 depletion in the diving turtle. Am. J. Physiol. 241:503, 1968.
15. Jöbsis, F.F.: Non-invasive monitoring of cerebral and myocardial oxygen sufficiency and circulatory parameters. Science 198:1264, 1977.
16. Keens, T.G., Chen, V., Patel, P., O'Brien, P., Levinson, H., and Ianuzzo, C.D.L.: Cellular adaptations of the ventilatory muscles to a chronically increased ventilatory load. J. Appl. Physiol. 44:905, 1978.
17. Lehninger, A.L.: The Mitochondrion: Molecular Basis of Structure and Function. Menlo Park, Calif., Benjaman, 1964.
18. Lincoln, D.R., Black, J.A., and Rittenberg, M.P.: The immunological properties of pyruvate kinase. II. The relationship of the human erythrocyte isozyme to the human liver isozymes. Biochim. Biophys. Acta 410:279, 1975.
19. Malan, A.: Hibernation as a model for studies on thermogenesis and its control in effectors of thermogenesis. In Girardier, L. and Seydoux, J. (eds.): Birkhauser, 1978, pp. 303–314.
20. Michenfelder, J.D., Milde, J., and Sundt, T.: Cerebral protection by barbiturate anesthesia. Arch. Neurol. 33:345, 1976.
21. Mitchell, P.: Chemiosmotic coupling in oxidative and photosynthetic phosphorylation. Biol. Rev. 41:445, 1965.
22. Nathan, C.F., Silverstein, S.C., Brukner, L.H., and Cohn, Z.: Extracellular cytolysis by activated macrophages and granulocytes. II. Hydrogen peroxidase mediator of toxicity. J. Exp. Med. 149:100, 1979.
23. Phillips, J.R., Theodore, J., and Robin, E.D.: Comparative enzymatic maturation of post natal rabbit brain and heart: relation to organ development and resistance to hypoxia. Clin. Res. 23:351, 1975.
24. Racker, E.: From Pasteur to Mitchell: a hundred years of bioenergetics. Fed. Proc. 39:210, 1980.
24a. Robin, E.D.: Dysoxia: Abnormal tissue O_2 utilization. Arch. Intern. Med. 137:905, 1977.
25. Robin, E.D.: Dysoxia and intrinsic mitochondrial diseases. In Robin, E.D. (ed.): Extrapulmonary Manifestations of Respiratory Disease. New York, Dekker, 1978, pp. 171–184.
26. Robin, E.D.: Overview: dysoxia—ab-

normalities of tissue O_2 use. In Robin, E.D. (ed.): Extrapulmonary Manifestations of Respiratory Disease. New York, Dekker, 1978, pp. 3-12.
27. Robin, E.D., Ensinck, J., Newman, A., Lewiston, N.J., Cornell, L., Davis, R., and Theodore, J.: Glucoregulation in diving mammals. (in press).
28. Robin, E.D., Lewiston, N.J., Newman, A., Simon, L.M., and Theodore, J.: Bioenergetic pattern of turtle brain and resistance to profound loss of mitochondrial ATP generation. Proc. Natl. Acad. Sci. 76:3922, 1978.
29. Robin, E.D., Simon, L.M., and Theodore, J.T.: Postadaptive cost: Impaired energy metabolism in hypoxically adapted lung cells upon return to normoxia. Trans. Assoc. Am. Physicians 61:388, 1978.
30. Robin, E.D., Vester, J.W., Murdaugh, H.B., Jr., and Millen, J.E.: Prolonged anaerobiosis in a vertebrate: Anaerobic metabolism in the freshwater turtle. J. Cell Comp. Physiol. 63:287, 1964.
31. Siesjo, B.K.: Brain Energy Metabolism. New York, Wiley, 1978.
32. Simon, L.M., Theodore, J., and Robin, E.D.: Regulation of biosynthesis/biodegradation of oxygen-related enzymes by molecular oxygen. In Robin, E.D. (ed.): Extrapulmonary Manifestations of Respiratory Disease. New York, Dekker, 1978.
33. Storey, K.B. and Hochachka, P.W.: Enzymes of energy metabolism from a vertebrate facultative anaerobe, *Pseudemys scripta*, turtle heart phosphofructokinase. J. Biol. Chem. 294:1417, 1974.
34. Susor, W.A. and Rutter, W.J.: Method for the detection of pyruvate kinase, aldolase, and other pyridine nucleotide-linked enzyme activities after electrophoresis. Anal. Biochem. 43:147, 1971.

II. Oxygen Affinity and Oxygen Unloading

7

Minimal P_{O_2} in Working and Resting Tissues

D.W. LÜBBERS

Introduction: Simulation of O_2 Supply by the Krogh Model

Oxygen is supplied to the tissues by convection, i.e., by blood perfusion, and by diffusion. The energy for convection is produced by the heart. The energy for diffusional transport originates from the O_2 gradient between capillaries and tissues which is built up by the oxygen consumption of the tissue oxidases (18). The local tissue P_{O_2} varies with both flow and O_2 transport capacity of the blood, as well as with the distance from the arterial supply and the O_2 consumption of the tissue. Thus, an oxygen pressure field develops around the capillary which characterizes the oxygen supply of the tissue. To understand the effect of the different parameters which influence the oxygen supply of the tissue, an analysis with a simplified model, the Krogh model, is useful (12,13,28,35) (Fig. 7-1). It considers a single capillary which supplies a cylindric space of constant radius and homogeneous oxygen consumption. In such a capillary the O_2 content decreases linearly from the arterial to the venous end. The steepness of the decrease depends on the oxygen transport capacity per time, $\dot{T}(O_2)$, and the tissue respiration, $\dot{A}(O_2)$. The oxygen consumed by the tissue, $\dot{A}(O_2)$, is taken from the oxygen content of the inflowing blood $[O_2]_a$. If the blood leaves the tissue with an oxygen content $[O_2]_v$,

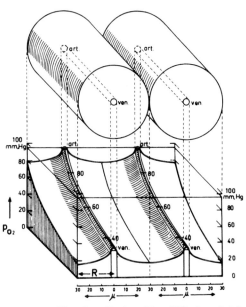

Fig. 7-1. Krogh cylinder (P_{O_2} diffusion field). P_{O_2} decreases from arterial to venous end of capillary. With homogeneous tissue O_2 consumption the decrease follows the O_2 dissociation curve. Within the tissue, P_{O_2} decreases perpendicularly to the capillary (Eq. 7-5). From Thews, G.: In Schadé, J.P. and McMenemy, W.H. (eds.): Selective Vulnerability of the Brain in Hypoxia. Oxford, Blackwell, 1963.

the product flow $\cdot ([O_2]_a - [O_2]_v)$ equals the tissue respiration.

$$\dot{A}(O_2) = ([O_2]_a - [O_2]_v) \cdot F$$
$$= [O_2]_{AVD} \cdot F \quad [7\text{-}1]$$

$$[O_2]_v = [O_2]_a - \frac{\dot{A}(O_2)}{F} \quad [7\text{-}2]$$

Equation 7-2 shows that the venous O_2 content is directly proportional to the difference between the arterial O_2 content and the quotient "tissue respiration divided by flow (F)." To maintain the venous oxygen content the following processes may occur: (1) if the respiration changes, the flow has to change by the same fractional amount; and (2) if the arterial O_2 content decreases, the quotient $\dot{A}(O_2)/F$ has to decrease or, with constant $\dot{A}(O_2)$, the flow has to increase correspondingly.

O_2 Pressure Field and Its Determinants

At first glance the situation seems to be very simple but it becomes more complicated when combining the convectional with the diffusive transport, because the oxygen transport within the tissue depends on the oxygen pressure and not on the oxygen concentration. The O_2 content depends on the P_{O_2} in the following way:

$$[O_2] = \alpha \cdot P_{O_2} \quad [7\text{-}3]$$

In blood α (Bunsen solubility coefficient) varies with the hemoglobin content and additionally, corresponding to the oxygen dissociation curve, with the P_{O_2}. Also, in tissue, α is different in different tissue constituents. Therefore, in such a system the oxygen flux, I_{O_2} (mol $\cdot$ cm^2 $\cdot$ s^{-1}), has to be written as

$$I_{O_2} = -D \frac{\Delta(\alpha \cdot P_{O_2})}{\Delta x} = -\alpha D \frac{\Delta P_{O_2}}{\Delta x} \quad [7\text{-}4]$$

(D, diffusion coefficient).

For an adequate oxygen supply of the tissue it is not the capillary oxygen content that must be sufficiently large, but the capillary oxygen pressure. For the Krogh model we obtain the following for the oxygen pressure difference that is necessary for a given tissue respiration (12):

$$\Delta P_{O_2} = P_{cO_2} - P_{tO_2} =$$
$$\frac{\dot{A}(O_2)}{2\alpha D} \cdot g(r_t, r_c, r_z) \quad [7\text{-}5]$$

with

$$g(r_t, r_c, r_z) = r_z^2 \cdot \ln\left(\frac{r_t}{r_c}\right) - (r_t^2 - r_c^2)$$

(r_c, radius of the capillary; r_t, distance from the capillary center; r_z, radius of tissue cylinder).

The P_{O_2} difference is directly proportional to the tissue O_2 consumption and inversely proportional to the oxygen conductivity, αD. A large conductivity decreases the necessary pressure difference (as, for example, for CO_2 transport where αD_{CO_2} is about 23 times larger than αD_{O_2}). g is a geometric factor which is mainly governed by the radii of the capillary and the supply area. The increase or decrease of the supply distance has a large effect, since in the equation the radii are nearly squared.

Compensatory Mechanism at Decreasing Arterial P_{O_2}

To maintain tissue P_{O_2} in case of a decrease in arterial P_{O_2}, the following compensatory mechanisms are possible:

1) An increase in hemoglobin content to increase the total amount of oxygen
2) A right shift of the hemoglobin binding curve to increase the venous P_{O_2}
3) An increase of flow to increase venous O_2 content and venous P_{O_2}
4) A decrease in the P_{O_2} gradient to allow a lower P_{VO_2}
    a) By a decrease of the supply distance (increased capillarity)
    b) By an increase of oxygen conductivity
    c) By facilitated diffusion (oxygen carrier)
5) A change of the mitochondrial kinetics

to increase the amount of ATP formed at low P_{tO_2}

The increase in hemoglobin content is limited by the increase in viscosity caused by the concomitant increase in the hematocrit value (31), otherwise flow and hemoglobin content can be varied independently. Figure 7-2 shows the P_{O_2} decrease in a capillary (Krogh cylinder) with different combinations of flow and hematocrit. The P_{O_2} decrease in the capillary blood from the arterial ($x = 0$) to the venous end ($x = 1.0$) is calculated for a brain capillary with normal blood flow, F_n, and a normal hematocrit, H, of 45% (30). An increase in the hematocrit value to 80% increases the venous P_{O_2}.

This increase can be approximately compensated by a 50% flow decrease. With 25% flow, a P_{O_2} of about 18 mmHg (2.3 kPa) at the venous end of the capillary can be maintained which is assumed to be sufficient to maintain brain metabolism (28). With a further decrease anoxic areas would occur. One can calculate that a smaller capillary distance would help to improve the O_2 supply situation, but the necessary increase of capillarity increases exponentially. If the capillary distance decreases from 30 to 25 μm, capillarity must increase from 1 to 1.44; for a decrease to 20 μm, from 1 to 2.25; for a decrease to 10 μm, from 1 to 9. The possible increase is strongly limited by the tissue geometry (30).

The variation of the oxygen dissociation curve with different P_{aO_2} shows that the right shift is only beneficial if the arterial P_{O_2} is sufficiently high. Using for simulation a right-shifted O_2 dissociation curve (pH = 7.2), the beneficial effect will be the highest with an arterial P_{O_2} of 100 mmHg (13.3 kPa) and then it gradually decreases. In the range of an arterial P_{O_2} of 40 mmHg (5.33 kPa) there is no longer any effect; below this value the right-shifted curve affects adversely the venous P_{O_2} (10,30,37). This special situation is explained by the fact that by the right shift of the O_2 dissociation curve the amount of oxygen bound at low oxygen pressures becomes too small.

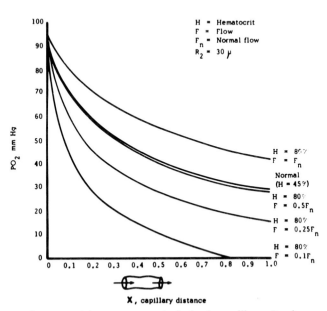

Fig. 7-2. P_{O_2} decrease from arterial to venous end of a brain capillary. P_{O_2} decrease is calculated for a Krogh cylinder (brain) and different combinations of flow, F, and hematocrit value, H. From Reneau, D.D. and Silver, I.A.: In Silver, I.A., Erecinska, M., and Bicher, H.I. (eds.): Oxygen Transport to Tissue. III. New York, Plenum Press, 1977.

To compensate the effect of arterial hypoxia as the main reaction, an increase in tissue blood flow is observed (Eq. 7-2) which is accompanied by the well-known increase in heart rate. However, the compensatory increase in flow is possible only if the blood volume is sufficiently high and the rheologic behavior of blood is not disturbed (31).

As mentioned before, calculations have shown the benefit of smaller capillary distances. In skeletal muscle all capillaries are opened and perfused in hypoxia. As an adaptation, higher capillarity can be produced by chronic tissue hypoxia (2,4,26,27, 34,36,38). It is not known whether an increase in oxygen conductivity can occur by altering the composition of the tissue.

Facilitated diffusion may occur in the presence of an oxygen carrier such as myoglobin (or hemoglobin) (11,43–45). In this case, a simultaneous diffusion of dissolved oxygen molecules, I_1, and oxygenated myoglobin molecules, I_2, can take place. The oxygen flux, I_{O_2}, is

$$I_{O_2} = I_1 + I_2 = \left(-D_{O_2}\, \alpha\, \frac{\Delta P_{O_2}}{\Delta x}\right)_1 + \left(D_{Mb}\, m\, c_{Mb}\, \frac{\Delta S}{\Delta x}\right)_2 \quad [7\text{-}6]$$

(Mb, myoglobin; m, number of binding sites; S, oxygen saturation of myoglobin).

The first term, I_1, describes the diffusion of oxygen molecules (Eq. 7-6), the second the diffusion of oxygenated myoglobin molecules, I_2. The fraction I_2 of the total flux increases with increasing concentration of myoglobin, c_{Mb}. It becomes larger at low P_{O_2} values corresponding to the hyperbolic dissociation curve of myoglobin. The transport conditions have been calculated for solutions and free movable myoglobin (7). However, there is no clear evidence as to the extent facilitated diffusion plays a physiologic role under hypoxic conditions.

Size of Hypoxic and Anoxic Zone

The P_{O_2} gradient within the tissue is produced by the local oxygen uptake of the cells and exactly mirrors the rate of oxygen flux through the tissue (17). As we heard in the foregoing presentation, about 80% of oxygen is used for the aerobic energy production achieved by the respiratory enzymes in the mitochondria. Experiments with isolated mitochondria and cells have shown (1) that with no further oxygen supply, oxygen is totally consumed and a complete anoxia is produced; and (2) that over a large range of O_2 concentration the respiratory rate is independent of the O_2 concentration of the medium, in other words follows a zero order kinetics (3.33).

Under certain conditions in isolated mitochondria, oxygen becomes rate limiting when the oxygen concentration of the medium is about 0.02 to 0.72 μM (3.33). These O_2 concentrations can be found at oxygen pressures of the medium of 0.012 to 0.072 mmHg (0.0016–0.0026 kPa). In the physiologic literature which analyzes O_2 transport, the pressure at which oxygen becomes rate limiting is called "critical mitochondrial P_{O_2}." Below this P_{O_2} the local respiration rate decreases. Compared with other biologic reactions, the oxygen concentration may not be very small, but considering the diffusion conditions, it is indeed extremely small. As pointed out above, the oxygen pressure difference is relevant for oxygen transport. Compared with a mixed venous P_{O_2} of 40 mmHg (5.33 kPa), the critical mitochondrial P_{O_2} is about 2000 times smaller, and thus in many cases negligible. Therefore, we can conclude that for oxygen transport mitochondrial respiration produces a perfect sink. The size of the gradient necessary for oxygen transport mainly depends on geometric factors, respiratory rate, and oxygen conductivity of tissue.

Figure 7-3 shows the P_{O_2} decrease perpendicular to a brain capillary in hypoxia (Krogh cylinder (6)). Assuming the energy need of the tissue remains constant, close to the capillary, we obtain a zone with constant O_2 consumption (normoxic zone). The limit of that zone is reached when the local P_{O_2} decreases to a value at which the oxygen concentration limits the respiratory

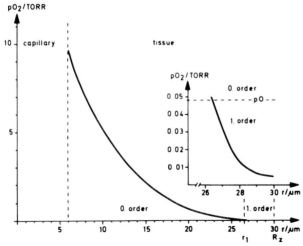

Fig. 7-3. Radial decrease in P_{O_2} in a Krogh cylinder in hypoxia. r_1, border of normoxic zone; r_z, radius of cylinder; $r_z - r_1$, width of hypoxic zone; *0.order,* zero order kinetics; *1.order,* first order kinetics. From Grossmann, U. and Lübbers, D.W.: In Silver, I.A., Erecinska, M., and Bicher, H.I. (eds.): Oxygen Transport to Tissue. III. New York, Plenum Press, 1977.

rate (critical P_{O_2}, in this case assumed to be 0.048 mmHg (6.4 Pa)). Beyond that distance the hypoxic zone begins, in which not enough oxygen is available to maintain the earlier energy need. Assuming a pseudo–first order kinetics (3), the local P_{O_2} decreases more slowly (Fig. 7-3, *insert*). The size of the hypoxic zone depends on the energy need of the tissue. Decrease in energy need increases its width. Its volume also depends on the steepness of the capillary P_{O_2} decrease: a slow decrease augments its volume (6).

Respiratory Enzymes at Low P_{O_2}

Experiments have shown that the respiratory rate is mainly determined by the breakdown of ATP, and in consequence by the local phosphate potential, ATP/ADP · P_i. It has been assumed that the reaction of oxygen with cytochrome oxidase was responsible for the zero order kinetics. Experiments of Wilson et al. (5,40–42), however, have shown that the redox state of different members of the respiratory chain changes at oxygen concentrations up to 100 μM. They concluded that in a large range the reaction kinetics of the respiratory chain are influenced by the oxygen concentration, but that the demand of ATP keeps the respiratory rate constant by producing an overall pseudo–zero order kinetics (5,40–42).

In cultured human fibroblasts (25) oxygen is able to regulate the enzymatic activity of the respiratory enzymes. It was found that in hypoxia the contents of cytochrome $a + a_3$, $c + c_1$, and b are decreased as compared with cells of well-oxygenated cultures. Likewise, high altitude produces some alterations of mitochondrial enzymes: succinic dehydrogenase increases (heart, liver, kidneys, brain) whereas cytochrome oxidase activity decreases. However, the turnover number of the cytochrome chain increases, i.e., at the same oxygen concentration a higher respiratory rate has been found (9,24,25,29,32). The physiologic consequences of these findings for tissue oxygen supply have still to be elucidated.

Flow Regulation in the Capillary Network at Decreasing Arterial P_{O_2}

Flow measurements in the brain cortex during hypoxia showed that despite an increase in total flow the simultaneous reactions of local flow can be different at different sites (14,22). For a single site the reaction pattern is rather fixed. Figure 7-4 shows simultaneous registration of microflow at four sites on the surface of the cat brain during hypoxia. Microflow is measured by the H_2-P_{H_2} clearance method in which hydrogen is generated electrochemically and measured polarographically (23). A decrease in local hydrogen pressure, P_{H_2}, means an increase in flow. In Fig. 7-4 the lowest trace shows an increase in flow, the second trace shows a small and the fourth trace a larger flow decrease, whereas the third follows passively the blood pressure change. At first glance, the decrease in microflow seems contradictory to the supply situation during tissue hypoxia; however, other experiments (8,22) have led us to the assumption that two families of capillaries may exist: (1) capillaries with high flow (flow capillaries); and (2) capillaries with functionally adapted flow (nutritional capillaries). In brain hypoxia, the flow capillaries probably restrict their flow, whereas in the other capillaries the flow increases (22). Kessler found similar conditions in the liver (8).

This redistribution of flow seems to be essential for maintaining tissue oxygenation during hypoxia. If this regulation is disturbed, an increase in blood flow may not be profitable for all regions of tissue, since the blood in the flow capillaries bypasses the nutritional capillaries. In such a capillary network the venous P_{O_2} does not mirror the O_2 supply of the tissue. Such situations can be detected by a tissue P_{O_2} histogram (8,16,19).

The structure responsible for the flow redistribution is not known. Our experiments with electric stimulation of capillaries (rabbit mesentery) have shown that at capillary branches a closure of the capillary can be obtained by this stimulus (1,20,21,39). It may be that this regulation at the capillary level plays an important role in blood flow regulation during hypoxia. We may add that a system with flow capillaries and nutritional capillaries needs a greater blood volume than does a simple capillary network. Blood volume loss can disturb this regulation heavily.

We have seen that arterial hypoxia limits the oxygen supply to the tissue, in other words, the amount of energy which can be used up by the tissue. If the energy need is higher than the supply, cell damage develops, since the cell needs a certain amount of energy to maintain its structure and functional integrity. It has been discussed whether the cell or the tissue has the potential to sacrifice a part of its functions to conserve energy (15,28). The initial shortage of oxygen could be detected by chemical reactions which depend over a large range on the local oxygen concentration. There are many oxidases which could act in this way. Wilson et al. (41,42) showed that mitochondria are able to sense the local oxygen concentration. But at the moment we do not know to what extent and in which manner an energy-conserving mechanism works.

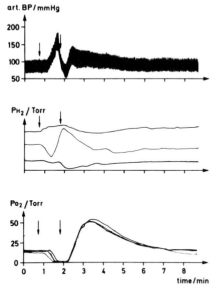

Fig. 7-4. Redistribution of flow during hypoxia. *art. BP/mmHg*, blood pressure in mmHg; P_{H_2}/*torr*, polarographically measured hydrogen pressure (multiwire element); P_{O_2}/*torr*, polarographically measured tissue P_{O_2} (multiwire element). From Leniger-Follert, E., Wrabetz, W., and Lübbers, D.W.: In Grote, J., Reneau, D., and Thews, G. (eds.): Oxygen Transport to Tissue. II. New York, Plenum Press, 1976.

Summary

The capillary models are useful to help understand the effect of the different parameters of convectional and diffusive transport. The arterial oxygen content, the hemoglobin content of blood, and the blood flow serve to decrease the arteriovenous O_2 difference in order to maintain the blood oxygen pressure. The blood oxygen pressure is mainly used to transport the O_2 into the tissue, i.e., to provide the necessary P_{O_2} gradient. The O_2 kinetics of the mitochondria make them an almost perfect oxygen sink. The capillary pattern with "flow" and "nutritional" capillaries shows that the actual supply situation in the tissue is more complicated and cannot be simulated by a simple model. Unfortunately, we do not know enough about the regulation within the microcirculation. Only if this regulation is intact does blood reach that part of the tissue which really needs it. I would think that a severe grade of hypoxia can be tolerated only if the redistribution system works properly and this necessitates a sufficiently large blood volume.

References

1. Addicks, K., Weigelt, H., Hauck, G., Lübbers, D.W., and Knoche, H.: Light- and electronmicroscopic studies with regard to the role of intraendothelial structures under normal and inflammatory conditions. In: Microcirculation in Inflammation. Bibl. Anat. 17:21, 1979.
2. Cassian, S., Gilbert, R.D., Bunnel, C.E., and Johnson, E.M.: Capillary development during exposure to chronic hypoxia. Am. J. Physiol. 220:448, 1971.
3. Chance, B., Schoener, B., and Schindler, F.: The extracellular oxidation reduction state. In Dickens, F. and Neil, E. (eds.): Oxygen in the Animal Organism. London, Pergamon Press, 1966, pp. 367–388.
4. Diemer, K. and Henn, R.: Kapillarvermehrung in der Hirnrinde der Ratte unter chronischem Sauerstoffmangel. Naturwissenschaften 52: 135, 1965.
5. Erecinska, M., Wilson, D.F., and Nishiki, K.: Homeostatic regulation of cellular energy metabolism: experimental characterization in vivo and fit to a model. Am. J. Physiol. 234:C82, 1978.
6. Grossmann, U. and Lübbers, D.W.: The size of the hypoxic zone at the border of an anoxic region within the tissue. In: Oxygen Transport to Tissue. III. Adv. Exp. Med. Biol. 94:655, 1977.
7. Hoofd, L. and Kreuzer, F.: Calculation of the facilitation of O_2 or CO transport by Hb or Mb by means of a new method for solving the carrier-diffusion problem. In: Oxygen Transport to Tissue. III. Adv. Exp. Med. Biol. 94:163, 1978.
8. Kessler, M.: Lebenserhaltende Mechanismen bei Sauerstoffmangel und bei Störungen der Organdurchblutung. München, Mitteilungen der Max-Planck-Gesellschaft, 1974, pp. 444–463.
9. Kinnula, V.L.: Rat liver mitochondrial enzyme activities in hypoxia. Acta Physiol. Scand. 95:54, 1975.
10. Kreuzer, F.: Influence of dissociation curve. *This volume*, 9.
11. Kreuzer, F.: Facilitated diffusion of oxygen and its possible significance; a review. Respir. Physiol. 9:1, 1970.
12. Krogh, A.: The number and distribution of capillaries in muscles with calculations of the oxygen pressure head necessary for supplying the tissue. J. Physiol. (Lond.) 52:409, 1919.
13. Krogh, A.: The rate of diffusion of gases through animal tissues with some remarks on the coefficient of invasion. J. Physiol. (Lond.) 52:391, 1919.
14. Leniger-Follert, E., Wrabetz, W., and Lübbers, D.W.: Local tissue pO_2 and microflow of the brain cortex under varying arterial oxygen pressure. In Grote, J., et al. (eds.): Oxygen Transport to Tissue. II. New York, Plenum Press, 1976, pp. 361–367.
15. Lowry, O.H.: Energy metabolism in brain and its control. In Ingvar, D.H. and Lassen, N.A. (eds.): Brain Work. Copenhagen, Munksgaard, 1975, pp. 49–63.
16. Lübbers, D.W.: The meaning of the tissue oxygen distribution curve and its measurement by means of Pt electrodes. In Kreuzer, F. (ed.): Oxygen Pressure Recording in Gases, Fluids and Tissues. Basel, Karger, 1969, pp. 112–123.

17. Lübbers, D.W.: Das O_2-Versorgungssystem der Warmblüterorgane. Jahrb. d. Max-Planck-Gesellschaft zur Förderung der Wissenschaften. München, 1974, pp. 87–112.
18. Lübbers, D.W.: Exchange processes in the microcirculatory bed. In Meessen, H. (ed.): Handbuch der allgemeinen Pathologie, III/7. Mikrozirkulation/Microcirculation. Berlin, Springer-Verlag, 1977, pp. 411–476.
19. Lübbers, D.W.: Quantitative measurement and description of oxygen supply to the tissue. In Jöbsis, F.F. (ed.): Oxygen and Physiological Function. Dallas, Professional Information Library, 1977, pp. 254–276.
20. Lübbers, D.W., Hauck, G., and Weigelt, H.: Reaction of capillary flow to electrical stimulation of the capillary wall and to application of different ions. In Betz, F. (ed.): Ionic Actions on Vascular Smooth Muscle. Berlin, Springer-Verlag, 1976, pp. 44–47.
21. Lübbers, D.W., Hauck, G., Weigelt, H., and Addicks, K.: Contractile properties of frog capillaries tested by electrical stimulation. In: Microcirculation in Inflammation. Bibl. Anat. 17:3, 1979.
22. Lübbers, D.W. and Leniger-Follert, E.: Capillary flow in the brain cortex during changes in oxygen supply and state of activation. In: Cerebral Vascular Smooth Muscle and its Control. Ciba Foundation Symposium 56. Amsterdam, Elsevier, 1978, pp. 21–47.
23. Lübbers, D.W. and Stosseck, K.: Quantitative Bestimmung der lokalen Durchblutung durch elektro-chemisch im Gewebe erzeugten Wasserstoff. Naturwissenschaften 57:311, 1970.
24. Mela, L., Goodwin, C.W., and Miller, L.D.: In vivo control of mitochondrial enzyme concentrations and activity by oxygen. Am. J. Physiol. 31(6):1811, 1976.
25. Mela, L., Goodwin, C.W., and Miller, L.D.: In vivo adaption of O_2 utilization to O_2 availability: comparison of adult and newborn mitochondria. In Jöbsis, F.F. (ed.): Oxygen and Physiological Function. Dallas, Professional Information Library, 1977, pp. 285–292.
26. Miller, A.T., Jr. and Hale, D.M.: Increased vascularity of brain, heart, and skeletal muscle of polycythemic rats. Am. J. Physiol. 219:702, 1970.
27. Opitz, E.: Increased vascularization of the tissue due to acclimatization to high altitude and its significance for oxygen transport. Exp. Med. Surg. 9:389, 1951.
28. Opitz, E. and Schneider, M.: Über die Sauerstoffversorgung des Gehirns und den Mechanismus von Mangelwirkungen. Ergebn. Physiol. 46:126, 1950.
29. Ou, L.C. and Tenney, S.M.: Properties of mitochondria from hearts of cattle acclimatized to high altitude. Respir. Physiol. 8:151, 1970.
30. Reneau, D.D. and Silver, I.A.: Some effects of high altitude and polycythaemia on oxygen delivery. In Silver, I.A., Erecinska, M., and Bicher, H.I. (eds.): Oxygen Transport to Tissue. III. New York, Plenum Press, 1977, pp. 245–253.
31. Schmid-Schönbein, H.: Blood rheology. *This volume*, 15.
32. Shertzer, H.G. and Cascarano, J.: Mitochondrial alterations in heart, liver, kidney of altitude acclimated rats. Am. J. Physiol. 223:632, 1972.
33. Starlinger, H. and Lübbers, D.W.: Polarographic measurement of the oxygen pressure performed simultaneously with optical measurements of the redox state of the respiratory chain in suspensions of mitochondria under steady-state conditions at low oxygen tensions. Pflügers Arch. 341:15, 1973.
34. Tenney, S.M. and Ou, L.C.: Physiological evidence for increased tissue capillarity in rats acclimatized to high altitude. Respir. Physiol. 8:137, 1970.
35. Thews, G.: Implications to physiology and pathology of oxygen diffusion at the capillary level. In Schadé, J.P. and McMenemy, W.H. (eds.): Selective Vulnerability of the Brain in Hypoxia. Oxford, Blackwell, 1963, pp. 27–35.
36. Turek, Z., Grandtner, M., and Kreuzer, F.: Cardiac hypertrophy, capillary and muscle fiber density, muscle fiber diameter, capillary radius and diffusion distance in the myocardium of growing rats adapted to a simulated altitude of 3500 m. Pflügers Arch. 335:19, 1972.
37. Turek, Z., Kreuzer, F., and Hoofd, L.J.C.:

Advantage or disadvantage of a decrease of blood oxygen affinity for tissue oxygen supply at hypoxia. A theoretical study comparing man and rat. Pflügers Arch. 342:185, 1973.
38. Valdivia, E.: Total capillary bed of the myocardium in chronic hypoxia. Fed. Proc. 21:221, 1962.
39. Weigelt, H., Addicks, K., Hauck, G., Lübbers, D.W.: Vital microscopic studies in regard to the role of intraendothelian reactive structures in the inflammatory process. In: Microcirculation in Inflammation. Bibl. Anat. 17:11, 1979.
40. Wilson, D.F., Erecinska, M., Drown, C., and Silver, I.A.: Effect of oxygen tension on cellular energetics. Am. J. Physiol. 233:C135, 1977.
41. Wilson, D.F., Erecinska, M., and Sussman, I.: Control of energy flux in biological systems. In: Energy Conservation in Biological Membranes. Berlin, Springer-Verlag, 1978, pp. 255–263.
42. Wilson, D.F., Owen, C.S., and Erecinska, M.: Regulation of mitochondrial respiration in intact tissue: a mathematical model. In Silver, I.A., Erecinska, M., and Bicher, H.I. (eds.): Oxygen Transport to Tissue. III. New York, Plenum Press, 1978, pp. 279–287.
43. Wittenberg, B.A., Wittenberg, J.B., and Caldwell, P.R.B.: Role of myoglobin in the oxygen supply to red skeletal muscle. J. Biol. Chem. 250:9038, 1975.
44. Wittenberg, J.B.: Myoglobin-facilitated oxygen diffusion: role of myoglobin in oxygen entry into muscle. Physiol. Rev. 50:559, 1970.
45. Wyman, J.: Facilitated diffusion and the possible role of myoglobin as a transport mechanism. J. Biol. Chem. 241:115, 1966.

8

Effects of High Altitude (Low Arterial P_{O_2}) and of Displacements of the Oxygen Dissociation Curve of Blood on Peripheral O_2 Extraction and P_{O_2}

JOCHEN DUHM

The rise in 2,3-diphosphoglycerate (2,3-DPG) content of human erythrocytes occurring at high altitude (caused by the rise in blood and red cell pH, respectively, and by the increased mean desaturation of hemoglobin) and the resulting right-hand shift of the oxyhemoglobin dissociation curve of blood serve to counterbalance the left-hand shift resulting from the hypoxia-induced respiratory alkalosis (mediated by the Bohr effect(s) of hemoglobin). Accordingly, *the main role of the 2,3-DPG change at high altitude (and also in acid-base disorders) is to maintain the oxygen dissociation curve of human blood at (or near) its original position.* This conclusion seems to be valid for man resting at altitudes up to 7000 m. The changes occurring at higher altitudes and during a rapid climb to a summit above 8000 m remain to be investigated. Also, the increase in body temperature at high altitude discussed by Dr. Zink at this symposium as well as the rise in temperature induced by physical exercise (14) need to be considered. Many reviews and original papers on the regulation of 2,3-DPG metabolism under physiologic and pathophysiologic conditions have been published (10–12,16–20,28,31,38,39,44,47). Reviews concerning the molecular mechanism of the interaction of 2,3-DPG with hemoglobin and general aspects of the physiologic function of hemoglobin have also appeared (7,8,43).

Whether the 2,3-DPG mechanism, compensating for the shift of the curve caused by the hypoxic respiratory alkalosis, is advantageous or disadvantageous for high altitude residents or newcomers can only be decided when the physiologic effects of displacements of the oxygen dissociation curve of blood are known. Many experimental studies on this topic have been performed (1,2,4–6,9,13,22,23,25–27,33–35, 41,42,48,51–56) and may serve as introduction into the several aspects of this highly controversial field. Displacements of the curve during exercise at sea level and moderate altitude are discussed by Dempsey et al. (14). Theoretical studies have also been performed (2,15,21,24,29,30,32, 36,37,44,46,49,50; Chapter 9, this volume).

However, despite extensive research, the physiologic consequences of displacements of the oxyhemoglobin dissociation curve of blood are still far from being clear and an enormous confusion exists in the literature, some authors paying much attention to a change in the P_{50} (O_2 pressure at

50% oxygenation of hemoglobin) by 1 or 2 mmHg, whereas others regard displacements of the oxygen dissociation curve of human blood to be of potential significance only when the P_{50} is altered by more than 10 mmHg.

In the present paper, an attempt is made to ascertain the areas where a real physiologic significance of a displacement of the oxyhemoglobin dissociation curve is to be expected. A theoretical analysis is performed on the basis that displacements of the oxyhemoglobin dissociation curve of blood can alter (1) the oxygen uptake in the lungs; (2) the peripheral P_{O_2} at a given oxygen extraction; and (3) the oxygen extraction at a given peripheral P_{O_2}.

To become independent of the hemoglobin content of blood and the blood flow (both parameters contributing to tissue oxygen supply) and independent of the oxygen demand and the absolute arteriovenous difference in O_2 content, the *oxygen extraction is ascertained as a percentage of total O_2 capacity*. This allows conclusions to be drawn which refer only to the position of the oxygen dissociation curve of blood (or intravasal hemoglobin) and which are not biased by variables such as blood flow, O_2 capacity, capillary density, and so on.

Methods and Definitions

The analysis employs the information contained in the five oxyhemoglobin dissociation curves shown in Fig. 8-1 which have been determined under standard conditions on human blood with normal erythrocytes or red cells of altered 2,3-DPG content. 2,3-DPG was increased or decreased experimentally (15) to yield P_{50} values of 15, 20, 27, 34, and 42 mmHg, respectively, the Hill coefficient, n, ranging between 2.6 and 2.8.

Since no computer program was used and the saturations and P_{O_2} values were read from the curves in Fig. 8-1 by eye, the absolute values given in Figs. 8-2–8-7 are subject to an error estimated not to exceed ± 5%.

The term "*critical O_2 extraction*" is defined as the percentage of total O_2 capacity released above which the increase in pe-

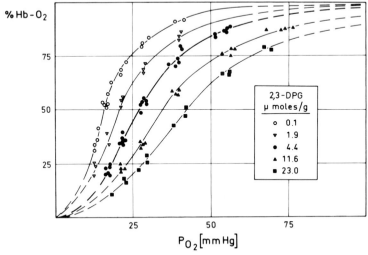

Fig. 8-1. Effect of 2,3-DPG content of erythrocytes on the oxygen dissociation curve of human blood under standard conditions (plasma pH, 7.4; 37°C; 40 mmHg CO_2). The cellular 2,3-DPG content has been altered as described earlier (15–17). The P_{50} values of the five curves are 15, 20, 27, 34, and 42 mmHg, respectively. From Duhm, J.: In Gerlach, E., et al. (eds.): Erythrocytes, Thrombocytes, Leukocytes. Recent Advances in Membrane and Metabolic Research. Stuttgart, Thieme, 1973.

ripheral P_{O_2}, seen upon a systemic right-hand shift of the oxyhemoglobin dissociation curve at lower extractions, turns into a decrease.

The "*critical peripheral (unloading)* P_{O_2}" is the capillary P_{O_2} below which the increase in O_2 extraction caused by a systemic right-hand shift of the curve occurring at higher capillary P_{O_2} values turns into a fall in O_2 extraction.

The "*arterial crossover* P_{O_2}" is the arterial P_{O_2} above which a right-hand shift and below which a left-hand shift of the curve is advantageous due to increased peripheral O_2 extraction or increased peripheral P_{O_2}.

"*Endcapillary*" means that part of a capillary (which is perfused with red cells or hemoglobin) with the lowest P_{O_2} along the capillary length.

"*Lethal corner*" means those cells in the tissue which are dependent on the endcapillary P_{O_2}.

Results and Discussion

Before going into details it is to be noted that the effects discussed below are independent of the manner in which a systemic displacement of the oxyhemoglobin dissociation curve is achieved, whether by changes in red cell organic phosphates, an abnormal hemoglobin structure, or systemic alterations in blood pH, P_{CO_2}, or temperature. "Local" effects, such as a decrease in O_2 affinity arising from a peripheral increase in H^+ or CO_2 concentration and temperature (and their reversal in the lung) are not considered in the present analysis, although they certainly will locally add to the systemic effects of a displacement of the oxygen dissociation curve of blood.

The parameters considered are (1) the arterial P_{O_2} at which the blood is oxygenated; (2) the O_2 extraction (percentage of O_2 capacity); and (3) the "venous" P_{O_2}. The term "venous" P_{O_2} can be replaced in the present theoretical context by the terms "peripheral blood P_{O_2}" or "(end) cap-

illary P_{O_2}" (although in reality they are not interchangeable due to inhomogeneities in perfusion, O_2 shunts, and so on).

Effects of Alterations of the P_{50} on Extraction

Figure 8-2 shows relations between the O_2 extraction and the arterial P_{O_2} with respect to the initial P_{50} and the value down to which the P_{O_2} has fallen at a certain point in the periphery (40, 25, and 10 mmHg in the bottom, middle, and top panel of Fig. 8-2, respectively).

Obviously, when the peripheral P_{O_2} is 40 mmHg (as in mixed venous blood at sea level and rest), each shift of the curve to the right (increase in P_{50}) leads to an increase in O_2 extraction, independent of the arterial P_{O_2}. However, when the peripheral P_{O_2} falls to 25 mmHg (as in the normal coronary sinus venous blood), an increase in P_{50} up to 42 mmHg is advantageous only at arteri-

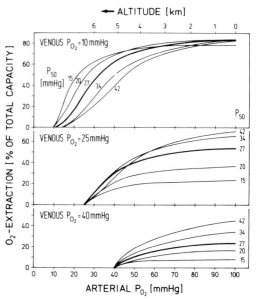

Fig. 8-2. Effects of displacements of the oxygen dissociation curve of blood and of the arterial P_{O_2} on O_2 extraction upon a fall of the P_{O_2} in the periphery to 40, 25, and 10 mmHg, respectively. The curves have been calculated from the data shown in Fig. 8-1.

al P_{O_2} values above 60 mmHg. Below an arterial P_{O_2} of 55 mmHg, less O_2 will be extracted with an initial P_{50} of 42 instead of 34 mmHg; and at an arterial P_{O_2} of about 45 mmHg, O_2 extraction with a P_{50} of 42 mmHg will be lower than with a P_{50} of 27 mmHg (but better than with a P_{50} of 20 or 15 mmHg).

When the peripheral P_{O_2} is only 10 mmHg (as in working skeletal muscle), the order seen in the lower panel of Fig. 8-2 is reversed and each right-hand shift will lead to a lowered O_2 extraction, especially at arterial P_{O_2} values below 40 mmHg. Thus, it becomes evident that the effect of displacements of the oxyhemoglobin dissociation curve on O_2 extraction is not uniform and can be either beneficial or detrimental, depending on (1) the initial P_{50}; (2) the arterial P_{O_2}: and (3) the value to which the P_{O_2} actually falls in the periphery.

Figure 8-3 shows the changes in O_2 extraction that result from an increase or decrease in P_{50} above or below its normal value of 27 mmHg in human blood for arterial P_{O_2} values to be expected at sea level and after a few days at altitudes of about 2500, 5000, and 6000 m, respectively.

No change in O_2 extraction means, by definition, that the same percentage is extracted at a particular peripheral P_{O_2} as with a P_{50} of 27 mmHg. These percentages are indicated by the small numbers at the zero lines in Fig. 8-3 (for instance, 24% of O_2 capacity is extracted with a P_{50} of 27 mmHg when the P_{O_2} falls from 100 mmHg at the arterial to 40 mmHg at the venous side).

At the arterial P_{O_2} of 100 mmHg, an increase in P_{50} from 27 to 34 or 42 mmHg causes a rise in O_2 extraction with a maximum gain of about 10 and 20% of O_2 capacity at peripheral P_{O_2} values of 30 and 40 mmHg, respectively. Conversely, a fall of the P_{50} to 20 or 15 mmHg leads to a decrease in O_2 extraction with a maximum loss of 20 or 30% of O_2 capacity at peripheral P_{O_2} values of about 30 and 25 mmHg (Fig. 8-3, left).

The pattern changes when the blood is oxygenated at lower P_{O_2} values. At an arterial P_{O_2} of 40 mmHg, for instance, an increase of the P_{50} to 42 mmHg decreases O_2 extraction. However, a left-hand shift of the curve is also disadvantageous under this condition as long as the peripheral P_{O_2} is above about 20 mmHg. Only at lower peripheral P_{O_2} values does O_2 extraction increase with a left-hand-shifted curve, as is expected from the improved oxygen uptake in the lungs. This is also seen at an arterial P_{O_2} of 30 mmHg.

Figure 8-4 is constructed similarly to

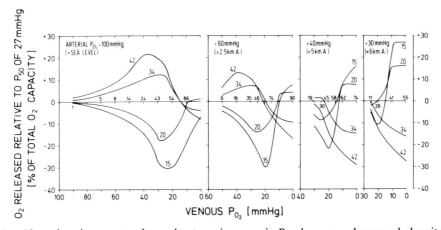

Fig. 8-3. Alterations in oxygen release due to an increase in P_{50} above or a decrease below its normal value of 27 mmHg. The O_2 released relative to a P_{50} of 27 mmHg is given in percentage of total O_2 capacity. The numbers of the curves indicate the P_{50}, and the small numbers at the zero line the percentage of O_2 released at each peripheral P_{O_2} from blood with a P_{50} of 27 mmHg.

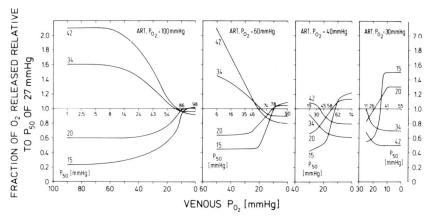

Fig. 8-4. Effects of displacements of the oxygen dissociation curve of blood on the fraction of O_2 released relative to a P_{50} of 27 mmHg. The small numbers at the zero line indicate the percentage of O_2 released from blood with a P_{50} of 27 mmHg.

Fig. 8-3, except that the fraction of O_2 released relative to a P_{50} of 27 mmHg is given at the ordinate instead of the percentage change in O_2 release. With a sea level arterial P_{O_2}, an increase in P_{50} to 34 or 43 mmHg causes a 1.6- or 2-fold increase in O_2 unloading at peripheral P_{O_2} values above 50 mmHg, whereas only up to 14% of O_2 capacity is released from normal blood. Below a peripheral P_{O_2} of 40 mmHg (i.e., with increasing O_2 extraction) the effect decreases monotonically to vanish at a capillary P_{O_2} of about 15 mmHg. When the blood is oxygenated at an arterial P_{O_2} of 30 mmHg (Fig. 8-4, *right*) an increase in P_{50} to 42 mmHg leads to a fall in O_2 extraction to one-half that seen with a normal curve. Conversely, lowering the P_{50} leads to an increase in O_2 extraction, yielding up to 1.5 times the amount of O_2 released with a P_{50} of 15 mmHg as compared to normal blood. This occurs, however, only at peripheral P_{O_2} values *below* about 15 mmHg.

The implications for tissue oxygenation evident from Figs. 8-3 and 8-4 may be summarized as follows:

1) At arterial P_{O_2} values of 100 to 80 mmHg every increase of the P_{50} from its normal value up to 42 mmHg increases O_2 extraction at all peripheral P_{O_2} values above 15 mmHg. This effect reaches its maximum at capillary P_{O_2} values between 40 and 30 mmHg. For cells which are supplied with oxygen from capillary regions with a P_{O_2} below about 15 mmHg, elevation of the P_{50} above 27 mmHg tends to decrease O_2 extraction, the disadvantageous effect, however, being small at sea level.

The capillary P_{O_2} "critical" with respect to oxygen unloading after a right-hand shift of the normal curve rises from 11 mmHg at sea level to 15 mmHg at 2500 m and further to 20 to 22 mmHg at 5000 to 6000 m (Fig. 8-6). The rise in critical capillary unloading P_{O_2} with falling arterial P_{O_2} is accompanied by an increase in magnitude of the detrimental effect of a right-hand shift on O_2 unloading below the critical P_{O_2} (Fig. 8-4).

2) At sea level arterial P_{O_2} values, a left-hand shift of the curve is disadvantageous with respect to O_2 unloading at any peripheral P_{O_2}.

3) A left-hand shift of the oxyhemoglobin dissociation curve becomes advantageous at high altitude only at capillary P_{O_2} values *below* 15 mmHg. At an arterial P_{O_2} of 30 mmHg, for instance, an organ will be able to extract three times the amount of oxygen from hemoglobin at an initial P_{50} of 15 mmHg as compared to an initial P_{50} of 42 mmHg—provided the organ properly functions at endcapillary P_{O_2} values of 8 to 10 mmHg. However, when the organ can

maintain its function at endcapillary P_{O_2} values of 20 mmHg, then the O_2 extraction will be the same with a P_{50} of 15 and 42 mmHg, but higher with intermediate values of the P_{50} (Fig. 8-4, *right*). Thus, the benefit from a shift of the curve to the left at high altitude can be far greater than that induced by the altitude-induced increase in total O_2 capacity of blood.

However, this potential benefit from a left-hand shift of the curve at high altitude is restricted to those tissues which can maintain their function at endcapillary P_{O_2} values *below* 15 to 10 mmHg. Such tissues may be skeletal muscle and the heart, but possibly not the human brain. Accordingly, *organs essential for vital function in organisms which will benefit from a decrease in the P_{50} below 27 mmHg at extreme altitudes should be constructed in such a way that each cell is sufficiently supplied with oxygen at endcapillary P_{O_2} values below about 10 mmHg*. This could be achieved either by a substantially lowered intercapillary distance, by a shortened capillary length, by extremely high local flow rates, by periodically alternating patterns of "microflow," or by a lowered oxygen demand due to an increased P/O ratio. It would be interesting to know, therefore, by which of these mechanisms such animals as the bar-hooded goose, which has been seen flying above the Himalayan mountains at altitudes as high as 9000 m above sea level, achieve a sufficient oxygenation of the brain (40).

4) There are regions in Figs. 8-3 and 8-4 where both an increase in P_{50} above and a decrease below its normal value of 27 mmHg lower O_2 extraction. This phenomenon is seen at peripheral P_{O_2} values between 20 and 25 mmHg with arterial P_{O_2} values of 40 to 30 mmHg. Such arterial P_{O_2} values are characteristic for the upper limit of altitude where man can live continuously (5000 to 6000 m). Accordingly, it would be attractive to speculate that the 2,3-DPG mechanism which maintains the curve close to its normal position serves to maintain high O_2 extraction at altitudes around 5000 m (both in residents and newcomers) in those organs which do not function properly at endcapillary P_{O_2} values below 20 mmHg. For instance, the human brain may be such an organ.

Effects of Alterations of the P_{50} on Peripheral P_{O_2}

Figure 8-5 shows relations between the arterial P_{O_2} at which the blood is oxygenated and the peripheral P_{O_2} which will result from an extraction of 10, 25, 50, and 75%

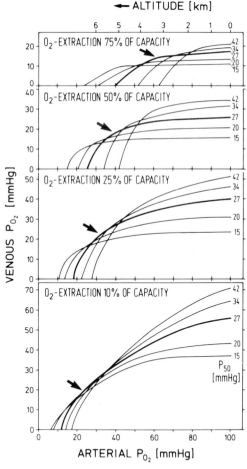

Fig. 8-5. Relations between arterial and peripheral P_{O_2} as influenced by displacements of the oxygen dissociation curve of blood at four levels of oxygen extraction. For explanation of the arrows see text.

of total O_2 capacity. The heavy line in each panel refers to the normal P_{50} of 27 mmHg.

At an arterial P_{O_2} of 100 mmHg, the peripheral P_{O_2} will be higher by about 32 mmHg following an increase in the P_{50} from 15 to 42 mmHg when only 10% of O_2 capacity is extracted (Fig. 8-5, *bottom*). With increasing O_2 extraction, this rise in peripheral P_{O_2} due to a right-hand shift progressively declines, being only 10 mmHg at 75% extraction. Apparently, *a right-hand shift of the curve is advantageous with respect to peripheral P_{O_2} even at the high O_2 extraction of 75%, provided the arterial P_{O_2} is high.*

The pattern changes when the blood is oxygenated at lower P_{O_2} values. For instance, at an arterial P_{O_2} of 40 mmHg, a right-hand shift slightly increases the peripheral P_{O_2} only at 10 and 25% extraction. At 50% extraction, an increase of the P_{50} above 27 mmHg is no longer advantageous, a left-hand shift, however, remaining disadvantageous. Accordingly, there are *regions where both an increase and a decrease in P_{50} will lower the peripheral P_{O_2}*. Such optima have also been reported by Neville (36,37).

Those regions in which a P_{50} of 27 mmHg is optimal are indicated by the arrows in Fig. 8-5. They are in the range of peripheral P_{O_2} values of 25 mmHg (10% extraction) and 15 mmHg (75% extraction) and shift to higher arterial P_{O_2} values with increasing O_2 extraction.

Furthermore, there are regions where the five curves in Fig. 8-5 come close together, i.e., *regions where a change in P_{50} between 15 and 45 mmHg has a minimal effect on peripheral P_{O_2}*. These regions are at arterial P_{O_2} values of about 30, 35, 45, and 75 mmHg when 10%, 25%, 50%, and 75% of O_2 capacity are released, respectively (Fig. 8-5). These regions are found at peripheral P_{O_2} values of 20 to 25 mmHg with 10% extraction and 10 to 15 mmHg with 75% extraction of O_2 capacity. This implies, for instance, that a shift of the curve will have almost no effect at 6000 m altitude when the flow in all organs is maintained at such a level that only 10% of O_2 capacity needs to be extracted.

From these considerations it is obvious that the effect of displacements of the oxygen dissociation curve of blood on peripheral P_{O_2} strongly depends on (1) the initial P_{50}; (2) the arterial P_{O_2}; and (3) the percentage of O_2 capacity to be extracted.

The Crossover Phenomena

The above discussion has shown that the effect of a right-hand displacement of the curve will turn from an advantageous into a disadvantageous result above an O_2 extraction "critical" with respect to peripheral P_{O_2} and below a peripheral P_{O_2} "critical" with respect to O_2 extraction. The values of critical O_2 extraction and of critical peripheral P_{O_2} depend on the arterial P_{O_2} and the initial P_{50} as shown in Figs. 8-6 and 8-7. The curves given in these figures describe the "arterial crossover" (or the "points of inversion"; Chapter 9, *this volume*), i.e., the arterial P_{O_2} below which a right-hand shift

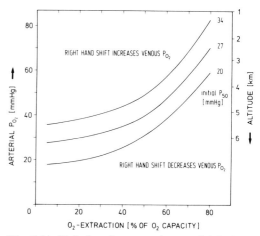

Fig. 8-6. Relations between the arterial "crossover" P_{O_2} and the "critical" O_2 extraction at different initial P_{50} values. In the areas above each of the three curves an increase in P_{50} above its initial value increases the peripheral P_{O_2}, whereas in the areas below each of the curves an increase in P_{50} lowers the peripheral P_{O_2}.

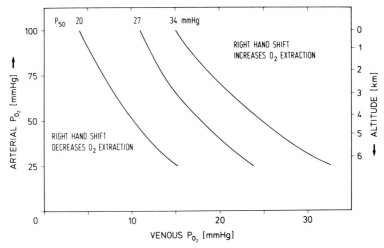

Fig. 8-7. Relations between the arterial "crossover" P_{O_2} and the "critical" peripheral P_{O_2} at three initial P_{50} values. In the areas above each of the three curves an increase in P_{50} will increase O_2 extraction, whereas in the areas below each of the three curves the opposite occurs.

becomes potentially disadvantageous due to a lowered peripheral P_{O_2} (depending on whether or not the lethal corners will sense the change in endcapillary P_{O_2}) (Fig. 8-6), or detrimental due to a lowered O_2 extraction (Fig. 8-7).

With increasing O_2 extraction higher arterial P_{O_2} values are a prerequisite in order for a right-hand shift of the curve to increase peripheral P_{O_2} (Fig. 8-6). Furthermore, the arterial crossover P_{O_2} is higher the greater the initial P_{50} (data similar to those shown in Fig. 8-6 have been derived by Turek and Kreuzer (49)).

Figure 8-7 shows that the critical peripheral P_{O_2} is higher with a lower arterial P_{O_2} and higher with a greater initial P_{50}.

The curves presented in Fig. 8-6 and 8-7 may have the practical purpose of enabling one to assess, for any particular arterial P_{O_2}, at which peripheral P_{O_2} value a right-hand shift of the oxygen dissociation curve will be advantageous or disadvantageous, respectively, for blood with an initial P_{50} of 20, 27, or 34 mmHg. Let us assume an initial P_{50} of 27 mmHg and an organ which properly functions down to an endcapillary P_{O_2} of 20 mmHg. From Fig. 8-7 it becomes evident that the critical arterial crossover P_{O_2} is 41 mmHg. Figure 8-6 shows that the critical O_2 extraction at this arterial P_{O_2} is 52% of total capacity. From these values it becomes apparent that a right-hand shift of the curve will be advantageous for the particular initial P_{50} and organ considered at any arterial P_{O_2} above 41 mmHg and any O_2 extraction below 52% of O_2 capacity. The magnitude of the advantageous effect will increase the higher the actual arterial P_{O_2}, and the lower the actual O_2 extraction.

These numbers (as well as the data shown in Figs. 8-2–8-7) are valid for each hematocrit or O_2 capacity; for each local flow rate, intercapillary distance, and length; for each microflow geometry and O_2 shunting; as well as for each rate of O_2 consumption, as long as the hematocrit is above zero in a particular capillary and the Hill parameter, n, of hemoglobin is about 2.7.

Another application of Figs. 8-6 and 8-7 is to consider which fraction of oxygen has already been extracted when blood arrives at the end of the capillaries. If we again assume a P_{50} of 27 mmHg and a situation where 30% of oxygen has been extracted, then Fig. 8-6 yields an arterial crossover P_{O_2} of 32 mmHg. This arterial crossover corresponds to a critical peripheral P_{O_2} of 22 mmHg (Fig. 8-7). Consequently, only when the arterial P_{O_2} lies above 32 mmHg will the size of the lethal corner (supplied with a capillary P_{O_2} of 22 mmHg) decrease upon a systemic right-hand shift of the curve when 30% of O_2 capacity needs to be extracted.

Conclusions

The position of the normal oxygen dissociation curve of human blood is optimal at altitudes of about 5000 to 6000 m as compared to a right- or left-hand-shifted curve with respect to (1) oxygen unloading at peripheral P_{O_2} values of 20 to 25 mmHg; and (2) maintaining peripheral P_{O_2} in the range of 15 mmHg at 50% O_2 extraction. Such optima have also right-hand-shifted curves, which, however, are to be found at lower altitudes and higher peripheral P_{O_2} values. With a left-hand-shifted curve, the optima occur at higher altitudes but lower peripheral P_{O_2} values.

The consequences of systemic displacements of the oxygen dissociation curve of blood are not uniform, but depend on (1) the arterial P_{O_2}; (2) the initial P_{50}; (3) the percentage of oxygen to be extracted; and (4) the value to which the P_{O_2} actually falls in the periphery.

Several implications may be drawn from these dependencies on the basis of the results decribed above. A shift of the curve which may be beneficial in one organ (e.g., in a heart which properly functions down to an endcapillary P_{O_2} of 10 mmHg) can be detrimental in another (e.g., in a brain which properly functions only down to an endcapillary P_{O_2} of 20 mmHg). Furthermore, an advantageous effect of a right-hand shift may become disadvantageous when the arterial P_{O_2} falls or the initial P_{50} is higher than in another case.

It is obvious, in addition, that *all factors that alter O_2 extraction will influence the result of a shift of the curve.* Such factors are, on one hand, the O_2 demand, and on the other, the oxygen supply, this being governed by several factors such as O_2 capacity, perfusion rate, intercapillary distance, capillary geometry, "microflow," and so on. However, all these factors exert their effects not *per se* but rather indirectly by altering O_2 extraction.

Addendum

The analyses made above are based on the tacit assumption that at every point of circulation there exist equilibria such as are shown in Fig. 8-1. This is certainly an oversimplification. However, as long as it is not known whether the nonequilibria occurring in blood with a normal dissociation curve are of significance, it is impossible to estimate the potential effects of reaction limitations in determining the consequences of displacements of the oxygen dissociation curve of blood. Admittedly, if there are drastic reaction limitations opposing those effects estimated for equilibrium conditions (which, however, appears unlikely to the author), all of the conclusions drawn in this paper may be wrong. Nevertheless, the statement may remain valid that the outcome of a displacement of the oxygen dissociation curve has some properties in common with politics, namely, that sometimes a right-hand shift and sometimes a left-hand shift will be of benefit, but that it is also sometimes better to remain at the original position, and, finally, that under certain conditions either shift will have almost no effect.

References

1. Aono, M., Harkness, D.R., Flores, A., and Gollan, F.: Oxygen transport and delivery in rabbits treated with cyanate. Life Sci. 15:1083, 1974.
2. Aste-Salazar, H. and Hurtado, A.: The affinity of hemoglobin for oxygen at sea level and at high altitudes. Am. J. Physiol. 142:733, 1944.
3. Bakker, J.C.: 2,3-Diphosphoglycerate, Haemoglobin and O_2 Release to Tissues.

Amsterdam, Aacademisch Proefschrift, 1977.
4. Bakker, J.C., Gortmaker, G.C., deVries-vanRossen, A., and Offerijns, F.G.J.: The influence of the position of the oxygen dissociation curve on oxygen-dependent functions of the isolated perfused rat liver. III. Studies at different levels of anaemic hypoxia. Pflügers Arch. 368:63, 1977.
5. Bakker, J.C., Gortmaker, G.C., and Offerijns, F.G.J.: The influence of the position of the oxygen dissociation curve on oxygen-dependent functions of the isolated perfused rat liver. II. Studies at different levels of hypoxia induced by decrease of blood flow rate. Pflügers Arch. 366:45, 1976.
6. Bakker, J.C., Gortmaker, G.C., Vrolijk, A.C.M., and Offerijns, F.G.J.: The influence of the position of the oxygen dissociation curve on oxygen-dependent functions of the isolated perfused rat liver. I. Studies at different levels of hypoxic hypoxia. Pflügers Arch. 362:21, 1976.
7. Bartels, H. and Baumann, R.: Respiratory function of hemoglobin. Int. Rev. Physiol. 14:107, 1977.
8. Bauer, C.: On the respiratory function of haemoglobin. Rev. Physiol. Biochem. Pharmacol. 70:1, 1974.
9. Bellingham, A.J.: Hemoglobins with altered oxygen affinity. Br. Med. Bull. 32:234, 1976.
10. Bellingham, A.J., Detter, J.C., and Lenfant, C.: Regulatory mechanisms of hemoglobin oxygen affinity in acidosis and alkalosis. J. Clin. Invest. 50:700, 1971.
11. Brewer, G.J. and Eaton, J.W.: Erythrocyte metabolism: interaction with oxygen transport. Science 171:1205, 1971.
12. Bromberg, P.A.: Cellular cyanosis and the shifting sigmoid: the blood oxygen dissociation curve. Am. J. Med. Sci. 260:1, 1970.
13. Charache, S. and Murphy, E.A.: Is placental oxygen transport normal in carriers of high affinity hemoglobins? In Labie, D., Poyart, C., and Rose, J. (eds.): Molecular Interactions of Hemoglobin. Paris, Editions INSERM, 1978, pp. 285-294.
14. Dempsey, J.A., Thompson, J.M., Forster, H.V., Cerny, F.C., and Chosy, L.W.: HbO_2 dissociation in man during prolonged work in chronic hypoxia. J. Appl. Physiol. 38:1022, 1975.
15. Duhm, J.: Effects of 2,3-diphosphoglycerate and other organic phosphate compounds on the oxygen affinity and intracellular pH of human erythrocytes. Pflügers Arch. 326:341, 1971.
16. Duhm, J.: 2,3-Diphosphoglycerate metabolism of erythrocytes and oxygen transport function of blood. In Gerlach, E., Moser, K., Deutsch, E., and Wilmanns, W. (eds.): Erythrocytes, Thrombocytes, Leukocytes. Recent Advances in Membrane and Metabolic Research. Stuttgart, Thieme, 1973, pp. 149-157.
17. Duhm, J.: 2,3-DPG-induced displacements of the oxyhemoglobin dissociation curve of blood: mechanisms and consequences. In Bicher, H.I. and Bruley, D.F. (eds.): Oxygen Transport to Tissue. Adv. Exp. Med. Biol. 37A:179, 1973.
18. Duhm, J.: Studies on 2,3-diphosphoglycerate: effects on hemoglobin, glycolysis, and on buffering properties of human erythrocytes. In Brewer, G. (ed.): Erythrocyte Structure and Function. Prog. Clin. Biol. Res. 1:167, 1975.
19. Duhm, J. and Gerlach, E.: On the mechanisms of the hypoxia-induced increase of 2,3-diphosphoglycerate in erythrocytes. Studies on rat erythrocytes in vivo and on human erythrocytes in vitro. Pflügers Arch. 326:254, 1971.
20. Duhm, J. and Gerlach, E.: Metabolism and function of 2,3-diphosphoglycerate in red blood cells. In Greenwalt, T. and Jamieson, G.A. (eds.): The Human Red Cell in Vitro. New York, Grune & Stratton, 1974, pp. 111-148.
21. Duvelleroy, M.A., Mehmel, H., and Laver, M.B.: Hemoglobin-oxygen equilibrium and coronary blood flow: an analog model. J. Appl. Physiol. 35:480, 1973.
22. Eaton, J.W., Brewer, G.J., and Grover, R.F.: Role of red cell 2,3-diphosphoglycerate in the adaptation of man to altitude. J. Lab. Clin. Med. 73:603, 1969.
23. Eaton, J.W., Skelton, T.D., and Berger, E.: Survival at extreme altitude: protective effect of increased hemoglobin-oxygen affinity. Science 183:743, 1974.
24. Garby, L. and Meldon, J.: The Respiratory Function of Blood. Topics in Hematology. New York, Plenum Medical Books, 1977.
25. Guy, J.T., Bromberg, P.A., Metz, E.N., Ringle, R., and Balcerzak, S.P.: Oxygen delivery following transfusion of stored

blood. I. Normal rats. J. Appl. Physiol. 37:60, 1974.
26. Hall, F.G., Dill, D.B., and Guzman Barron, E.S.: Comparative physiology in high altitudes. J. Cell. Comp. Physiol. 8:301, 1936.
27. Hebbel, R.P., Eaton, J.W., Kronenberg, R.S., Zanjani, R.D., Moore, L.G., and Berger, E.M.: Human llamas. Adaptation to altitude in subjects with high hemoglobin oxygen affinity. J. Clin. Invest. 62:593, 1978.
28. Jacobasch, G., Minakami, S., and Rapoport, S.M.: Glycolysis of the erythrocyte. In Yoshikawa, H. and Rapoport, S.M. (eds.): Cellular and Molecular Biology of Erythrocytes. Tokyo, University of Toyko Press, 1974, pp. 55–92.
29. Lenfant, C.: Red-cell function: theoretical considerations and physiological aspects. In Chaplin, H., Jr., Jaffe, E.R., Lenfant, C., and Valeri, C.R. (eds.): Preservation of Red Cells. Washington, D.C., Nat. Acad. Sciences, 1973, pp. 57–66.
30. Lenfant, C. and Sullivan, K.: Adaptation to altitude. N. Engl. J. Med. 284:1298, 1971.
31. Lenfant, C., Torrance, J., English, E., Finch, C.A., Reynafarje, C., Ramos, J., and Faura, J.: Effect of altitude on oxygen binding by hemoglobin and organic phosphate levels. J. Clin. Invest. 47:2652, 1968.
32. Lenfant, C., Torrance, J.D., and Reynafarje, C.: Shift of the O_2-Hb dissociation curve at altitude: mechanism and effect. J. Appl. Physiol. 30:625, 1971.
33. Lichtman, M.A., Cohen, J., Young, J.A., Whitbeck, A.A., and Murphy, M.: The relationship between arterial oxygen flow rate, oxygen binding by hemoglobin, and oxygen utilization after myocardial infarction. J. Clin. Invest. 54:501, 1974.
34. Mehmel, H.C., Duvelleroy, M.A., and Laver, M.B.: Responses of coronary blood flow to pH-induced changes in hemoglobin-O_2 affinity. J. Appl. Physiol. 35:485, 1973.
35. Mondzelewski, J.P., Guy, J.T., Bromberg, P.A., Metz, E.N., and Balcerzak, S.P.: Oxygen delivery following transfusion of stored blood. II. Acidotic rats. J. Appl. Physiol. 37:64, 1974.
36. Neville, J.R.: Theoretical analysis of altitude tolerance and hemoglobin function. Aviat. Space Environ. Med. 48:409, 1977.
37. Neville, J.R.: Altered haem-haem interaction and tissue-oxygen supply: a theoretical study. Br. J. Haematol. 35:387, 1977.
38. Oski, F.A. and Delivoria-Papadopoulos, M.: The red cell, 2,3-diphosphoglycerate, and tissue oxygen release. J. Pediatr. 77:941, 1970.
39. Oski, F.A. and McMillan, J.A.: Clinical significance of 2,3-diphosphoglycerate in hematology. In Gordon, A.S., Silber, R., and LoBue, J. (eds.): The Year in Hematology, 1977. New York, Plenum Medical Books, 1977, pp. 104–130.
40. Petschow, D., Würdinger, I., Baumann, R., Duhm, J., Braunitzer, G., and Bauer, C.: Causes of high blood O_2 affinity of animals living at high altitude. J. Appl. Physiol. 42:139, 1977.
41. Rand, P.W., Norton, J.M., Barker, N.D., Lovell, M.D., and Austin, W.H.: Responses to graded hypoxia at high and low 2,3-diphosphoglycerate concentrations. J. Appl. Physiol. 34:827, 1973.
42. Riggs, T.A., Shafer, A.W., and Guenter, C.A.: Acute changes in oxyhemoglobin affinity. Effects on oxygen transport and utilization. J. Clin. Invest. 52:2660, 1973.
43. Rørth, M.: Hemoglobin interactions and red cell metabolism. Ser. Hematol. 5:1, 1972.
44. Shappell, S.D. and Lenfant, C.J.M.: Adaptive, genetic, and iatrogenic alterations of the oxyhemoglobin-dissociation curve. Anesthesiology 37:127, 1972.
45. Shappell, S.D. and Lenfant, C.: Physiological role of the oxyhemoglobin dissociation curve. In Surgenor, D. MacN. (ed.): The Red Blood Cell. Vol. 2. New York, Academic Press, 1975, pp. 841–871.
46. Shappell, S.D., Murray, J.A., Bellingham, A.J., Woodson, R.D., Detter, J.C., and Lenfant, C.: Adaptation to exercise: role of hemoglobin affinity for oxygen and 2,3-diphosphoglycerate. J. Appl. Physiol. 30:827, 1971.
47. Thomas, H.M., Lefrak, S.S., Irwin, R.S., Fritts, H.W., and Caldwell, P.R.B.: The oxyhemoglobin dissociation curve in health and disease. Am. J. Med. 57:331, 1974.
48. Torrance, J.D., Lenfant, C., Couz, J., and Marticorena, E.: Oxygen transport mechanisms in residents at high altitude. Respir. Physiol. 11:1, 1971.
49. Turek, Z. and Kreuzer, F.: Effect of a shift of the oxygen dissociation curve on myocardial oxygenation at hypoxia. In Grote, J., Reneau, D., and Thews, G. (eds.): Oxygen Transport to Tissue. II. Adv. Exp. Med. Biol. 75:657, 1976.

50. Turek, Z., Kreuzer, F., and Hoofd, L.J.C.: Advantage or disadvantage of a decrease of blood oxygen affinity for tissue oxygen supply at hypoxia. A theoretical study comparing man and rat. Pflügers Arch. 342:185, 1973.
51. Turek, Z., Kreuzer, F., and Reginalda, B.E.M.: Blood gases at several levels of oxygenation in rats with a left-shifted oxygen dissociation curve. Pflügers Arch. 376:7, 1978.
52. Turek, Z., Kreuzer, F., Turek-Maischeider, M., and Reginalda, B.E.M.: Blood O_2 content, cardiac output, and flow to organs at several levels of oxygenation in rats with a left-shifted blood oxygen dissociation curve. Pflügers Arch. 376:201, 1978.
53. Valeri, C.R.: Oxygen transport function of preserved red blood cells. In Valeri, C.R. (ed.): Blood Banking and the Use of Frozen Blood Products. Cleveland, CrC Press, 1976, pp. 141–174.
54. Woodson, R.D., Wranne, B., and Detter, J.C.: Effect of increased blood oxygen affinity on work performance of rats. J. Clin. Invest. 52:2717, 1973.
55. Wranne, B., Nordgren, L., and Woodson, R.D.: Increased blood oxygen affinity and physical work performance in man. Scand. J. Clin. Lab. Invest. 33:337, 1974.
56. Yhap, E.O., Wright, C.B., Popovic, N.A., and Alix, E.C.: Decreased oxygen uptake with stored blood in the isolated hindlimb. J. Appl. Physiol. 38:882, 1975.

9
Influence of the Position of the Oxygen Dissociation Curve on the Oxygen Supply to Tissues

F. KREUZER AND Z. TUREK

Until 10 years ago our understanding of the physiologic implications of the blood oxygen dissociation curve for oxygen transport in the organism was fragmentary and fraught with contradictory experimental results and a lack of a more fundamental theoretical approach (10). A displacement of the oxygen dissociation curve to the right was supposed to enhance unloading of oxygen to the tissues and thus to represent an effective adaptational mechanism at high altitude, in the view of Aste-Salazar and Hurtado (1). After 1967 an important role of the organic phosphates, particularly 2,3-diphosphoglycerate (2,3-DPG), in the adaptation to hypoxia was invoked. More recent work, however, has suggested that this effect may be less important than originally presumed (11,15,21).

If a shift of the oxygen dissociation curve to the right (e.g., by 2,3-DPG) indeed involved an efficient adaptation to hypoxia, animals with genetically lower affinity of hemoglobin for oxygen should be better suited for high altitude than those with higher affinity. Surprisingly, however, most animals native to high altitude show a rather high oxygen affinity (3,5–9). Furthermore, studies in rats with relatively low oxygen affinity did not demonstrate any advantage from a further decrease of oxygen affinity after exposing the animals to simulated altitude for a prolonged period of time (4,16,20).

These discrepancies prompted us to reinvestigate this problem with particular reference to the oxygenation of the tissues. A theoretical study comparing man and rat (18) convinced us that a shift of the oxygen dissociation curve to the right or left will not have a uniform effect on tissue oxygenation at all levels of ambient P_{O_2}. We deduced theoretically that a shift of the oxygen dissociation curve to the right will have a beneficial effect by increasing tissue oxygen pressure at normoxia and mild hypoxia, but an adverse effect by decreasing tissue oxygen pressure at severe hypoxia, and conversely with a shift to the left. This concept has been confirmed experimentally in various situations (2,19).

Subsequently we examined (17) the effect of a right shift of the oxygen dissociation curve by 3 torr on myocardial oxygenation at hypoxia, again comparing man (control P_{50}, 26.6 torr) and rat (control P_{50}, 37 torr). When plotting myocardial tissue oxygen pressure at the periphery of

Krogh's tissue cylinder against capillary length for three values of arterial oxygen pressure, this tissue oxygen pressure was higher after the shift at 70 torr Pa_{O_2} in both species, still higher in man but lower in the rat at 50 torr Pa_{O_2}, and lower in both species, particularly in the rat, at 30 torr Pa_{O_2} (Fig. 9-1). Thus, the point of inversion, where a favorable effect turns into an adverse situation, is located differently and furthermore depends on various factors such as the original position of the oxygen dissociation curve (the lower the oxygen affinity of the blood, the higher the inversion point), the blood oxygen capacity (whose increase decreases the oxygen pressure at the inversion point), and the arteriovenous oxygen difference, i.e., oxygen consumption/blood flow (whose increase increases the oxygen pressure at the point of inversion). It may be concluded that the higher the oxygen pressure at the point of inversion, the sooner the situation for tissue oxygenation will worsen with a right shift of the oxygen dissociation curve when arterial oxygen pressure decreases.

When subsequent calculations provided some unexpected results we further investigated the influence of the original position of the oxygen dissociation curve on the effect of a shift over a wider range of values of P_{50}, and also included the effect of a varying Hill slope, n, of the oxygen dissociation curve in the Hill plot according to Neville (13,14). This author showed that changes in the shape of the oxygen dissociation curve, also expressed as the cooperativity or heme-heme interaction (Hill's n), could substantially modify oxygen transport; a review of published observations suggested that hemoglobin cooperativity is not normally invariant. In these recent calculations we also compared the effect of P_{50} on mixed-venous oxygen pressure as well as mixed-venous oxygen content.

These recent considerations were prompted and will be exemplified by data obtained during an Italian high altitude expedition to Peru in the spring of 1978, where native guinea pigs were compared with sea level animals and with animals having stayed in a low pressure chamber

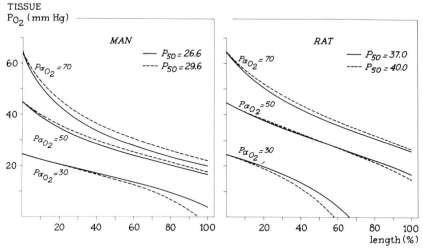

Fig. 9-1. Myocardial tissue oxygen pressure (P_{O_2}) at the periphery of Krogh's tissue cylinder in man (**A**) and in the rat (**B**) before (*solid line*) and after (*broken line*) a shift of the oxygen dissociation curve to the right by 3 torr for three values of the arterial oxygen pressure (Pa_{O_2}). Blood O_2 capacity, 20 ml O_2/100 ml blood; myocardial O_2 consumption, 0.15 ml O_2/min·g; coronary blood flow, 1.5 ml/min·g; arteriovenous O_2 difference, 10 ml O_2/100 ml blood; arteriovenous pH difference, 0.04. From Turek, Z. and Kreuzer, F.: In Grote, J., et al.: Oxygen Transport to Tissue. II. Adv. Med. Exp. Biol. 75:657, 1976. New York, Plenum Press, 1976.

Table 9-1. Most Relevant Data in Guinea Pigs Native to High Altitude (4100 m) Measured in Morococha (4560 m), in Guinea Pigs Born at Sea Level and Exposed to Simulated High Altitude (4200 m; LPC) for 4–5 Weeks, and in the Respective Sea Level Controls Measured at Hypoxia or Normoxia.[a]

Factor	Guinea pigs measured at hypoxia				Guinea pigs measured at normoxia	
	Morococha	Control I	LPC	Control II	LPC	Control
P_{aO_2} (torr)	43.1 ± 7.6	44.2 ± 8.9	55.7 ± 5.5	47.2 ± 9.6	94.7 ± 4.8	89.6 ± 12.7
Hematocrit	47.3 ± 3.5	39.3 ± 1.0	46.0 ± 2.2	40.4 ± 1.1	47.2 ± 1.8	39.4 ± 2.2
	(10)	(11)	(8)	(7)	(6)	(19)
pH_a	7.521 ± 0.054	7.494 ± 0.073	7.517 ± 0.056	7.464 ± 0.059	7.466 ± 0.039	7.460 ± 0.071
	(10)	(11)	(8)	(7)	(6)	(19)
$(a-v)_{O_2}$ (ml/100 ml)	8.37 ± 1.83	8.08 ± 1.98	9.36 ± 2.23	7.69 ± 2.34	9.35 ± 1.74	10.41 ± 2.25
	(10)	(11)	(8)	(8)	(6)	(19)
$P_{\bar{v}O_2}$ (torr)	26.8 ± 6.3	21.5 ± 3.4	26.8 ± 2.8	24.6 ± 4.4	34.7 ± 6.7	27.1 ± 6.0
	(10)	(11)	(8)	(7)	(6)	(19)

[a] Mean values ± S.D. All guinea pigs weighed about 390 g. Numbers in parentheses are the numbers of animals studied.

for 4 to 5 weeks. First, the most relevant data concerning the adaptation of these animals to high altitude will be demonstrated (Table 9-1). Hematocrit (and oxygen capacity, not listed) increases at altitude, but the altitude hematocrit of 46 to 47 is not much higher than the normal values of rat or man at sea level, the sea level values of the guinea pig being markedly lower. The red blood cells of the guinea pig are of similar size or slightly smaller than those of man and intermediate between those of man and rat.

The arterial pH values are high throughout but there is no important difference between high altitude and sea level; this might suggest a deficient renal compensation of respiratory alkalosis in the guinea pig, contrary to the classically known situation in man native to high altitude. Arteriovenous oxygen difference and cardiac output of the Morococha group are not much different from those of their controls. Apparently the only important difference is an increase in mixed-venous oxygen pressure at high altitude which, however, has to be compared with a remarkably low value in the normoxic controls (about 27 torr). This increase of mixed-venous oxygen pressure at high altitude might be due to an increased slope of the physiologic oxygen dissociation curve, a straight line connecting the arterial and venous points of oxygen content (not shown graphically here). The value of the in vitro standard P_{50} is slightly increased in the natives (much more so in man and rat), whereas n is decreased (Table 9-2). Thus, on the whole the guinea pig seems to be closer to most Andean camelids (llama, vicuña, alpaca) and rodents (chinchilla, viscacha) that show both a low P_{50} related to body weight and a modest or absent polycythemia at high altitude (12), than to animals with a high P_{50} and a marked polycythemia such as man and rat.

In view of these discrepancies between various animal species we wanted to extend these considerations to make them applicable to a wider range of animals, i.e., to evaluate the effect of the original P_{50} on mixed-venous oxygen pressure in general. For this purpose we plotted mixed-venous oxygen pressure as well as mixed-venous oxygen content against P_{50} over a wide range of P_{50} values (10–55 torr) (Fig. 9-2A and B, respectively). For this plot we chose the data from the Morococha guinea pigs (arterial oxygen pressure, 43.1 torr; arteriovenous oxygen difference, 8.37 ml/100 ml), but such a plot may be constructed for any other situation to illustrate the general tendencies. The parameters for these plots were two different values of the oxygen capacity (20.35 and 17.75 ml/100 ml, respectively) and two different values of the Hill number, n (3.021 and 2.358, respectively).

In Fig. 9-2A the mixed-venous oxygen pressure as a function of P_{50} starts from low values, passes through a rather flat top, and further declines with increasing P_{50}. Thus, apparently it depends on the starting P_{50} whether a shift of the oxygen dissociation curve to the left or right can increase mixed-venous oxygen pressure. A higher oxygen capacity implies a higher mixed-venous oxygen pressure (as ex-

Table 9-2. Values of P_{50} and Hill Number n for Standard Oxygen Dissociation Curves Estimated In Vitro and In Vivo for Guinea Pigs Native to High Altitude (Morococha), Exposed to Simulated High Altitude (LPC), and Sea Level Controls.

	In vitro			In vivo		
	P_{50}	n	Number of animals	P_{50}	n	Number of animals
Morococha	30.6[a]	2.36[a]	10	31.0	3.02	8
LPC	29.0	3.14	8	30.2	3.08	8
Control	28.4	3.01	17	29.0	2.89	16

[a]Significantly different from control value ($p < 0.01$).

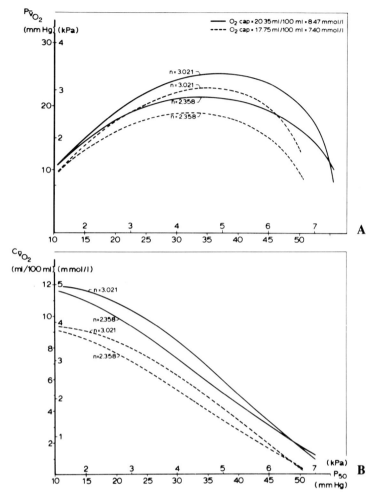

Fig. 9-2. Mixed-venous oxygen pressure (**A**) and mixed-venous oxygen content (**B**) as a function of P_{50} calculated from arterial oxygen pressure (43.1 torr), arterial and mixed-venous pH, and arteriovenous oxygen difference (8.37 ml O_2/100 ml) of the Morococha group of guinea pigs (Table 9-1), using as parameters two values of oxygen capacity (20.35 and 17.75 ml O_2/100 ml from the Morococha and sea level control group, respectively) and two values of Hill number, n (2.358 and 3.021 from the in vitro and in vivo determinations of the Morococha group, respectively) (Table 9-2). From Turek, Z. et al.: Pfluegers Arch. 384:109, 1980.

pected) as well as a higher P_{50} at the maximum. A higher value of n also induces a higher mixed-venous oxygen pressure and a higher P_{50} at the maximum. Thus, when altering P_{50}, the effect on mixed-venous oxygen pressure will depend not only on the original P_{50} but also on oxygen capacity and the value of n.

The plot of mixed-venous oxygen content against P_{50} in Fig. 9-2**B**, on the other hand, shows that an increase in P_{50} invariably decreases mixed-venous oxygen content. Again the values of the mixed-venous oxygen content are higher with increased oxygen capacity and n.

Further calculations demonstrated the effects of the arterial oxygen pressure and of the arteriovenous oxygen difference on the position of the maximum in the curves of mixed-venous oxygen pressure versus P_{50} (Fig. 9-3). It appears that the maximum occurs at lower values of P_{50} with decreasing arterial oxygen pressure and with increasing arteriovenous oxygen difference.

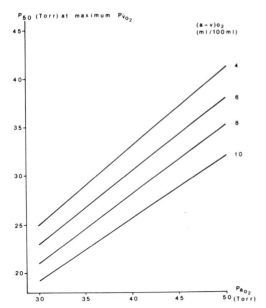

Fig. 9-3. Values of P_{50} (torr) at maximum mixed-venous oxygen pressure ($P\bar{v}_{O_2}$) in the plot of mixed-venous oxygen pressure against P_{50} (Fig. 9-2A) plotted against arterial oxygen pressure (Pa_{O_2}, torr) for four values of arteriovenous oxygen difference [(a-v)O_2, ml/100 ml], the oxygen capacity being 20 ml/100 ml.

Let us compare the three species, i.e., guinea pig (P_{50}, 29 torr), rat (P_{50}, 37 torr), and man (P_{50}, 27 torr) under the conditions of Fig. 9-2A, using an oxygen capacity of 20.35 ml/100 ml and an n of 3.021 as parameters. The respective values of mixed-venous oxygen pressure would be about 24, 25, and 23 torr. Both in the guinea pig and in man a right shift of the oxygen dissociation curve would increase, and a left shift would decrease mixed-venous oxygen pressure, whereas in the rat both a right and a left shift would decrease mixed-venous oxygen pressure, since the starting P_{50} is located near the maximum, although to a minor extent due to the flatness of the maximum.

The curves of Fig. 9-2 enable us to determine the result if the P_{50} of the guinea pig were the same as that of the rat (about 37 torr). Apparently mixed-venous oxygen pressure would not be much different irrespective of which value of n were used. On the other hand, mixed-venous oxygen content would be much lower. Maintaining the mixed-venous oxygen content as at the original P_{50} of about 29 torr obviously necessitates a further increase in blood oxygen capacity, which indeed happens in the rat. Thus, the lower P_{50} of the guinea pig, as compared with the rat, enables the guinea pig to avoid the need of increasing the oxygen capacity.

In conclusion, an evaluation of the effect of a shift of the oxygen dissociation curve must take into account all the parameters involved, and must consider the specific species and individual situation in terms of these parameters.

References

1. Aste-Salazar, H. and Hurtado, A.: The affinity for oxygen at sea level and at high altitudes. Am. J. Physiol. 142:733, 1944.
2. Bakker, J.C., Gortmaker, G.C., Vrolijk, A.C.M., and Offerijns, F.G.J.: The influence of the position of the oxygen dissociation curve on oxygen-dependent functions of the isolated perfused rat liver. Pflügers Arch. 362:21, 1976.
3. Banchero, N., Grover, R.F., and Will, J.A.: Oxygen transport in the llama (*Lama glama*). Respir. Physiol. 13:102, 1971.
4. Baumann, R., Bauer, C., and Bartels, H.: Influence of chronic and acute hypoxia on oxygen affinity and red cell 2,3-diphosphoglycerate of rats and guinea pigs. Respir. Physiol. 11:135, 1971.
5. Bullard, R.W., Broumand, C., and Meyer, F.R.: Blood characteristics and volume in two rodents native to high altitude. J. Appl. Physiol. 21:994, 1966.
6. Chiodi, H.: Oxygen affinity of the hemoglobin of high altitude mammals. Acta Physiol. Lat. Am. 12:208, 1962.
7. Chiodi, H.: Comparative study of the blood gas transport in high altitude and sea level camelidae and goats. Respir. Physiol. 11:84, 1970/1971.
8. Hall, F.G.: Minimal utilizable oxygen and the oxygen dissociation curve of blood of rodents. J. Appl. Physiol. 21:375, 1966.
9. Hall, F.G., Dill, D.B., and Barron, E.S.G.:

Comparative physiology in high altitudes. J. Cell. Comp. Physiol. 8:301, 1936.
10. Kreuzer, F.: Transport of O_2 and CO_2 at altitude. In Margaria, R. (ed.): Exercise at Altitude. Amsterdam, Excerpta Medica Foundation, 1967, pp. 149–158.
11. Lenfant, C., Torrance, J.D., and Reynafarje, C.: Shift of the O_2-Hb dissociation curve at altitude: mechanism and effect. J. Appl. Physiol. 30:625, 1971.
12. Monge, C. and Whittembury, J.: High altitude adaptations in the whole animal. In Bligh, J., Cloudsley-Thompson, J.I., and Macdonald, A.G. (eds.): Environmental Physiology of Animals. Oxford, Blackwell Scientific Publishers, 1976, pp. 289–308.
13. Neville, J.R.: Altered haem-haem interaction and tissue-oxygen supply: a theoretical analysis. Br. J. Haematol. 35:387, 1977.
14. Neville, J.R.: Theoretical analysis of altitude tolerance and haemoglobin function. Aviat. Space Environ. Med. 48:409, 1977.
15. Torrance, J.D., Lenfant, C., Cruz, J., and Marticorena, E.: Oxygen transport mechanisms in residents at high altitude. Respir. Physiol. 11:1, 1970/1971.
16. Turek, Z., Grandtner, M., Ringnalda, B.E.M., and Kreuzer, F.: Hypoxic pulmonary diffusing capacity for CO and cardiac output in rats born at a simulated altitude of 3500 m. Pflügers Arch. 340:11, 1973.
17. Turek, Z. and Kreuzer, F.: Effect of a shift of the oxygen dissociation curve on myocardial oxygenation at hypoxia. In Grote, J., Reneau, D., and Thews, G. (eds.): Oxygen Transport to Tissue. II. Adv. Exp. Med. Biol. 75:657, 1976.
18. Turek, Z., Kreuzer, F., and Hoofd, L.J.C.: Advantage or disadvantage of a decrease of blood oxygen affinity for tissue oxygen supply at hypoxia. A theoretical study comparing man and rat. Pflügers Arch. 342:185, 1973.
19. Turek, Z., Kreuzer, F., and Ringnalda, B.E.M.: Blood gases at several levels of oxygenation in rats with a left-shifted blood oxygen dissociation curve. Pflügers Arch. 376:7, 1978.
20. Turek, Z., Ringnalda, B.E.M., Hoofd, L.J.C., Frans, A., and Kreuzer, F.: Cardiac output, arterial and mixed-venous O_2 saturation, and blood O_2 dissociation curve in growing rats adapted to a simulated altitude of 3500 m. Pflügers Arch. 335:10, 1972.
21. Weiskopf, R.B. and Severinghaus, J.W.: Lack of effect of high altitude on hemoglobin oxygen affinity. J. Appl. Physiol. 33:276, 1972.

10

Carbon Dioxide and Oxygen Dissociation Curves During and After a Stay at Moderate Altitude

D. BÖNING, F. TROST, K.-M. BRAUMANN, H. BENDER, AND K. BITTER

Most investigations on oxygen deficiency and altitude adaptation have been carried out under rather extreme conditions to obtain clear and large effects. However, the great mass experiment performed every winter by millions of people who enjoy skiing for some weeks in the mountains also bears some importance for medicine. What happens in the body during such a short stay with moderate exercise intensity? Do possible changes persist more than a few days after return? The main subject of this study concerns the effects on blood gas binding and their possible relevance for exercise performance.

Methods

The investigations were performed in a group of eight male and three female physical education students (age 26 ± 3 years) taking part in a 2-week ski course. During the year preceding the experiments all but two had not performed intense physical training and nobody was a top level athlete. Three were moderate smokers who ceased to smoke 36 h before each test. During the ski course the group lived for 14 to 15 days at 2208 m altitude (Livigno, Italy) and changed daily between 1800 and 2900 m.

During the 3 weeks preceding ascent two exercise tests in sitting position with stepwise increasing load (every third minute + 50 W, 65 rpm, Monark Bicycle Ergometer) until exhaustion were performed at Hannover. The first test served only for familiarization. Oxygen uptake and respiration were measured by use of a closed spirometer system (Meditron Magnatest 610); heart rate was measured by use of an electrocardiograph (Hellige Simpliscriptor EK 31).

During the second test heparinized blood was sampled from a cubital vein (determination of blood gas status and concentrations of different blood constituents) and a hyperemized ear lobe (determination of arterial pH) after 20 min of recumbency. During exercise only venous lactate concentrations and arterial pH were measured at the end of each step.

The tests began at approximately 9 A.M., 12 noon, or 3 P.M. A light meal was allowed up to 1 h before arrival at the laboratory. The tests were repeated at identical times for each subject during the first, second, and third weeks after descent. At altitude blood could only be sampled in the morning before breakfast. After having arisen for washing, the subjects were placed in the supine position for at least 20 min. No exercise test took place.

Immediately after sampling pH, P_{CO_2}, P_{O_2} and S_{O_2} were measured at 37°C by use of a Radiometer Blood Gas Analyser (BGA1) and a Radiometer Oxygen Saturation Meter (OSM2). Red cell pH values were obtained after freeze-thaw hemolysis. After equilibration at appropriate gas tensions the CO_2 dissociation curve (CDC) and the arterial P_{CO_2} were determined by the Astrup technique (10); the oxygen dissociation curve (ODC) was determined by a mixing method (5).

After correction to whole blood pH 7.4 or red cell pH 7.2 with saturation dependent Bohr coefficients (1), half saturation pressure P_{50} and slope n at 50% S_{O_2} could be read from a Hill plot. The hematocrit value (microhematocrit centrifugation) and the concentrations of hemoglobin (cyanhemoglobin method), 2,3-diphosphoglycerate (DPG; photometrically by Sigma Kit), adenosine triphosphate (ATP; enzymatically by Boehringer test kit) and lactic acid (Lac; enzymatically by Boehringer test kit) were measured in whole blood; and the concentration of inorganic phosphate (molybdenum blue formation) was measured in plasma. Statistical testing was performed by analysis of variance or the Student test (11).

Results and Discussion

Blood Investigations

The altitude effect was most clearly demonstrated by the decreased O_2-saturation, whereas whole blood pH values exhibited the expected alkalosis only slightly after 1 week at altitude (Fig. 10-1). Since altitude measurements were performed only in the morning when pH is relatively low (2), the real extent of alkalosis might have been underestimated by 0.01 to 0.02 units. In contrast to whole blood, red cell pH even decreased in the mountains and reached initial values only in the third week after return. As usually observed, the P_{aCO_2} decrease at altitude was followed by a compensatory decrease of base excess (BE) which disappeared only slowly after return.

An unexpected result is the change of the slope of the CDC in the log P_{CO_2} – pH nomogram at the end of the ski course which implies a better buffering against CO_2. An increase or decrease in blood P_{CO_2} from 20 to 60 torr or vice versa changes pH by

$$\Delta pH = \frac{\pm(\log 60 - \log 20)}{-1.539} = \pm 0.310$$

units before altitude. After 2 weeks in the mountains the pH change will be only 0.253 (−0.057; −18.4%); during the first week after descent, only 0.272 (−0.037). Together with the improved buffering of deoxygenated blood this factor stabilizes arterial and venous blood pH. Similar results were obtained by Lefrancois (7) in subjects living at La Paz (3660 m). The cause for the increased slope of the CDC is difficult to explain; a left shift of the equilibration line exerts such an effect (10), but to a much less extent than observed in our experiments.

After maximal exercise the CDC was left shifted because of a BE decrease of 9.5 to 11.5 mmol/liter; the slope rose (−2.0 to −2.2) without significant differences among pre- and postaltitude measurements.

Buffering against lactate (Lac) formed during exercise can roughly be estimated from $\Delta[Lac]/\Delta pH$ if the P_{CO_2} effect on pH is subtracted by use of the postexercise slope of the CDC. Changes in oxygen saturation were small and therefore not corrected for. During the first week after return the buffer value for fixed acid was improved (Table 10-1). The cause is probably the same as that for the altered blood buffering against CO_2; tissue buffers, however, might also play a role.

The DPG concentration increased by

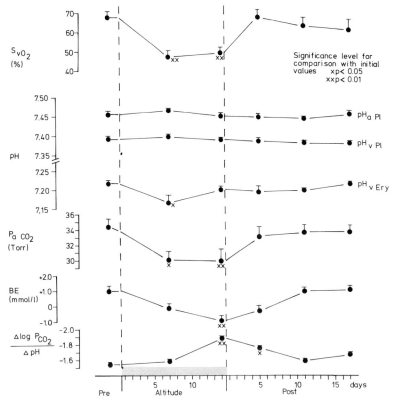

Fig. 10-1. Blood gases and acid-base status during and after a ski course at 2208 m of altitude. Means and standard errors of 11 subjects. *a*, arterial (ear lobe); *v*, cubital venous blood.

about 25% at altitude (Fig. 10-2). Usually this increase is explained by intracellular alkalosis at altitude but this does not exist, at least at the moment of blood sampling. After descent [DPG] fell to initial values during the first week, but showed a slight renewed increase during the following days. [ATP] in red cells decreased from 3.72 ± 0.12 (SE) to 3.37 ± 0.12 μmol/g Hb ($p < 0.01$) at altitude, whereas inorganic phosphate concentration in plasma exhibited no significant changes. The whole blood hemoglobin concentration showed little alteration during the sojourn at altitude and afterwards, but mean cellular hemoglobin concentration decreased at altitude by 1.5 g% ($p < 0.01$), obviously caused by osmotic effects of DPG.

The half saturation pressure, P_{50}, for standard conditions in plasma (pH 7.4; Bohr coefficient, -0.48) showed the expected increase at altitude and an unexplainable decrease in the first week after return to Hannover (Fig. 10-2). When P_{50} is calculated for a red cell pH of 7.2 (Bohr coefficient, -0.59) the altitude increase to-

Table 10-1. Ratio of Lactate Concentration and pH Changes (After Subtraction of P_{CO_2} Effects) in Venous Blood After Maximal Exercise.[a]

	Prealtitude	Postaltitude		
		5th Day	11th Day	17th Day
ΔLac/ΔpH	44.4 ± 2.5	55.5 ± 3.8[b]	44.2 ± 3.8	44.3 ± 2.1

[a] Mean $\pm$ SE.
[b] $p < 0.05$, compared with prealtitude value.

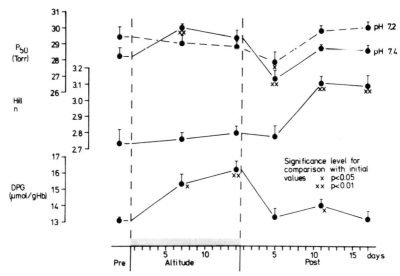

Fig. 10-2. Half saturation pressure (P_{50}) and slope (n) in the Hill plot of the oxygen dissociation curve (whole blood pH, 7.4; red cell pH, 7.2; 37°C). 2,3-Diphosphoglycerate concentration (DPG) in fresh blood. Means and standard errors of 11 subjects.

tally disappears. This means that the change in P_{50} at altitude depends more on the intracellular acidosis caused by DPG than on its direct binding to hemoglobin. A second essential parameter of the ODC is its slope in the Hill plot (cooperativity). Astonishingly, n increased markedly in the second and third weeks after descent. We have observed similar effects in athletes (400-1500 m runners) performing relatively intense anaerobic exercise (3). According to Edwards and Rigas (6) and Neville (9) young erythrocytes, the percentage of which might be augmented by altitude and physical training, show a steepened slope of the ODC.

Exercise Testing

After return from altitude, improvements in performance could be observed, but they were small. Figure 10-3 shows the venous blood lactate concentrations as a function of the work rate (data presented are only for the male participants because of the known sex differences in performance). Postaltitude concentrations tend toward lowered values at submaximal loads ($p < 0.05$). Furthermore, some subjects reached higher loads than initially or exercised longer on the last step. Table 10-2 presents parameters of performance. Maximal oxygen uptake, physical work capacity 170 (PWC 170), and total work were slightly but significantly increased during at

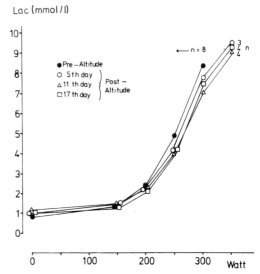

Fig. 10-3. Lactate (*Lac*) concentrations in cubital venous blood at different work loads. Means of 8 males.

Table 10-2. Parameters of Physical Performance (Means and Standard Errors) for the Male Subjects (n = 8).

	Prealtitude	Postaltitude		
		5th Day	11th Day	17th Day
$\dot{V}_{O_2}$ max	3.98 ± 0.14	4.03 ± 0.14	4.14 ± 0.14[a]	3.97 ± 0.12
PWC_{170} (W)	248.5 ± 11.5	262.2 ± 11.7	260.4 ± 11.9	271.9 ± 13.0[b]
Total work (kW · s)	166.2 ± 9.2	176.7 ± 8.8	181.9 ± 10.2[a]	179.9 ± 11.0

[a] $p < 0.05$, compared with prealtitude value.
[b] $p < 0.01$, compared with prealtitude value.

least 1 week of the postaltitude period. Liesen and Hollmann (8) observed similar changes after sojourn at altitude (1950 m above sea level) including intense physical training which had begun 5 months before ascent. Buskirk et al. (4) could not detect any synergistic effect of exercise plus hypoxia after 7 weeks at 4000 m. Whether the improvement after the ski course results from altitude adaption, physical activity, or both cannot be decided.

One might speculate that the augmented slope of the ODC contributes to the increased performance. The total gain in P_{O_2} between 95 and 20% S_{O_2} amounts to 3–4 torr. Low saturations are reached in blood perfusing the heart and skeletal muscles. Therefore, high n values might affect the anaerobic threshold during exercise by improved oxygen diffusion to tissue cells besides improved arterial oxygenation.

Acknowledgment

Supported by Deutsche Forschungsgemeinschaft Bo 360/4.

References

1. Böning, D., Draude, W., Trost, F., and Meier, U.: Interrelation between Bohr and temperature effects on the oxygen dissociation curve in men and women. Respir. Physiol. 34:195, 1978.
2. Böning, D., Schweigart, U., and Kunze, M.: Diurnal variations of protein and electrolyte concentrations and of acid-base status in plasma and red cells of normal man. Eur. J. Appl. Physiol. 32:239, 1974.
3. Braumann, K.-M., Böning, D., and Trost, F.: Oxygen dissociation curves in trained and untrained subjects. Eur. J. Appl. Physiol. 42:51, 1979.
4. Buskirk, E.R., Kollias, J., Akers, R.F., Prokop, E.K., and Reategui, E.P.: Maximal performance at altitude and on return in conditioned runners. J. Appl. Physiol. 23:259, 1967.
5. Edwards, M.J. and Martin, R.J.: Mixing technique for the oxygen-haemoglobin equilibrium and Bohr effect. J. Appl. Physiol. 21:1898, 1966.
6. Edwards, M.J. and Rigas, D.A.: Electrolyte-labile increase of oxygen affinity during in vivo aging of hemoglobin. J. Clin. Invest. 46:1579, 1967.
7. Lefrancois, R.: Pouvoir tampon du sang humain à haute et basse altitude. J. Physiol. (Suppl. 3) 62:401, 1970.
8. Liesen, H. and Hollmann, W.: Der Einfluss eines 2 wöchigen Höhentrainings auf die Leistungsfähigkeit im Flachland, gemessen an spiroergometrischen und metabolischen Parametern. Sportarzt Sportmed. 8:157, 1972.
9. Neville, J.R.: Erythrocyte age and shape of the oxygen dissociation curve. Proc. Intern. Physiol. Sci. 23:548, 1977.
10. Siggaard-Andersen, O.: The Acid-Base Status of the Blood, Copenhagen, Munskgaard, 1974.
11. Winer, B.J.: Statistical Principles in Experimental Design. New York, McGraw-Hill, 1971.

III. Hypoxia and Anaerobic Metabolism

11
Ventilatory, Circulatory, and Metabolic Mechanisms During Muscular Exercise at High Altitude (La Paz, 3500 m)

M. Paz Zamora, J. Coudert, J. Arnaud, E. Vargas, J. Ergueta Collao, N. Gutierrez, H. Spielvogel, G. Antezana, and J. Durand

The scientific study of the natives of the Andes in South America is no longer an isolated or exotic undertaking (9). The Andean peoples are undergoing cultural, social, and economic development unequaled in their histories. The agricultural and mining potential makes the Andean area one of vital importance. However, a whole series of health problems has emerged from the requirements of integration and migration to new economic development centers.

Bolivia extends over approximately 1,000,000 km² with permanent settlements between 500 and 5000 m above sea level. The climate ranges from tropical to cold. Its labor force amounts to approximately 2,500,000 people, of which approximately 70% live above altitudes of 2000 m, working in agriculture, metal, mining, smelting, and general services. La Paz, the political and administrative capital, is a city with 800,000 inhabitants and is situated at 3500 m above sea level.

Since all developmental activities are based on physical labor, knowledge of the repercussions of physical exertion on the human organism, the organism's physiologic limits, and the effects of environmental factors is of prime importance. It is useful to know the performance of the respiratory, cardiovascular, and metabolic systems during light, medium, and maximum levels of exertion with respect to the native resident population living at high altitude as well as population groups living at low altitude, since these groups may have to change altitude temporarily or permanently.

It has been shown that in the investigation of sports-related exercise, many useful parameters may be obtained which may then be applied to other fields (3,6,7,11,23).

Normal Biologic Parameters (3500 m)

Studies carried out at the Instituto Boliviano de Biología de Altura (IBBA) show that the most common blood group is O, that men have from 5,200,000 to 5,600,000 and women from 5,100,000 to 5,400,000 red blood cells, with a hematocrit from 49 to 53% in men and from 47 to 50% in women. Men have 15.5 to 17.5 g% hemoglobin and women 15.0 to 16.5 g%. Respiratory values do not show a significant difference in comparison to sea level values (minute ventilation, 7–9 liters/min BTPS; respiratory frequency, 16–18/min), with the exception of a higher vital capacity and a larger residual volume. Lung compliance is not significantly different from that of natives of Santa Cruz (400 m); neither are inspiratory and expiratory resistances (181 ml/cm H_2O, 4.2 cm H_2O/liter/s, and 3.9 cm H_2O/liter/s; and 175 ml/cm H_2O, 4.0 cm H_2O/liter/s, and 3.8 cm H_2O/liter/s, respectively, in both groups). Lung diffusing capacity for carbon monoxide at steady state is increased: for a person measuring 1.70 m and aged 30, 28.98 ml/min/mmHg were measured.

Gas transport is different at high altitude in comparison to sea level because lower PI_{O_2} (93 mmHg) secondary to lower P_B (490 mmHg) causes a Pa_{O_2} of 66 mmHg and a Pa_{CO_2} of 30 mmHg. Pa_{O_2} is 58–60 mmHg; Pa_{CO_2}, 29–30 mmHg; Sa_{O_2}, 88–90%; standard bicarbonate, 19 mEq; and pH 7.40, which does not change.

In cardiovascular circulation the systemic pressures are the same as at sea level. Pulmonary artery pressure is higher for individuals of the same age and sex (P_s, 29 ± 6.6 mmHg; P_d, 13 ± 3.8 mmHg; and $\bar{P}$, 21 ± 4.2 mmHg) without significant differences in pulmonary capillary pressure (mean PCP, 9 ± 2.7 mmHg); cardiac output (6.43 ± 1.69 liter/min); and cardiac index (3.91 ± 1.09 liter/min/m²). Total pulmonary vascular resistance and pulmonary arteriolar resistance calculated in La Paz are 265 ± 80 and 148 ± 43 dyn/s/cm^{-5}, respectively. Skin blood flow and blood volume are also lowered in sea level natives during acclimatization to high altitude (4,14–17). Finally, cerebral blood flow (2,7,10,20) is decreased (32.6 ml/min/100 g), whereas neither oxygen consumption (3.18 ± 2.8 ml/min/100 kg) nor the respiratory quotient changes.

The VIIIth Bolivarian games took place in La Paz, Bolivia, at an altitude of 3500 m in October of 1977. It was therefore very important to investigate the biological responses to acute high altitude exposure and the physical capacity of as many participants as possible. In La Paz we observed an increase in the respiratory and metabolic function of the erythrocyte, which contained 3.13 ± 0.5% methemoglobin and 79.85 mg/100 ml red blood cell for reduced glutathione (GSH). ATP is also increased (1968 nmol/ml red blood cell), as well as 2,3-DPG (6193 nmol/ml red blood cell). Total lipids are 772 ± 160 mg%; triglycerides and total cholesterol are 135 ± 9 mg% and 182 ± 32 mg%, respectively (5,13,21).

During maximal muscular exercise, for female sprinters and basketball players maximal heart rate and maximal oxygen consumption are 169.0 ± 12.50 beats/min and 51.59 ± 10.07 ml/min/kg, and 175.0 ± 15.04 beats/min and 48.06 ± 4.14 ml/min/kg, respectively. In men, for long distance running, cycling, boxing, and soccer, the values are 174.0 ± 9.20 beats/min and 68.68 ± 7.71 ml/min/kg, 174.0 ± 12.90 beats/min and 63.73 ± 6.96 ml/min/kg, 170 ± 13.35 beats/min and 49.30 ± 2.41 ml/min/kg, and 175.0 ± 6.0 beats/min and 48.40 ± 11.0 ml/min/kg, respectively (Tables 11-1 and 11-2).

Maximal Muscular Exercise (3500 m)

Fifteen hundred athletes from Bolivia, Colombia, Ecuador, Panama, Peru, and Venezuela gathered for the VIIIth Bolivarian Games. It was expected that the high altitude of La Paz would be a source of concern, as was the case during the Olympic Games held in Mexico City in 1968. Consequently, the respective competing teams took compensatory action. Some athletes arrived in La Paz 2 weeks before the opening day; others spent some time in high altitude regions in their own countries to achieve some degree of acclimatization. In addition, the First Pre-Bolivarian Symposium of Sports Medicine was organized and held 3 months before opening day. During this symposium various topics were discussed, including age, sex, nutrition, doping, and the relation of high altitude to health. It was concluded that the effects of

Table 11-1. Biologic Data of Athletes Born and Residing at La Paz (3500 m) During Muscular Exercise.

	Women	
	Sprinters	Basketball
Hct (%)	43.33 ± 2.89	48.70 ± 2.63
Hb (g%)	14.33 ± 0.29	16.40 ± 0.53
HR (min)	169.0 ± 12.50	175.0 ± 15.04
pH	7.24 ± 0.05	7.27 ± 0.04
Pa_{O_2} (mmHg)	62.6 ± 4.77	61.25 ± 3.01
Pa_{CO_2} (mmHg)	22.8 ± 3.08	20.98 ± 3.64
Lac. Ac. (nmol/liter)	6.68 ± 0.83	7.98 ± 0.84
$\dot{V}_{O_2max}$ (liter/min) STPD	2.40 ± 0.35	3.01 ± 0.18
$\dot{V}_{O_2max}$ (ml/min/kg) STPD	51.59 ± 10.07	48.06 ± 4.14

high altitude would depend on the athletes' health, their degree of training, and the intensity and duration of each competition.

Actually, of the 1528 medical consultations that were necessary for the participants, only 41 had a direct relation to the high altitude. These were motivated by slight signs of high altitude sickness such as headache, insomnia, nausea, epistaxis, and dryness of the lips and conjunctiva.

Regarding the performance, 36 Bolivarian records were broken, as many or more than in any other previous game. Most of these were in events of short duration, as expected. However, records were also broken in some long-lasting events requiring sustained effort.

One of the participating countries sent its swimming team to La Paz almost 4 weeks before the beginning of the competitions. This team did acclimatization and training in La Paz. On the 11th day after arrival a 200-m competition in five styles was held. Evaluation of the scores indicates the beneficial and positive effects of acclimatization and the correct training system.

Of the 16 swimmers, men and women, who competed in the earlier races and almost 20 days later in the games, 14 improved their scores (87.5%) and ten of them (71.4%) even received medals.

Evaluation of scores in sports such as weight lifting, shooting, boxing, fencing, gymnastics, judo, and wrestling, all requiring intense effort and great skillfulness, proves that they can be performed at high altitude and that records can even be improved.

We are convinced that it is completely possible to do muscular exercise of the

Table 11-2. Biologic Data of Athletes Born and Residing at La Paz (3500 m) During Muscular Exercise.

	Men			
	Long distance runners	Cycling	Boxing	Soccer
Hct (%)	51.08 ± 1.10	49.0 ± 5.66	51.0 ± 2.08	53.2 ± 4.0
Hb (g%)	17.12 ± 0.46	15.50 ± 0.71	17.14 ± 0.48	16.87 ± 0.65
HR (min)	174.0 ± 9.20	174.0 ± 12.99	170.0 ± 13.35	175.0 ± 6.0
pH	7.24 ± 0.08	7.27 ± 0.06	7.28 ± 0.05	7.34 ± 0.095
Pa_{O_2} (mmHg)	66.84 ± 12.33	58.33 ± 3.06	62.83 ± 2.94	64.59 ± 4.64
Pa_{CO_2} (mmHg)	25.30 ± 3.17	23.93 ± 1.44	4.83 ± 3.80	25.25 ± 3.98
Lac. Ac. (nmol/liter)	7.63 ± 0.75	5.32 ± 0.01	6.87 ± 0.83	20.08 ± 3.21
$\dot{V}_{O_2max}$ (liter/min) STPD	—	4.35 ± 0.35	3.37 ± 0.38	3.22 ± 3.07
$\dot{V}_{O_2max}$ (ml/min/kg) STPD	68.68 ± 7.71	63.73 ± 6.96	49.30 ± 2.41	48.4 ± 11.0

highest competitive level at the altitude of La Paz, without significant risk for the health of the participants. The 36 new records achieved in the Bolivarian Games at La Paz are the best proof of this statement. The overall high performance at the Games must also be mentioned (1,8,19). However, there is a difference between the events of short duration with intense effort and the long-lasting events with sustained effort. The latter require a longer period of acclimatization, i.e., about 3 weeks (22).

Research has shown that the maximum oxygen uptake is decreased at high altitude and improves slightly in the first 2 weeks. We have been able to demonstrate this in 18 top soccer players, born and living at sea level, who were subjected to spiroergometric studies, as well as to determination of hematocrit, arterial oxygen and carbon-dioxide tension, and lactate levels on the first and ninth days of stay in La Paz. These values were compared to those obtained from 18 resident top athletes (Table 11-3). The comparison showed that the values for $\dot{V}_{O_2max}$ of the first group did not change within the first nine days and that they were lower than those of the resident group (37 compared to 49 ml/kg/min ($p < 0.02$). Hematocrit and arterial oxygen tension did not show any differences. Arterial CO_2 tension and levels of lactate were significantly higher in the resident soccer players. We concluded that the capacity of O_2 transport to the periphery is comparable in high altitude and sea level athletes, that the limiting factor of O_2 utilization is located at the tissue level, and that racial factors, in addition to acclimatization, play an important role in performance at high altitude (12,18).

Accordingly, a group of cyclists who were born at sea level showed a smaller $\dot{V}_{O_2max}$ without significant improvement between the second and tenth days after arrival in La Paz, than did a similar group of high altitude natives.

The Bolivarian Games also gave us the opportunity to perform a further study. It has been reported (2a) that the oxygen affinity of hemoglobin is decreased in people born and living at high altitude. This is also observed in people born at sea level and transferred to high altitude (13a), and is presumed to be an adaptation mechanism. Samaja was unable to prove this observation. In view of this finding we studied this mechanism in lowlanders staying at high altitude.

Eight top male athletes, members of a cycling team, all born at sea level and whose age ranged from 19 to 25 years (average, 22 years), arrived in La Paz 2 weeks before the opening day of the games. Venous blood was drawn 5 h after arrival, and on the second, fifth, and 12th days. In each sample hematocrit, hemoglobin, and intraerythrocyte concentration of 2,3-diglycerolphosphate (2,3-DPG) and adenosine triphosphate (ATP) were determined. A dissociation curve for oxyhemoglobin was obtained using the method of tonometry at standard conditions (P_{CO_2}, 40 mmHg; pH, 7.4; T, 37°C). Starting with the stabilized oxyhemoglobin dissociation curve, P_{50} was evaluated and the oxygen tension (P_{O_2}) at 50% oxygen saturation (S_{O_2}) was determined.

The following results were obtained: Hematocrit increased from 46.7 ± 2.6 to 50.6 ± 2.9% (Fig. 11-1). Hemoglobin increased from 16.8 ± 0.9 to 17.8 ± 1.1 g%. P_{50} increased from 26.8 ± 1 mmHg on the second day to 29.0 ± 0.5 mmHg on the

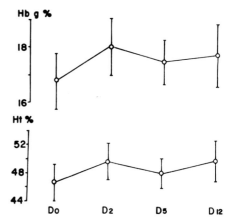

Fig. 11-1. Hemoglobin variation during 12 days of acclimatization to high altitude.

Table 11-3. Biometric Data of the Two Groups Studied in La Paz.

	Number	Age	Weight	Height	Sup. corp. (m²)
Group A (Athletes living at sea level)	18	26 (±7)	74 (±7)	177 (±8)	1.89 (±0.14)
Group B (Athletes living at high altitudes)	18	25 (±3)	67 (±6)	171 (±7)	1.78 (±0.11)

	Group A			Group B	
Parameter	3rd day high altitude	p	9th day high altitude	p	Athletes living permanently at high altitude
$\dot{V}_E$: liter/min BTPS	59.5 ± 9.2	NS	121.3 ± 20.3	$0.01 < p <$	103.0 ± 18.7
liter/min STPD	59.5 ± 10.6	NS	62.7 ± 10.3	$p < 0.001$	53.3 ± 10.3
ml/min STPD	2741 ± 141	NS	2641 ± 880	$0.01 < p < 0.02$	3228 ± 30.7
$\dot{V}_{O_2}$: ml/min/kg STPD	1360 ± 236	NS	1325 ± 83	$p < 0.005$	1851 ± 350
ml/min/kg STPD	37 ± 2	NS	36.7 ± 3	$0.001 < p < 0.01$	48.4 ± 11
HR:min⁻¹	179 ± 12	NS	175 ± 15	NS	175 ± 6
Ht(%)	53.0 ± 4.6	NS	52.0 ± 3.1	NS	53.2 ± 4
Hb(%)	17.35 ± 2.01	NS	17.16 ± 1.03	NS	16.87 ± 0.65
Pa_{O_2}(mmHg)	67.86 ± 4.67	NS	66.83 ± 5.36	NS	64.59 ± 4.64
Pa_{CO_2}(mmHg)	21.95 ± 8.68	NS	21.84 ± 2.85	$0.01 < p < 0.02$	25.25 ± 3.98
pH_a	7.134 ± 0.243	NS	7.288 ± 0.100	$0.05 < p < 0.10$	7.384 ± 0.095
Lac. Ac. (mmol/liter)	12.81 ± 2.64	NS	14.84 ± 3.04	$0.01 < p < 0.02$	20.08 ± 3.21

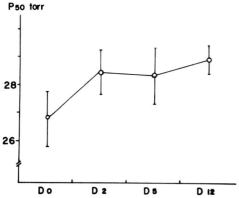

Fig. 11-2. Variation of P_{50} during 12 days of acclimatization to high altitude.

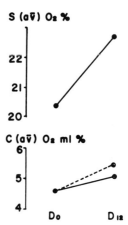

Fig. 11-4 Concentration of O_2 during 12 days of acclimatization to high altitude.

12th day (Fig. 11-2). This change parallels the increase in 2,3-DPG (0.73 ± 0.05 to 1.12 ± 0.08 mol/mol Hb). ATP also increased (0.23 ± 0.02 to 0.36 ± 0.02 mol/mol Hb; Fig. 11-3).

The decrease in oxygen affinity of hemoglobin benefits the organism mostly at the level of the arteriovenous difference, either expressed in saturation or content of oxygen ($\triangle S$ $a\text{-}v_{O_2}$ or $\triangle C$ $a\text{-}v_{O_2}$). With the increase in P_{50} alone the arteriovenous oxygen gradient increases by 11%. This allows the tissues at rest to extract 5 ml more O_2 per liter of blood (Fig. 11-4). The rapid acclimatization process at the erythrocyte level benefits the newcomer athlete when the competitions are held.

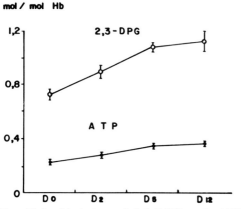

Fig. 11-3. Variation of 2,3-DPG and ATP during 12 days of acclimatization to high altitude.

References

1. Alexander, J.K., Hortleg, L.H., Modelski, M., and Grover, R.F.: Reduction of stroke volume during exercise in man following ascent to 3100 m altitude. J. Appl. Physiol. 23:849, 1967.
2. Arnaud, J. and Gutierrez, N.: Funcion respiratoria del globulo rojo en la altura (La Paz, 3500 m). Rev. Inst. Boliviano Biol. Altura 5(4):19, 1974.
2a. Aste Salazar, H., and Hurtado, A.: The affinity of hemoglobin for oxygen at sea level and at high altitudes. Inst. de Biol. Andina. Am. J. Physiol. 142:733, 1944.
3. Balke, B., Nagle, F.J., and Daniels, J.: Altitude and maximal performance in work and sports activity. J.A.M.A. 194:6, 1965.
4. Durand, J., Martineaud, J.P., and Seroussi, S.: Effect of altitude on resistance and capacitance of the skin vessels. Aspen, Colorado, International Symposium on Altitude and Cold, 1968.
5. Durand, J., Raynaud, J., Martineaud, J.P., Varene, P., and Coudert, J.: Les echanges energetiques au cours de l'exercice musculaire a l'altitude. C.N.R.S. Definition et analyse biologique des populations amerindiennes. Etude de leur environnement. R.C.S.P. 87:153, 1973.
6. Funahashi, A. and Kobayashi, Y.: Bibliographical study on the effects of high altitude on aerobic work capacity; mainly, maximal O_2 uptake and heart rate. Natural Sci. 22(11):13–19 1, 1973.
7. Hagerman, F., Addington, W., and

Gaensler, E.: Severe steady exercise at sea level and altitude in Olympic oarsmen. Med. Sci. Sports 7(4): 275, 1975.
8. Hebbelick, M. and Dirix, A.: A physiological study on Belgian cyclists during the tour of Mexico. J. Sports Med. Phys. Fitness 8(2):1, 23-30, 1968.
9. Hurtado, A.: Influencia de la altura sobre el hombre. WHO, 1970.
10. Ignazi, G. and Paz-Zamora, M.: Dimensions du thorax et mesures spirometriques dans une serie de jeunes aymaras de l'altiplano bolivien. Cah. Anthropol. 4:57, 1975.
11. Knuttgen, H.G.: Oxygen debt, lactate, pyruvate, and excess lactate after muscular work. J. Appl. Physiol. 2:4, 1962.
12. Lahiri, S., Kao, F.F., Velasquez, T., Martinez, C., and Pezzia, W.: Irreversible blunted respiratory sensitivity to hypoxia in high altitude natives. Respir. Physiol. 6:360, 1969.
13. Lefrancois, R., Pasquis, P., and Gautier, H.: Mecanisme de l'adaptation respiratoire a la haute altitude. Anthropologie des populations andines. INSERM 63:259, 1976.
13a. Lenfant, C., Torrance, J.D., English, E., Finch, C.A., Reynaforge, C., Ramos, J., and Faura, J. Effect of altitude on oxygen binding by hemoglobin and on organic phosphate levels. J. Clin. Invest. 47:2652, 1968.
14. Lockhart, A., Vargas, E., Bisseliches, F., Antezana, G., and Paz-Zamora, M.: Effects hemodynamiques de l'orthostatisme en altitude. J. Physiol. 63:70, 1971.
15. Paz-Zamora, M., Coudert, J., Freminet, A., Gascard, J.P., and Durand, J.: Aerobic and anaerobic capacity at 3750 m in resident athletes and sea level athletes during acclimatization. In: Sport in the Modern World. Changes and Problems. Munich, Organizing Committee for the Games of the XXth Olympiad, 1972.
16. Paz-Zamora, M., Coudert, J., and Lefrancois, R.: Sensibilidad ventilatoria al oxigeno con relacion al tiempo de estadia en la altura. Rev. I.B.B.A. 4(17):1, page 5, 1972.
17. Paz-Zamora, M., Ergueta Collao, J., Antezana, G., Vargas, E., Coudert, J., Gutierrez, N., Haftel, W., and Valot, J.L.: Parámetros Biologicos Normales (La Paz, 3500 m). Special publication of Comite Organizador de los VIII Juegos Bolivarianos. Comite Olimpico de Bolivia, 1974.
18. Rasmussen, S.A.: Exercise physiology at the cellular level. J. Sports Med. Phys. Fitness 12(2): 197, 1972.
19. Scano, A. and Venerando, A.: Studi sull acclimatazione degli atleti italiani a citta del Messico. Dal Coni Scuola Centrale dello Sport. Rome, Inst. di Medicin delle Sport, 1968.
20. Soren, C., Sorensen, S.C., Lassen, N., Severinghaus, J.W., Coudert, J., and Paz-Zamora, M.: Cerebral glucose metabolism and cerebral blood flow in high altitude residents. J. Appl. Physiol. 37(3):1, 1974.
21. Sorensen, S.C. and Severinghaus, J.W.: Irreversible respiratory insensitivity to acute hypoxia in man born at high altitude. J. Appl. Physiol. 25:217, 1968.
22. Strauzenberg, E.: Short outline of the problem of competition at altitude above 2000 m. J. Sport Med, 1976.
23. Velasquez, T.: Aspects of physical activity in high altitude natives. Am. J. Physiol. Anthropol. 32:251, 1970.

12

The Effects of Hypoxia on Maximal Anaerobic Alactic Power in Man

P. E. di Prampero, P. Mognoni, and A. Veicsteinas

Many types of physical exertion are characterized by a baseline of moderate activity interrupted by sudden bursts of violent effort. In these conditions the muscle energy expenditure varies over a very large range in a practically instantaneous way. Indeed the whole body energy expenditure can increase in a few tenths of a second by a factor of 40, or even more in athletes; i.e., from rest (approx. 4 ml · min^{-1} · kg^{-1} O_2 consumption) to an energy requirement of 160–220 ml · min^{-1} · kg^{-1}, if expressed in O_2 equivalent, during maximal effort (2,4,7).

Therefore the assessment of maximal power (a) provides useful practical information for athletes, coaches, and sportsmen in general; and (b) allows deeper insight to be gained into the processes of muscle energetics in man. It thus becomes apparent why the development of reliable and easy methods for the assessment of maximal power has been given constant attention in the field of exercise and environmental physiology (5,7).

To our knowledge, however, no measurements of maximal anaerobic power in man during acute or chronic hypoxia have been reported. Therefore, we set out to study the effects of hypoxia on muscle anaerobic metabolism during short bursts of exhausting exercise.

Principle

At the beginning of muscular exercise, O_2 consumption and lactate (La) production do not contribute significantly to the energy requirement of the working muscles. It is well known that these two energy-yielding mechanisms are relatively sluggish in comparison with the mechanical events of the contraction (2), these last being entirely dependent on ATP splitting. As a consequence, during the first few seconds of exercise, the energetics of muscular work can be described as follows:

$$\dot{W} = n \cdot \dot{Al}$$

where $\dot{W}$ is the mechanical power output; $\dot{Al}$ the net utilization, per unit time, of alactic anaerobic energy sources (ATP and phosphocreatine splitting); and n is the mechanical efficiency of the exercise, as defined by the ratio between external mechanical power and rate of energy expenditure. Obviously, $\dot{W}$ and $\dot{Al}$ must be expressed in the same units. Throughout this study the following equivalents will be utilized: 1 ml O_2 = 5 cal = 20.9 J = 2.135 kgm (at RQ = 0.96).

Equation 12-1 can be utilized to assess the maximal anaerobic alactic power ($\dot{A}l_{max}$) in man, from measurements of $\dot{W}_{max}$ during short-term exercise (several seconds) of extremely high intensity ("all-out" efforts), the efficiency of which is known. If these conditions are met, the resulting $\dot{A}l_{max}$ is a measure of the maximal rate of high energy phosphates splitting in man. This principle has originally been proposed and exploited by Margaria et al. (7) to determine $\dot{A}l_{max}$ from the vertical component of the velocity in subjects running up a normal flight of stairs at top speed.

Methods

In this study $\dot{A}l_{max}$ was determined from the peak power output during an "all-out" cycloergometric effort as originally proposed by Ikuta and Ikai (5).

The experiments were performed on eight nonathletic subjects whose physical characteristics are reported in Table 12-1. The exercises were performed on a Monark cycloergometer. The load allowing maximal absolute power output to be reached had been previously determined: it ranged from 4.9 to 6.0 kg. The subjects were then asked to perform an all-out effort of 7 s duration. The time for each pedal revolution was recorded throughout the test by means

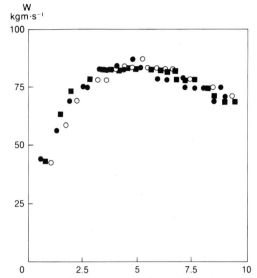

Fig. 12-1. Time course of mechanical power output ($\dot{W}$, kgm·s^{-1}) over a 10-s "all-out" effort on the cycloergometer on one subject. Average of three trials at sea level (■), at 3000 (●), and 4500 (○) m simulated altitude (FI_{O_2} = 0.145 and 0.121), respectively.

of a microswitch on the crankshaft. The time course of the mechanical power output could then be calculated (Fig. 12-1).

The following sets of conditions were studied: (a) PB = 750 torr while breathing (a) 100% O_2, (b) 14.5% O_2 in N_2 (corresponding to an altitude of 3000 m a.s.l.), and (c) 12.1% O_2 in N_2 (corresponding to an altitude of 4500 m a.s.l.); and (b) after 3 weeks acclimatization at 4540 m a.s.l. on the Peruvian Andes (Morococha).

In all these conditions $\dot{W}_{max}$ was determined from the mean of the three highest power values (over 1 cycle) observed in three separate trials. In all cases the differences between separate $\dot{W}_{max}$ measurements in the same experimental conditions did not exceed 2%.

In addition, for the exercises performed when breathing pure O_2, the blood La concentration was determined in samples drawn from the antecubital vein immediately before the exercise and in the following recovery period, up to the tenth minute, at 1-min intervals.

Table 12-1. Physical Characteristics of Subjects.

Subject	Age (years)	Height (cm)	Weight (kg)
P.M.	42	192	87
C.L.	25	176	63
A.M.	24	172	67
A.V.	34	170	55
L.B.	27	178	77
F.B.	28	165	57
M.C.	28	171	74
C.S.	28	176	78

Results

The external mechanical power output ($\dot{W}$, kgm · s^{-1}) is indicated in Fig. 12-1 as a function of the time from the onset of the test for one subject in the three experimental conditions at PB = 750 torr. $\dot{W}$ increases during the first 3 s of exercise to a maximal level which is maintained up to the sixth second and declines thereafter.

In Table 12-2 $\dot{W}_{max}$, average of three trials, has been indicated in all experimental conditions together with the load at which $\dot{W}_{max}$ was attained. $\dot{W}_{max}$ amounted on the average ($\pm$ SD) to 1.11 $\pm$ 0.12, 1.13 $\pm$ 0.10, 1.12 $\pm$ 0.10, 1.10 $\pm$ 0.09 kgm · kg^{-1} · s^{-1} in the four experimental conditions of this study: Aa, Ab, Ac, and B, respectively.

In all cases the pedal frequency was of the order of 150–160 min^{-1}. At this frequency the average mechanical efficiency of the exercise, as determined from the ratio between mechanical work performed and overall O$_2$ consumed (above resting) during the exercise and in the following 15 min of recovery, is 0.20.

Assuming $n = 0.20$, $\dot{A}l_{max}$ can therefore be calculated from Eq. 12-1: it amounted on the average to 159 $\pm$ 15, 162 $\pm$ 14, 160 $\pm$ 12, and 156 $\pm$ 13 ml · kg^{-1} · min^{-1} in the four experimental conditions of this study (Table 12-2).

The lactic acid concentration in blood increased significantly in the recovery period. The peak La concentration in blood, attained 5 to 7 min after the end of the exercise, amounted on the average to 2 mM ($n = 16$) above resting (range 1 to 4).

Discussion

The absolute $\dot{W}_{max}$ values reported in Table 12-2 are slightly lower than those obtained on four sprinters by Murase et al. (10) with the same method, which amounted on the average to 1.37 kgm · kg^{-1} · min^{-1}. They compare favorably, however, with those reported in the original paper by Ikuta and Ikai (5). Also the Al_{max} values, as calculated in Table 12-2, are substantially equal to those obtained by Margaria's staircase method on nonathletic subjects of similar age (7).

It appears from Table 12-2 that the maximal anaerobic power, as measured, is not affected by the experimental conditions of this study. Thus, it seems legitimate to conclude that the energetic mechanism underlying this type of exercise (mainly ATP and PC splitting) is not affected by acute or chronic hypoxia, at least within the investigated range. This finding is consistent with Knuttgen and Saltin's (6) data, which show that the ATP and PC concentrations are not affected by acute hypoxia, at least up to a simulated altitude of 4000 m a.s.l. (462 torr).

It was stated in the introduction that the energetic processes tested by this procedure are supposed to involve the splitting of high energy phosphates ($\sim$P). The aim of the discussion that follows is to show that this statement is true within reasonable approximation, and that therefore $\dot{A}l_{max}$ is indeed a measure of the maximal rate of $\sim$P splitting.

The overall increase in blood La concentration as an effect of the exercise (2 mM) is equivalent to an O$_2$ consumption of 6.7 ml · kg^{-1} (the increase of 1 mM in blood La concentration, according to Margaria et al. (8), being equivalent to an O$_2$ consumption of 3.3 ml · kg^{-1}). On the other hand, the overall mechanical work performed during the 7-s period amounted on the average to 7 kgm · kg^{-1}, i.e., on the basis of an efficiency of 0.20, to 18.3 ml O$_2$ · kg^{-1}.

On this basis alone, therefore, the La contribution to the energy requirement would amount to 6.7/18.3 = 0.37. A similar figure is arrived at if, instead of overall work and La, mechanical power and rate of La accumulation in blood are considered.

This figure applies to the overall energy balance at the fifth minute of recovery; it is presumably in excess, however, if the actual maximal power output during the test itself is considered. This is so because (a) it

Table 12-2. $\dot{W}_{max}$ (kgm · kg^{-1} · s^{-1}), $\dot{A}l_{max}$ (ml O_2 · kg^{-1} · s^{-1}), and Load (kg) on All Eight Subjects in the Four Experimental Conditions of this Study: (A) PB, 750 torr, While Breathing (a) 100% O_2, (b) 14.5% O_2, (c) 12.1% O_2; and (B) after 3 Weeks at 4540 m a.s.l. (PB, 430 torr). Each Value is the Mean of Three Determinations. Average and SD are also reported (n = 24).

Subject	100% O_2			14.5% O_2			12.1% O_2			After 3 weeks at 4540 m a.s.l.		
	Load (kg)	$\dot{W}_{max}$	$\dot{A}l_{max}$	Load (kg)	$\dot{W}_{max}$	$\dot{A}l_{max}$	Load (kg)	$\dot{W}_{max}$	$\dot{A}l_{max}$	Load (kg)	$\dot{W}_{max}$	$\dot{A}l_{max}$
P.M.	5.97	1.036	147.99	5.70	1.061	151.57	5.75	1.049	151.0	5.75	1.029	147.0
C.L.	6.00	1.260	182.28	5.50	1.249	176.44	5.15	1.189	171.0	5.15	1.169	167.0
A.M.	5.05	1.308	186.85	5.50	1.320	188.57	4.90	1.283	182.4	4.90	1.263	180.4
A.V.	5.35	1.308	168.57	5.50	1.201	171.57	5.45	1.180	167.7	5.45	1.159	165.6
L.B.	5.10	0.928	132.57	6.00	0.942	134.57	5.00	0.900	136.6	5.00	0.872	124.6
F.B.	5.00	1.163	166.14	5.00	1.232	176.00	5.00	1.155	173.0	5.00	1.176	168.0
M.C.	5.00	1.044	149.14	5.50	1.081	154.42	5.50	1.060	154.4	5.60	1.039	148.4
C.S.	5.45	0.971	136.71	5.60	1.032	147.42	5.55	1.080	155.7	5.45	1.062	151.7
Mean ± SD	5.36 ± 0.35	1.107 ± 0.12	159.03 ± 15.4	5.45 ± 0.20	1.131 ± 0.10	161.57 ± 14.5	5.28 ± 0.20	1.120 ± 0.10	160.4 ± 12.0	5.28 ± 6.20	1.10 ± 0.09	156.4 ± 13

has been shown (3) that a certain fraction, up to two-thirds, of the overall La found in blood at the end of exhausting exercise, is produced in the recovery after the exercise is over: "anaerobic recovery" (1); and (b) $\dot{W}_{max}$ in these conditions is attained within 3 s from the onset of exercise (Fig. 12-1), a time too short for the rate of La production to rise substantially (9).

The experiments of Margaria et al. (9) during treadmill running at 18 km·h^{-1} (5 m·s^{-1}) at 20% incline indicate in fact that no accumulation of La in blood takes place for exercise durations up to 4 s. This type of exercise is comparable to ours both in terms of external power output (1.0 kgm·kg^{-1}·s^{-1}) and in terms of exhaustion time (approx. 10 s). It can then reasonably be assumed that also in our experimental conditions no substantial La production occurs during the first seconds of exercise, i.e., when $\dot{W}_{max}$ is attained.

In the above experiments of Margaria et al. (9) the La determinations have been made on blood samples, a procedure that may be questioned on several grounds. However, similar conclusions can be arrived at from the data of Saltin and Essén (11) during intermittent cycloergometer exercise at 2400 kgm·min^{-1} (0.60 kgm·kg^{-1}·s^{-1}) load. In these experiments muscle La and PC concentrations were determined immediately after working periods of 10, 20, 30, and 60 s duration. The La and PC concentrations in muscle tissue are indicated in Fig. 12-2 as a function of the exercise duration. It appears from this figure that (a) La accumulation in the muscle is very limited in the first few seconds of exercise (indeed, linear extrapolation of the La vs time function indicates that no La production occurs for $t < 6$ s); and (b) PC utilization is proportionately greater in shorter exercises.

These experiments therefore confirm the hypothesis that La production in the first few seconds of exercise is indeed negligible. It must be pointed out, however, that the work intensity in Saltin and Essén's ex-

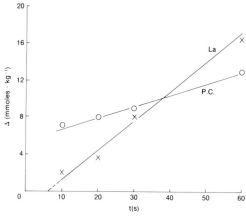

Fig. 12-2. Net lactic acid accumulation and net PC breakdown in the muscle ($\triangle$ mmol·kg^{-1}) as a function of time (s) during cycloergometer exercise. La and PC concentrations were determined by muscle biopsies in man immediately after the end of working periods of 10, 20, 30, and 60 s, during a study on intermittent exercise by Saltin and Essén (11) (rest/work ratio = 2 in all cases). Work load = 40 kgm·s^{-1}. From the data of Saltin and Essén (11).

periments was lower than in our case (0.6 vs 1.1 kgm·kg^{-1}·s^{-1}).

In conclusion, measurements of the maximal power developed during short "all-out" bursts of cycloergometric exercise allow assessment of the maximal rate of $\sim$ P splitting in man ($\dot{Al}_{max}$). This amounts on the average to 160 ml O$_2$·kg^{-1}·min^{-1} in oxygen equivalents and is not affected by acute or chronic (3 weeks) hypoxia, at least up to 4500 m a.s.l.

References

1. Cerretelli, P., Ambrosoli, G., and Fumagalli, M.: Anaerobic recovery in man. Eur. J. Appl. Physiol. 34:141, 1975.
2. di Prampero, P.E.: Energetics of muscular exercise. Rev. Physiol. Biochem. Pharmacol. 89:143, 1981.
3. di Prampero, P.E., Peeters, L., and Margaria, R.: Alactic O$_2$ debt and lactic acid production after exhausting exercise in man. J. Appl. Physiol. 34:628, 1973.
4. di Prampero, P.E., Piñera-Limas, F., and

Sassi, G.: Maximal muscular power (aerobic and anaerobic) in 116 athletes performing at the XIX Olympic Games in Mexico. Ergonomics 13:665, 1970.
5. Ikuta, K. and Ikai, M.: Study on the development of maximum anaerobic power in man with bicycle ergometer. Res. J. Phys. Ed. (Japan). 17:151, 1972.
6. Knuttgen, H.G. and Saltin, B.: Oxygen uptake, muscle high energy phosphates, and lactate in exercise under acute hypoxic conditions in man. Acta Physiol. Scand. 87:368, 1973.
7. Margaria, R., Aghemo, P., and Rovelli, E.: Measurement of muscular power (anaerobic) in man. J. Appl. Physiol. 21:1662, 1966.
8. Margaria, R., Aghemo, P., and Sassi, G.: Lactic acid production in supramaximal exercise. Pflügers Arch. 326:152, 1971.
9. Margaria, R., Cerretelli, P., and Mangili, F.: Balance and kinetics of anaerobic energy release during strenuous exercise in man. J. Appl. Physiol. 19:623, 1964.
10. Murase, Y., Hoshikawa, T., Yasuda, N., Ikegami, Y., and Matsui, H.: Analysis of the changes in progressive speed during 100-m dash. In Komi, P.V. (ed.): Biomechanics V-B. Baltimore, University Park Press, 1976, pp. 200-207.
11. Saltin, B. and Essén, B.: Muscle glycogen, lactate, ATP and CP in intermittent exercise. In Pernow, B. and Saltin, B. (eds.): Muscle Metabolism During Exercise. New York, Plenum Press, 1971, pp. 419-424.

13

Anaerobic Metabolism at High Altitude: The Lactacid Mechanism

P. CERRETELLI, A. VEICSTEINAS, AND C. MARCONI

The effects of acute and chronic hypoxia on maximal aerobic power have been the object of several extensive studies (4,5). By contrast, the characteristics of maximal anaerobic performance of man at altitude are not well documented.

Acute hypoxia does not cause a reduction of blood maximal lactic acid concentration, $[La_b max]$, after supramaximal exercise and therefore does not affect the maximal lactacid capacity of the subject (3). Because plasma bicarbonate levels drop following hypoxic hypocapnia, a reduction of $[La_b max]$ may be expected, on the contrary, in both altitude natives and acclimatized lowlanders. The aim may be to prevent intolerable changes of blood and tissue pH.

Actually, the results of the pioneering study of Edwards (7) indicate a progressive decrease of $[La_b max]$ in acclimatized lowlanders exercising at higher altitudes. The few measurements carried out by Pugh et al. (18) on some members of the Scientific Himalayan Expedition, by Hansen et al. (9), and by Lahiri et al. (13) seem to confirm this finding. On the other hand, resting blood lactate concentration $[La_b rest]$ data are rather controversial, ranging from 0.8 (the normal sea level value) to about 2 $mmol \cdot liter^{-1}$ (4,13).

In the course of various expeditions to the Khumbu Valley of Nepal at altitudes above 3850 m and to Morococha in the Peruvian Andes (4540 m), an extensive study of anaerobic (lactacid) metabolism was carried out on a large number of natives and acclimatized Caucasian lowlanders. The aim of the study was to assess one or more of the following variables:

1) Resting blood lactate $[La_b rest]$ and/or pyruvate $[Pa_b rest]$.
2) Maximal blood lactate $[La_b max]$, i.e., an indicator of the maximal anaerobic lactacid capacity of the subject, and in some instances, maximal pyruvate $[Pa_b max]$, following continuous or intermittent supramaximal exhaustive work loads of variable duration carried out at increasing altitudes, while breathing air or pure oxygen at ambient pressure.
3) The relationship between blood hydrogen ion concentration $[H^+]$ and $[La_b]$ after supramaximal efforts of varying duration.

4) The kinetics of lactate (La) disappearance from blood during recovery following supramaximal loads while breathing both air and oxygen.

Material and Methods

The measurements were carried out on a total of 41 male subjects. These included 6 Sherpas, 9 Peruvian Indians, and 18 Caucasians, all acclimatized over periods ranging between 3 and 16 weeks to altitudes ranging between 3850 and 7500 m; and 3 Sherpas resident at sea level and 5 nonacclimatized Caucasians carrying out exhaustive work loads in a decompression chamber. The exercise was usually carried out on a bicycle ergometer (Monark, Varberg AB, Sweden); occasionally, the effort consisted in uphill running, since several subjects, particularly the Sherpas, were unable to pedal. Lactate was assessed enzymatically (Test-Combination, Boehringer, Mannheim) from blood samples drawn from the antecubital vein. The [La_b] attained between the 6th and the 12th minute following a supramaximal exhaustive load, when equilibrium between the tissues and the body-fast compartments is usually reached, was taken as La_bmax. Venous blood pH (pH_v) was determined by microelectrodes (Radiometer, Copenhagen) on the same samples drawn for lactate. The effects of O_2 breathing on [La_b], both at rest and at work, were studied while the subject was breathing on a closed circuit system.

Results

Hematologic and Metabolic Characteristics of the Subjects

Mean values, obtained for representative subject subgroups, of erythrocyte counts (RBC), hemoglobin concentration (Hb), hematocrit (Hct), blood gases (Pa_{O_2}, Pa_{CO_2}), venous blood pH (pH_v) both at rest and after an exhaustive anaerobic load, and maximal aerobic power ($\dot{V}_{O_2max}$), appear in Table 13-1. These values are in agreement with most data from the literature (2). In acclimatized lowlanders, pH_v, after very heavy work loads, was found occasionally to drop as low as 7.10, i.e., to levels equal to those usually attained at sea level.

The Effect of Altitude Exposure on Blood Lactate and Pyruvate

Table 13-2 summarizes the data on blood lactate and pyruvate obtained for different groups of subjects in various experimental conditions. Both in air and in O_2, [La_brest] is approximately the same as at sea level; by contrast, [La_bmax] appears to be considerably reduced in both cases. The somewhat higher [La_bmax] values found during O_2 breathing are explained by the fact that the subjects could carry out a given supramaximal work load over a longer period of time. [La_bmax] drops with increasing altitude, as shown in Fig. 13-1. Results from other authors, along with data from Table 13-2, have been included in Fig. 13-1 (9,13,18), as well as the results of measurements obtained on three Sherpas living at sea level and five Caucasians exposed to acute hypoxia. The reduction of maximal lactacid capacity with increasing altitude, as reflected by the drop of [La_bmax], is most likely a function of the reduced plasma bicarbonate concentration consequential to decreased Pa_{CO_2}. Should Pa_{CO_2} at the summit of Mt. Everest drop to 10 torr, the size of the maximal lactacid O_2 debt would be practically nil. The subject could rely only on oxidative energy sources with only minor contributions from alactic sources during unsteady state conditions.

Blood pyruvate concentrations, both at rest and at work, are within the limits found at sea level (14).

Table 13-1. Blood and Metabolic Characteristics of the Subjects.

Subjects	RBC (c/mm³)	Hb (g%)	Hct (%)	Pa_{O_2} (torr)	Pa_{CO_2} (torr)	pH_v (rest)	pH_v (exhaustive exercise)	$\dot{V}_{O_2max}$ (ml · kg⁻¹ · min⁻¹)
Caucasians								
5350 m (PB, 390 torr)	7.0	23	66.3	41	20	7.36	7.15	36.8
3850 m (PB, 470 torr)	5.3	18	49.4	65	24	7.39	7.21	37.4
Sherpas								
5350 m	5.0	20	62.7	42	20	7.38	7.22	39.7
3850 m	4.7	17	53.3	65	27	7.30	7.19	40.6
Peruvian Indians								
4540 m (PB, 440 torr)	5.5	17.5	54	—	—	7.35	7.19	40.5

Table 13-2. Lactic (La) and Pyruvic Acid (Pa) Concentrations in Blood (Mean ± SE), at Rest and Exhaustion, Breathing Air or Pure O_2 at the Indicated Altitudes.

Metabolic conditions	Breathing conditions	Altitude (m)	Subjects		La (mmol · liter^{-1})	Pa (mmol · liter^{-1})
Rest	Air	5350	C		0.97 ± 0.07	0.02 ± 0.1
		5350	Sh		0.92 ± 0.12	0.1
		4540	P.I.		1.09 ± 0.04	
	O_2	5350	C		0.94 ± 0.09	
Exhaustive exercise	Air	5350	C		5.93 ± 0.35	0.6 ± 0.1
		5350	Sh		6.52 ± 1.19	0.6
		4540	P.I.	Continous	5.84 ± 0.10	
				Intermittent	6.63 ± 0.11	
	O_2	5350	C		6.93 ± 0.53	

C, Caucasians; Sh, Sherpas; P.I., Peruvian Indians. From Ceretelli, P.: In West, J.B. (ed.): Pulmonary Gas Exchange. Vol. II. New York, Academic Press, 1980.

Relationship Between Blood Lactate and H$^+$ Concentration

In Fig. 13-2 the relationship between increase in blood H$^+$ ion concentration, $\Delta[H^+]$, as calculated from the drop in pH,

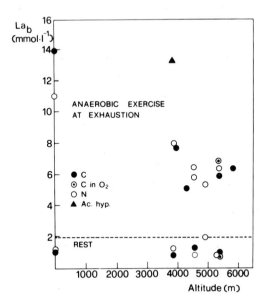

Fig. 13-1. Blood lactate [La$_b$] as a function of altitude. Resting values appear below the dashed line. Maximal exercise values appear above the dashed line. Caucasians breathing air (●), breathing O_2 (⊙), and during acute hypoxia (▲). Natives (Sherpas and Peruvian Indians) breathing air (○).

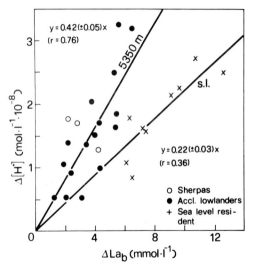

Fig. 13-2. Changes in blood H$^+$ concentration, [H$^+$], as a function of the increase in blood lactate concentration, Δ[La$_b$], in altitude natives (○), lowlanders acclimatized to 5350 m (●), and sea level residents (×). (From ref. 6.)

and the corresponding increase in blood lactic acid concentration, Δ[La$_b$], is shown for a group of sea level residents and for a number of subjects acclimatized to altitudes above 5350 m. Each point was calculated as the difference (Δ) between [H+] and [La$_b$] immediately before and after the end of a supramaximal work load carried out either until exhaustion or over progressively longer fractions of the exhaustion time. Linear regressions through the origin calculated by the least squares method for the group of sea level dwellers (Milan) and for acclimatized individuals (5350 m) allow the calculation of buffer values, β, for the whole body. These values for the two conditions under study are given by the reciprocal of the slope of the lines appearing in Fig. 13-2. At 5350 m, β is approximately half that of the sea level control value. From Fig. 13-2 it also appears that the highest concentration of H^+ attained is approximately the same for the two groups, independent of $[La_b]$. Thus, the maximal lactacid capacity of the subject seems to be reduced at altitude in direct proportion to the drop in the body buffer value and its limit is imposed by some variable apparently related to blood $[H^+]$. It may be hypothesized that lowering of intracellular pH might interfere with the function of a rate-limiting enzyme along the glycolytic sequence, probably phosphofructokinase (16), thus acting as a negative feedback mechanism aimed at preventing further acidification. This mechanism seems to become operational when blood pH falls to about 7.1, independent of other conditions.

Changes of the Body Buffer Value During Altitude Acclimatization

Recently, Piiper (17) has calculated the drop in blood pH in man at sea level for a given lactic acid production and a particular modality of H^+ ion buffering. A similar approach will be extended here to an acclimatized subject.

For Piiper's calculations (Fig. 13-3A) the following assumptions were made:

1) A muscle mass of 28 kg with an intracellular space of 14 liters H_2O and an extracellular space of 7 liters H_2O.
2) A lactate accumulation of 560 mEq.
3) A limitation of H^+ ion buffering to the muscle intra- (ICS) and extracellular (ECS) space and to blood (5 liters, containing 4.1 liters H_2O).
4) A nonbicarbonate buffer value for in-

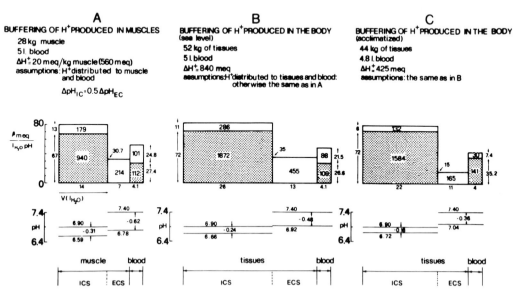

Fig. 13-3. Buffering of H^+ ions formed in muscle. Unshaded areas, bicarbonate; shaded areas, nonbicarbonate. **Upper panel** Abscissa: volume, V. (liters H_2O). Ordinate: buffer value, β; Area: buffer capacity. **Lower panel** pH changes. ICS, intracellular space; ECS, extracellular space. **A** Sea level conditions: La formed is buffered by muscles and blood only (from ref. 17). **B** Sea level conditions: La is buffered by all tissues except bones. **C** Altitude acclimatization: same as in **B**.

tracellular muscle water of 67 mEq · pH^{-1} · kg H$_2$O (11,12).
5) A Pa_{CO_2} of 40 torr.
6) A distribution of H$^+$ between the intra- and extracellular spaces of the muscle so that $(-\Delta pH_i / -\Delta pH_e) = 0.5$ (10).

The estimated blood pH drop was 0.62 of a unit.

Following Piiper's approach (Fig. 13-3B and C) a comparative estimate was attempted of the blood pH drop expected in a sea level dweller and in an acclimatized lowlander, for assumed amounts of lactic acid (ratio 1 to 0.5) produced as an effect of a supramaximal effort.

Piiper's assumptions were modified as follows:

1) The overall amount of lactic acid produced at exhaustion was 840 and 425 mEq for the sea level resident and for the acclimatized lowlander, respectively.
2) The buffering of H$^+$ was extended to all tissues with the exception of the bones, i.e., 52 and 44 kg, respectively, for the sea level control and the acclimatized subject.
3) The average tissue buffer value was increased to 72 mEq · pH^{-1} · kg H$_2$O (Heisler, personal communication, 1978).
4) The Pa_{CO_2} of the acclimatized subject was 20 torr.
5) Blood hemoglobin concentration in the acclimatized subject was 22 g%.

The calculated blood pH changes were 0.48 of a pH unit in the sea level control and 0.36 in the acclimatized subject. These values, very close to those observed experimentally, are compatible with a large drop, even though not to half its original value, of the body buffer value, β, at altitude. The discrepancy between the observed (Fig. 13-2) and the calculated (Fig. 13-3) reduction of β depends obviously on the error of some of the assumptions.

From Fig. 13-3 it may also be seen that most of the drop of β can be ascribed to the reduction of bicarbonate buffers, particularly in the ECS. The increase in Hb concentration plays only a minor role in the overall buffer capacity of the body.

The Effects of Intermittent Exercise on Lactacid O$_2$ Debt at Altitude

An alternate explanation for the lower lactacid capacity found in acclimatized lowlanders and altitude natives could be a reduction in muscle glycogen stores due to chronic hypoxia. Although muscle mass is certainly reduced after exposure to low oxygen (8), probably because of a partial mobilization of muscle proteins, to our knowledge no measurements have been made of either muscle glycogen concentration of glycogen mobilization during muscular work in man exposed to prolonged hypoxia. A functional estimate of the amount of glycogen available for glycolysis was recently attempted on a group of seven Peruvian Indians native to Morococha. The subjects were asked to perform supramaximal efforts (1300–1600 kgm · min^{-1}) on a bicycle ergometer following two different procedures: (1) a 2- to 3-min continuous performance; and (2) a series of four exercise bursts each lasting 80–100 s and followed by a 90 s pause. It was shown by Benadé and Heisler (1) that the kinetics of diffusion of H$^+$ ions from the muscle cells is faster ($t_{1/2}$, 39 s) than that of lactate ions ($t_{1/2}$, 9 min). Thus, a possible blockage of a rate-limiting enzyme by increased [H$^+$] in the cytosol could be temporarily released during the resting phase following each exercise period. As a consequence, more glycogen, if available, could be converted into lactic acid as an effect of repeated working periods rather than during a single exhaustive work load. [La$_b$max] determined during the recovery phase following the two described exercise modes was found to be 5.8 mmol · liter^{-1} (after continuous) vs 6.6 mmol · liter^{-1} (after repeated)

(Table 13-2). Unfortunately, the difference is too small to determine whether the size of available glycogen stores may be the limiting factor to anaerobic exercise at altitude.

The Kinetics of Lactate Disappearance from Blood During Recovery

The pattern of La_b disappearance from blood during recovery in acclimatized lowlanders is rather peculiar. At sea level, as is well known (15), $[La_b]$ drops exponentially shortly after the end of exercise: a variable delay (0–6 min) observed at the onset of the recovery phase depends on the intensity of the preceding work load and/or on the absolute level of $[La_b]$. By contrast, at altitude, a long delay (8–20 min) is observed before $[La_b]$ starts dropping in spite of its relatively low concentration, particularly during oxygen breathing (Fig. 13-5). Following the delay, La_b disappears at a rate similar to that observed at sea level (Figs. 13-4 and 13-6). This indicates that lactate utilization as a fuel and/or gluconeogenesis from lactate are unaffected by altitude. To explain the long delay observed in the La_b recovery curve following supramaximal exercise it has been hypothesized that at altitude the nutritional perfusion of the muscle during and after the effort, particularly when breathing oxygen, is impaired, thus reducing the rate of lactate release into blood. The observed "plateau" in $[La_b]$ would then be the result of the interplay between rates of release and removal of La from the tissues and may not represent the arterial La_bmax, which therefore in Table 13-2 and Fig. 13-1 might be somewhat underestimated. The possibility that the nutritional circulation of exercising muscles of acclimatized lowlanders is impaired both when breathing ambient air and pure O_2 has already been discussed elsewhere in this volume (Chapter 2, *this volume*).

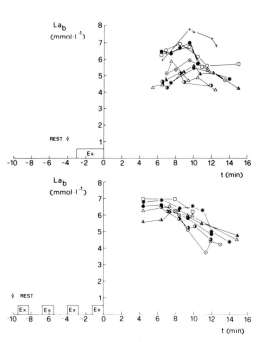

Fig. 13-4. Blood lactate $[La_b]$ during recovery following continuous (2–3 min) or intermittent (four periods of 80–100 s work interrupted by three 90-s intervals) anaerobic exercises.

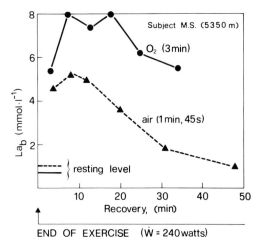

Fig. 13-5. Blood lactate concentration $[La_b]$ during recovery following exercise performed at 5350 m. *Dashed line*, breathing air; *solid line*, breathing O_2. Work load, 240 W; time in parentheses indicates duration of work to exhaustion. (From ref. 5.)

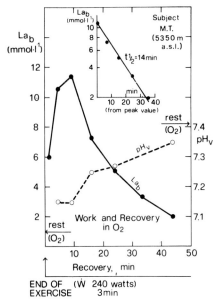

Fig. 13-6. Blood lactate [La_b] and pH_v following a 3-min exercise (240 W) at 5350 m. It may be observed that at the end of exercise, after a 10-min rise in [La_b], lactate is removed according to an exponential function with a halftime of 14 min (see inset at top). Correspondingly, venous blood pH (pH_v), after the initial drop, tends to resume its resting level.

Summary and Conclusions

High altitude exposure impairs the maximal anaerobic (lactacid) performance of man. The reduction of the energy released by way of anaerobic glycolysis parallels the drop of bicarbonate in plasma and in the extracellular space. At 5350 m, along with a reduction of the body buffer value to about half its sea level control value, the maximal lactacid capacity of the subject is halved. The extent to which anaerobic glycolysis operates appears to be limited by a negative feedback mechanism probably triggered by intracellular [H^+] when blood pH falls to about 7.1.

The rate of lactate disappearance from blood after supramaximal efforts at altitude is the same as at sea level ($t_{1/2} = 15-18$ min). However, lactic acid diffusion from the muscle into blood seems to be impaired by changes in microcirculation occurring during and/or after exercise.

References

1. Benadé, A.J.S. and Heisler, N.: Comparison of efflux rates of hydrogen and lactate ions from isolated muscles in vitro. Respir. Physiol. 32:369, 1978.
2. Biology Data Book, Vol. 3, 2nd ed.: Bethesda, Federation of American Societies for Experimental Biology, 1974.
3. Cerretelli, P.: Lactacid O_2 debt in acute and chronic hypoxia. In Margaria, R. (ed.): Exercise at Altitude. Amsterdam, Excerpta Medica, 1967, pp. 58–64.
4. Cerretelli, P.: Limiting factors to oxygen transport on Mount Everest. J. Appl. Physiol. 40:658, 1976.
5. Cerretelli, P.: Metabolismo ossidativo ed anaerobico nel soggetto acclimatato all' altitudine. Minerva Aerosp. 67:11, 1976.
6. Cerretelli, P.: Gas exchange at high altitude. In West, J.B. (ed.): Pulmonary Gas Exchange. Vol. II. New York, Academic Press, 1980.
7. Edwards, H.T.: Lactic acid in rest and work at high altitude. Am. J. Physiol. 116:367, 1936.
8. Gold, A.J., Johnson, T.F., and Costello, L.C.: Effects of altitude stress on mitochondrial function. Am. J. Physiol. 224:946, 1973.
9. Hansen, J.E., Stelter, G.P., and Vogel, J.A.: Arterial pyruvate, lactate, pH, and P_{CO_2} during work at sea level and high altitude. J. Appl. Physiol. 23:243, 1975.
10. Heisler, N.: Intracellular pH of isolated rat diaphragm muscle with metabolic and respiratory changes of extracellular pH. Respir. Physiol. 23:243, 1975.
11. Heisler, N. and Piiper, J.: The buffer value of rat diaphragm muscle tissue determined by P_{CO_2} equilibration of homogenates. Respir. Physiol. 12:169, 1971.
12. Heisler, N. and Piiper, J.: Determination of intracellular buffering properties in rat diaphragm muscle. Am. J. Physiol. 222:747, 1972.
13. Lahiri, S., Milledge, J.S., Chattopadhyay, H.P., Bhattacharyya, A.K., and Sinha,

A.K.: Respiration and heart rate of Sherpa highlanders during exercise. J. Appl. Physiol. 23:545, 1967.
14. Margaria, R., Cerretelli, P., di Prampero, P.E., Massari, C., and Torelli, G.: Kinetics and mechanism of oxygen debt contraction in man. J. Appl. Physiol. 18:371, 1963.
15. Margaria, R., Edwards, H.T., and Dill, D.B.: The possible mechanism of contracting and paying the oxygen debt and the role of lactic acid in muscular contraction. Am. J. Physiol. 106:689, 1933.
16. Newsholme, E.A. and Start, C.: Regulation in Metabolism. London, Wiley, 1973.
17. Piiper, J.: Buffering of lactic acid produced in exercising muscle. Proc. Symp. Onset of Exercise. Toulouse, 1971, pp. 175–185.
18. Pugh, L.G.C.E., Gill, M.B., Lahiri, S., Milledge, J.S., Ward, M.P., and West, J.B.: Muscular exercise at great altitudes. J. Appl. Physiol. 19:431, 1964.

14

Oxygen Deficit and Debt in Submaximal Exercise at Sea Level and High Altitude

J. Raynaud and J. Durand

This study was designed to follow $\dot{V}_{O_2}$ changes at onset and offset of exercise in order to compare by indirect method the relative role played by the two creditors of complementary energy, creatine phosphate and anaerobic glycolysis, in lowlanders studied at sea level (SL) and after translocation to high altitude (HA).

Subjects and Methods

Three subjects were studied comparatively at SL (Paris) and after 3 weeks in La Paz (3800 m). Protocols and equipment were identical in the two places. A 30-min rest period preceded exercise performed on a bicycle at a pedaling frequency of 60 rpm, at three different work loads for two durations, 10 and 25 min. Each test was performed twice; $\dot{V}_{O_2}$ was measured by open circuit method every minute during the transient phases at the beginning and end of exercise and over longer periods of time when the changes expected were not so abrupt. Initial O_2 deficit was computed by adding the sequential differences between steady state V_{O_2} ($V_{O_2 \, s \cdot st}$) and actual $\dot{V}_{O_2}$ during the initial transient phase. Final debt was computed as the sum of the differences between actual sequential $\dot{V}_{O_2}$ measured during recovery and rest $\dot{V}_{O_2}$. The first minute of deficit and debt was considered as representative of the fast components. The slow component was estimated as the difference between total value and the value of the first minute. Blood samples were drawn repeatedly for determinations of plasma lactic acid concentration, [La], during the 25-min exercise performed at the highest work load. Maximal O_2 uptake ($\dot{V}_{O_2 \, max}$) was determined by Maritz's indirect method (8); the maximal heart rate at HA was taken as equal to that at SL minus 10 (7).

Results

Steady State Data

$\dot{V}_{O_2 \, s \cdot st}$ is linearly related to the mechanical work and the slope is the same at SL and HA. $\dot{V}_{O_2 max}$ is decreased at HA by about 15%.

Transient Phases Data

The first minute deficit follows a linear relationship with $\dot{V}_{O_2\,s\cdot st}$. The slope has the dimension of time and has the same value at SL and HA (SL, 53 min; HA, 54 min). Similarly, the actual values of the first minute debt plotted against $\dot{V}_{O_2\,s\cdot st}$ describe a slope identical at SL and HA, but steeper (60 min) than that described by the first minute deficit. There is no significant difference in the slopes made with 10- or 25-min exercises. Thus, the first minute debt is independent of the duration of exercise and inspired P_{O_2} (Fig. 14-1). The slow component of the deficit increases linearly with relative work load; the intercept suggests that anaerobic threshold appears at about 45% of $\dot{V}_{O_2max}$ at SL and HA.

When the values of total debt at SL are plotted against relative work load, the mean

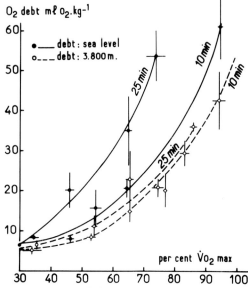

Fig. 14-2. Total debt at SL and HA in 10- and 25-min exercises. Debt increases with duration of work and is smaller at HA. Differences between the SL and HA curves are much more obvious in 25-min than in 10-min exercise.

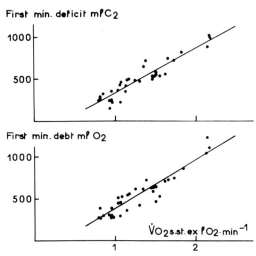

Fig. 14-1. Actual values of first minute deficit and debt, which are assumed to be an index of the fast component of deficit and debt, are plotted against $\dot{V}_{O_2}$ at steady state of exercise. These slopes are defined by data obtained at SL and the corresponding equations are:

y (first min deficit) = 0.53 × −198; r = 0.94
y (first min debt) = 0.60 × −221; r = 0.95
At HA, the following equations are obtained:
y (first min deficit) = 0.54 × −200; r = 0.95
y (first min debt) = 0.60 × −215; r = 0.90

Slope and intercept at HA are identical to those at SL.

points describe a curvilinear relationship. Moreover, the points corresponding to 25-min exercise are always situated above those corresponding to 10-min exercise. Therefore, total debt increases not only with intensity but also duration of exercise. At HA, these findings are noted again, but for given conditions of work load and duration, total debt is always smaller at HA than at SL (Fig. 14-2).

Peak values of [La] reached during exercise are higher at HA, but the time course of [La] during exercise and recovery follows the same pattern at HA and SL.

Discussion

$\dot{V}_{O_2\,s\cdot st}$ at a given work load shows no difference at HA and SL despite a larger increase in cardiac and ventilatory rates. Thus, gross efficiency is independent of physical fitness and is unaffected by altitude.

Oxygen deficit is always larger at HA

than at SL. The mean effect of HA is to reduce the P_{O_2} gradient from air to the tissues. Oxygen conductance would have to increase at HA in order to deliver the same amount of O_2 to the working muscles as at SL. However, the necessary improvement in conductance is not attained and O_2 transport is impaired in its time course and reduced in its maximal value: O_2 deficit increases and $\dot{V}_{O_2max}$ decreases. However, the deficit remains related to relative workload by the same relationship at SL as at HA.

Oxygen debt was calculated by subtracting rest $\dot{V}_{O_2}$ from sequential $\dot{V}_{O_2}$ during recovery. Rest $\dot{V}_{O_2}$ seems a better reference for determination of O_2 debt than the "asymptotic" value reached during recovery. The term "asymptotic" demonstrates that debt is not over and consequently a part of the studied phenomenon is disregarded. Debt as a whole is always larger than deficit at HA and SL. However, conditions such as work intensity, duration, and environment modify the discrepancy.

It was arbitrarily assumed that the lack or the excess of anaerobic energy represented by the first minute of deficit and debt was mainly related to the splitting and the restoration of the fast creditor. The step of 1 min was chosen in the light of conclusions presented by authors (9,10) who have studied the time course of muscular creatine phosphate concentration in dog and man and have shown that split creatine phosphate is proportional to load and complete splitting and replenishment are accomplished within 2 min for a given work load (6). Additionally, it was assumed that anaerobic glycolysis comes into play late and its role is insignificant during the first minute of work. Keeping in mind these limitations, the slopes can be interpreted as time constants of the fast processes of exponential character. They are identical (0.54 min) at SL and HA. This result is consistent with data previously reported (9,10). Since Lohman's reaction is independent of P_{O_2}, it was expected that the deficit fast component kinetics be unchanged at HA; this hypothesis was experimentally verified. Such a result was already suggested by Cerretelli (3). The time constant of the fast component of debt is the same at SL and HA. Therefore, replenishment of the stores is affected neither by altitude nor by duration of exercise.

The slow component of the deficit theoretically estimates the energy released by anaerobic glycolysis. It is larger at HA compared to SL and this increase is confirmed by higher concentration of La at HA. It is linearly related to $\dot{V}_{O_2max}$ wherever the exercise is performed. Hermansen and Saltin (5) also report that when blood [La] is related to relative work load, all values from different acute altitudes fall on the same line.

The interpretation of the slow component of the debt is more questionable. There was no attempt to find two or three slow components in extra $\dot{V}_{O_2}$ during recovery, since fitting more than one exponential curve to scattered experimental data is unreliable. Therefore, after subtracting the fast component, independent of duration and environment as seen above, from the total debt, the whole remaining extra O_2 volume was considered as a single slow component. Classically, a part of extra $\dot{V}_{O_2}$ during recovery is supposed to oxidize La. However, [La] attains peak value around the tenth minute of exercise. The disappearance rate is not affected by termination of exercise; it would therefore be risky to attribute the slow component of the debt entirely to La metabolism, all the more because a higher [La] occurs at HA simultaneously with a smaller debt slow component. A similar finding was observed by Reynafarje and Velasquez (11). Consolazio et al. (4) have reported too that a significant decline in O_2 debt occurred at HA. This challenging result has given rise to another series of experiments at HA (12). It was shown that in subjects translocated to HA the body shell cools off during exercise and for about 30 min of recovery due to decreased cutaneous blood flow and greater evaporative rate. On the contrary, at SL the shell warms up during exercise and skin temperature returns slowly to rest value during recovery. Thus, one-fifth of body mass is not in the same thermal conditions at HA as at SL.

In conclusion, these results argue in favor of a role of temperature in the long-lasting increased $\dot{V}_{O_2}$ after exercise. Cain (2) has pointed out the interaction of temperature with other more specific factors related to exercise and environment on O_2 consumption during exercise. On the other hand, Brooks et al. (1) have emphasized the striking effect of temperature on mitochondrial functions. However, the results reported here do not allow quantification of the contribution of the temperature factor to the magnitude of O_2 debt.

References

1. Brooks, G.A., Hittelman, K.J., Faulkner, J.A., and Beyer, R.E.: Temperature, skeletal muscle mitochondrial function, and oxygen debt. Am. J. Physiol. 220:1053, 1971.
2. Cain, S.M.: Exercise O_2 debts of dogs at ground level and at altitude with and without β-block. J. Appl. Physiol. 30:838, 1971.
3. Cerretelli, P.: Lactacid O_2 debt in acute and chronic hypoxia. In Margaria, R. (ed.): Exercise at Altitude. Amsterdam, Excerpta Medica, 1967, pp. 58-64.
4. Consolazio, C.F., Nelson, R.A., Matoush, L.O., and Hansen, J.E.: Energy metabolism at high altitude (3475 m). J. Appl. Physiol. 21:1732, 1966.
5. Hermansen, L. and Saltin, B.: Blood lactate concentration during exercise at acute exposure to altitude. In Margaria, R. (ed.): Exercise at Altitude. Amsterdam, Excerpta Medica, 1967, pp. 48-53.
6. Hultman, E., Bergström, J., and Anderson, N. McL.: Breakdown and resynthesis of phosphorylcreatine and adenosine triphosphate in connection with muscular work in man. Scand. J. Clin. Lab. Invest. 19:56, 1967.
7. Kollias, J., Buskirk, E.R., Akers, R.F., Prokop, E.K., Baker, P., and Picon-Reategui, E.: Work capacity of long time residents and newcomers to altitude. J. Appl. Physiol. 24:792, 1968.
8. Maritz, J.S., Morrison, J.F., Peter, J., Strydom, N.B., and Wyndham, C.H.: Practical method of estimating an individual maximum oxygen intake. Ergonomics 4:97, 1961.
9. Piiper, J., di Prampero, P.E., and Cerretelli, P.: Oxygen debt and high energy phosphates in gastrocnemius muscle of the dog. Am. J. Physiol. 215:523, 1968.
10. Piiper, J. and Spiller, P.: Repayment of O_2 debt and resynthesis of high-energy phosphates in gastrocnemius muscle of the dog. J. Appl. Physiol. 28:657, 1970.
11. Reynafarje, B. and Velasquez, T.: Metabolic and physiological aspects of exercise at high altitude. Fed. Proc. 25:1397, 1966.
12. Varene, P., Jacquemin, C., Durand, J., and Raynaud, J.: Energy balance during moderate exercise at altitude. J. Appl. Physiol. 34:633, 1973.

IV. Flow Distribution and Oxygen Transport

15

Blood Rheology in Hemoconcentration

H. SCHMID-SCHÖNBEIN

Theories on the O_2-transport function of blood have frequently neglected the complex fluid dynamic conditions of the blood perfusing the exchange blood vessels in the lung and the peripheral tissues. In the mammalian circulatory and respiratory systems, not only the transport of the blood in arteries and veins but even more so the distribution of the blood into, and its passage through, the terminal ramifications of the vasculature are greatly influenced by the rheologic behavior of the blood cells, primarily the erythrocytes, but also the granulocytes.

The highly variable macroscopic viscosity of mammalian blood—a property unique to suspensions of nonnucleated cells—has been the subject of a large number of recent reviews (1,3–5,7,9,18,19). Due to passive participation of red cells in flow (via membrane rotation and intracellular flow of hemoglobin) the apparent blood viscosity of rapidly flowing normal blood is unusually *low*, especially at high hematocrit levels. This behavior is most unusual, as all other known suspensions and/or emulsions have an extremely high viscosity at volume fractions above 0.4 and cease to flow at volume fractions above 0.60. In contrast, mammalian blood, subjected to the normal shear stresses, is a fluid even at a hematocrit value above 0.90. Its viscosity is then only about 60 cP (mPa · s) at 37°C, similar to that of olive oil (Fig. 15-1).

Recent microrheologic analysis of the blood has revealed that the nonnucleated mammalian red cells assume unique flow properties by adapting themselves passively to the forces of flow, due to properties similar to those of fluid droplets.[1] As a consequence, the disturbance to the flow of the continuous phase (plasma) is minimized, while at the same time an intracellular mixing of the red cell content (e.g., dissolved oxygen or oxyhemoglobin) is induced (25).

The rheologic behavior of *slowly flowing* blood is entirely different. In the absence of adequate flow forces, the cells are aggregated in rouleaux and thence united into larger elastic networks. With increasing hematocrit (45–65%) the density of these networks—and hence their mechanical stability as well as their influence on the flow of blood—progressively increases. Such aggregated blood can easily assume

---

[1] A high-speed film showing fluid droplike behavior of mammalian erythrocytes was presented during the conference.

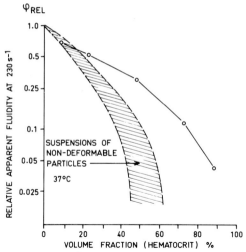

Fig. 15-1. Influence of the volume fraction (hematocrit) on relative apparent fluidity of red cells suspended in plasma. For comparison, relative apparent fluidities of nondeformable particles in various media are shown. In all such suspensions, fluidity falls to zero as the volume fraction exceeds 0.6. In contrast, the fluidity of packed red cell suspensions is maintained even at hematocrit values above 90%.

the properties of a viscoelastic solid below its yield shear stress; i.e., it can withstand finite flow forces without flowing. The recent development of appropriate macroscopic rheometers (12) has made it possible to measure a finite yield shear stress directly—a quantity previously only extrapolated or measured *indirectly* (7,18).

Blood Fluidity as a Consequence of Red Cell Fluidity

The "flow anomaly" of blood in macrovessels is already remarkable; however, in microscopic blood vessels it is even more pronounced. At rapid flow, the viscosity is far lower; at low flow, it is very much higher than in macrovessels. Recent work by Gaehtgens et al. (11) has shown that the flow viscosity of blood in microvessels is associated with typical asymmetric deformations (10) of the erythrocytes. Nondeformable avian erythrocytes have a

much higher viscosity in microvessels (Gaehtgens, personal communication).

The Concept of Red Cell Fluidity

The complex microrheologic behavior of blood is better understood if discussed in terms of apparent blood fluidity, i.e., the inverse value of apparent blood viscosity. As described in detail elsewhere (20,21) the relative apparent fluidity of blood (i.e., $\phi_{blood}/\phi_{plasma} = \eta_{plasma}/\eta_{blood}$) is not significantly different from the ideal value of 1.0 in the majority of blood vessels perfused with normally high driving pressure (Fig. 15-2). Furthermore, at rapid flow the relative apparent fluidity is almost unaffected by the hematocrit level (up to about 55%; ref. 1). A value of $\phi_{rel} = 0.8-1.0$ means that

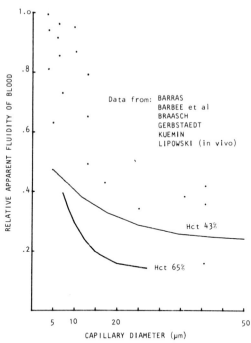

Fig. 15-2. Influence of capillary diameter on relative apparent fluidity of mammalian blood due to the pronounced Fahraeus-Lindqvist effect. Relative apparent blood fluidity raises to 1.0 in microvessels. Furthermore, the hematocrit has a smaller effect on apparent blood fluidity in microvessels.

the flow of plasma remains practically unaffected by the introduction of up to 50% additional red blood cells, which on the other hand increase the oxygen transport capacity by a factor of 70.

The rapid unimpeded flow of concentrated cell suspensions through exchange vessels is a unique transport phenomenon, based on the fluidity of the mammalian denucleated red cell, which is ontogenetically a product of an incomplete division of an erythroblast so overloaded with hemoglobin that mitosis results in two erythroblast fragments (17). The first one, containing the pyknotic nucleus, becomes phagocytosed in the bone marrow. The second one, the reticulocyte, becomes borne in the circulation, provided it is disc shaped, i.e., endowed with a favorable ratio of surface area to volume, and can thus be deformed without straining the membrane.

As a result of the favorable surface area to volume ratio, the flexibility of the membrane, and the fluidity of the cytoplasm, the reticulocyte and even more so the mature erythrocyte can easily pass through narrow slits in venular walls of the splenic microcirculation, where red cells devoid of this combination of properties are sequestered and subsequently phagocytosed. Consequently, all red cells found circulating in the peripheral blood have been highly selected with respect to deformability, and thus are ideally capable of promoting flow and distribution of blood to all microvessels.

This behavior is based on the following flow features: in response to the distribution of shear stresses found in a perfused vessel and in vessel intersections, the fluid red cells are stationarily deformed, elongated, and oriented in flow; thus the membrane is driven into a rotation around the cell content (Fig. 15-3). Furthermore, the cells rapidly migrate to the axial core, producing a marginal lubricating layer of low viscosity plasma. In narrow capillaries and in intersections, where asymmetric flow forces are dragging the erythrocyte

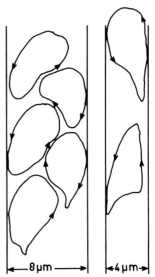

Fig. 15-3. Schematic representation of membrane rotation ("tanktreading") in the shear field of arterioles and capillaries. Whenever asymmetric flow forces exert asymmetric viscous drag on the membrane its rotation around the fluid cell content is enforced.

membrane, not a stationary shape (bullet or parachute), but rather a frequent shape change is induced which is also caused by a motion of membrane relative to its content (Fig. 15-3). In intersections and at confluences with a sudden lateral deviation of flow forces, the membrane or rather the whole cell is rotating, permitting a smooth passage past these critical traffic locations. Finally, the fluidity of red cells is responsible for their rapid axial migration even at high hematocrit, which not only produces a lubricating plasma layer but also a reduction of dynamic hematocrit. Both effects contribute to the Fahraeus and the Fahraeus-Lindqvist effect, i.e., the strong increase of relative apparent blood fluidity in narrow capillaries.

As an added dividend of red cell fluidity, a rapid and highly effective intracellular mixing is induced, since the constant or intermittently occurring membrane motion induces complex intracellular laminar flow of hemoglobin, oxyhemoglobin, and dissolved oxygen (25). Consequently, the diffusive transport of O_2 into or out of red cells is augmented by a convective motion in moving cells. By this simple mechanism, nature has "squared the transport circle" for O_2, producing both intracellular and extracellular (bolus flow) vortices (2), while at the same time producing maximally laminar flow with a minimum of pressure loss ($\phi_{rel} = 0.9 - 1.0$). As still another benefit, the mammalian capillaries, which literally grow around the preexisting flexible red cells, are much smaller than those of species with nucleated RBCs (23), consequently, the exchange area for any given capillary volume is much higher and the diffusion distance from the center of the exchange capillary to the wall is smaller. Finally, the volume occupied by capillaries is much smaller than that in species with nondeformable red cells, as deduced theoretically by Burton (2) and demonstrated by Stöckenius (22), who showed that birds have a much higher cerebral capillary density than do mammals at the cost of neuronal tissue, especially of dendrites and axons.

The extremely high fluidity of blood passing through the microcirculation depends on the presence of adequate flow forces; in their absence, the blood undergoes a transition from a highly fluid emulsion to a very viscous and viscoelastic, reticulated suspension with functional properties of a solid. The blood then progressively loses its fluidity—which may be zero if the hematocrit and/or the fibrinogen level is elevated above normal (Fig. 15-4).

Disadvantages and Risk Factors of Blood Rheology

Although there are obvious advantages in a miniature O_2-exchange system in mammals

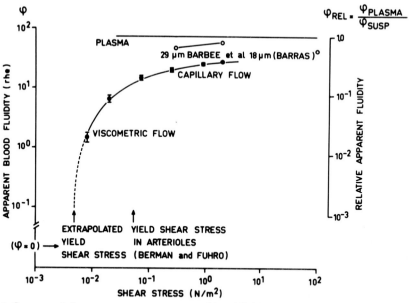

Fig. 15-4. Influence of shear stresses on apparent blood fluidity in macrovessels and microvessels. Due to the Fahraeus-Lindqvist effect apparent blood fluidity in the majority of microvessels is much higher than in macrovessels or in viscometric flow (rotational viscometers).

Whenever the shear stresses fall (due to any flow retardation) the fluidity of blood is drastically reduced. A yield shear stress on the order of 10^{-1} has been measured in vivo and in vitro (13). When the shear stresses acting in the microvessels fall below the yield shear stresses, blood behaves like a solid in the absence of coagulation.

and in man their normal operation requires a certain minimum of flow forces; in the case of a generalized or localized reduction in driving force, elastic cell-cell and cell-wall interactions interfere with the high fluidity, producing a reticulated suspension of elastic particles with very low fluidity. In the absence of flow forces, the cells are not only not deformed, but they aggregate into typical rouleaux and networks of rouleaux. This reversible formation of a structure reversibly increases the apparent viscosity of blood, or reduces its fluidity, for several reasons (Fig. 15-4): (1) the aggregates immobilize plasma and thereby functionally decrease the volume fraction of plasma (or increase the effective hematocrit level); and (2) the networks reach across planes of shear and thereby impede the relative motion of fluid lamellae. More importantly, the fluidity may be fully abolished whenever the rouleaux straddle bifurcations or completely fill the lumen of a tapering blood vessel or a capillary with small diameter (15). All these phenomena are strongly influenced by the hematocrit level: below 35% they have only negligible effects; above 55% they become extremely effective.

The extent of aggregation is primarily a function of the concentration of high molecular weight plasma proteins, such as fibrinogen, α_2-macroglobulin, and IgM (22), which also very strongly elevate plasma viscosity and thence further jeopardize blood fluidity. Again, the effect of aggregation and plasma viscosity on apparent blood fluidity is primarily a function of the hematocrit. It is immediately obvious that, merely due to reasons of strength, aggregates at 0.55 hematocrit have a much more severe effect than the same aggregates at 0.35 hematocrit, where even the aggregates are "lubricated" by about twice their volume of plasma.

Due to the formation of elastic aggregates, blood in the microvessels exhibits strong viscoelastic properties and hence behaves like a solid below its yield point (these materials have been called Bingham bodies or Casson bodies in the hemorheologic literature[2]).

It is intuitively clear that in vessels with diameters similar, equal to, or even smaller than the cell aggregates, the measured yield shear stresses are much higher than in macroscopic rheometers. Indeed, we have recently succeeded in measuring yield shear stresses between 0.1 and 0.5 Pa in models of a simple vascular network (13) with two branches of significantly different length. These experiments support a rheologic concept explaining the well-known stasis phenomenon on the basis of the non-newtonian blood fluidity and localized inhomogeneity of vascular lengths and thus shear stresses in the microcirculation. As discussed in detail elsehwere (19), this theory predicts that the blood flow in vessels parallel to main arteriovenous pathways can come to a complete stop despite a finite arteriovenous pressure gradient; i.e., it behaves like a solid by "collateral blood viscidation" *before* being clotted (15). More importantly, the fluidity even before full stagnation may be so severely jeopardized that the creeping red cell masses are deoxygenated long before they leave the exchange vessels. By their very presence they impede the further supply of fresh, oxygenated blood and the removal of metabolites. Thus, the viscoelasticity of creeping blood leads to "stagnation anoxia" due to localized or generalized loss of blood fluidity.

In the microcirculation, the blood rheology is further strongly affected by local changes in hematocrit and vasomotion under myogenic, metabolic, nervous, and

---

[2] For good reasons, a number of authors have denied or modified previous claims, i.e., that whole blood in macroscopic rheometers exhibits a yield shear stress, below which it has infinite viscosity or zero fluidity. In such instruments—and probably also in large-bore natural blood vessels—at normal hematocrit levels (45%) the formation of static aggregates is accompanied by a lubricating film of plasma (a newtonian fluid), which allows a relative motion of fluid lamellae near the wall and thus the institution of a finite shear rate no matter how small the shear stress.

hormonal control. It is well known that the hematocrit in the microcirculation is generally much lower than in the macrocirculation, a phenomenon caused by axial migration as well as by a so-called screening effect, i.e., a fluid dynamic mechanism that keeps the red cells away from capillaries, since cells tend to prefer the channel with the faster flow. On the other hand, under pathologic conditions the microvascular hematocrit can be as high or higher than the central venous hematocrit.

Recent work by Klitzman and Duling (16) shows that the microvascular hematocrit is under vasomotor control. Using pharmacologic vasodilatation and vasoconstriction, they demonstrated a decrease in capillary hematocrit (capillary hematocrit = $\frac{1}{8}$ central venous hematocrit) under vasoconstriction, as well as an increase in hematocrit under vasodilatation (capillary hematocrit = central venous hematocrit). Unpublished observations (Schmid-Schönbein, et al.) tend to support this finding: in myogenic and metabolic autoregulation we observed striking phasic rises in capillary hematocrit with dilatation of arterioles, and falls with arteriolar constriction.

Although the mechanism of this local vasomotor influence on the main determinant of the flow properties of blood remains to be clarified, the following fact can be accepted: microvascular hematocrit can vary between values well below 10% (where blood behaves practically like a newtonian fluid) and a value identical to central venous (or arterial) hematocrit. Under conditions of local endothelial damage and thus increased microvascular wall permeability, local hemoconcentration can occur, leading to a preferential increase in high molecular weight plasma proteins, thus in plasma viscosity, and in red cell aggregation, complicated by a further increase in hematocrit. It must be stressed in this context that a given increase in permeability leads to both faster and more pronounced hemoconcentration if blood with a higher than normal hematocrit is lodged in or perfusing the affected vessel. A higher tendency to become stagnant is easily understandable since not only is the apparent fluidity for any given shear stress lower or yield shear stress higher, but the dependence of apparent fluidity on shear stress is far more nonlinear.

The highly variable blood fluidity and the equally dynamic vasomotor control certainly interact in a very complex, as yet poorly understood fashion. It is obvious, however, that the pronounced autoregulatory capacity of the microcirculation will not only compensate an arterial hypotension by arteriolar dilatation and therefore local increase in vascular conductance ($\pi r^4 \cdot 1^-$), but will also tend to increase local driving pressure (ΔP) and thus shear stresses ($\Delta P \cdot r/2 \cdot 1$).[3] Consequently, autoregulation of blood flow also functions as a safety device that maintains blood fluidity.

On the other hand, the vasomotor dilatational reserve is *limited*; once it is recruited fully, or checked by strong vasoconstrictor influences, the viscosity (or fluidity) of blood becomes a limiting factor of perfusion. Measurements in our laboratories have shown indeed that the autoregulatory capacity is distinctly impaired at high hematocrit and improved at low hematocrit (8). More work is essential to understand the interaction between vasomotor factors and flow anomalies of blood in tissue perfusion. Such interaction must be investigated by experiments capable of measuring the vasomotor reserve, rather than the steady state perfusion at rest or during work.

We have recently seen in an isolated skeletal muscle preparation that the combination of low driving pressure, elevated hematocrit, and maximum vasodilatation bears the risk of complete and often irreversible stagnation (Schmid-Schönbein and Johnson, unpublished observations). As

---

[3] ΔP = driving pressure, r = vessel radius, l = vessel length

shown previously, this combination of conditions is characterized by an extremely sluggish recovery from induced stagnation (13), which may be related to the slow rise in perfusion during skeletal muscle exercise (see Gaehtgens).

Summary and Consequences for High Altitude Physiology

1) The fluidity of blood is a complex function of the hematocrit, the plasma composition, the tube diameter, and most of all the incident driving pressure. At high driving pressures, the fluid drop-like features of red cells allow relatively unimpeded flow of blood even at extremely high hematocrit levels, and in the presence of high molecular weight plasma proteins. However, the mere reduction of driving pressures seriously impedes or even abolishes blood fluidity due to extreme crowding and the formation of rouleaux.

2) The structural properties of blood manifest themselves primarily in the microvessels. They can, however, be compensated by the active components of the vascular bed, e.g., by elevated contractile forces of the right ventricle for the maintenance of pulmonary perfusion, and by peripheral vasodilatation and/or increased cardiac contraction in the major circulation.

3) Whenever the compensatory mechanisms are fully recruited or exhausted, the fluidity of blood becomes a limiting factor for perfusion. Under such conditions, the advantage of a high hematocrit turns into a serious disadvantage, especially if the true increase in red cell mass by stimulated erythropoiesis is complicated by an absolute or relative hypovolemia with dehydration and further increase in the concentration of high molecular weight plasma proteins. If fluidity is locally reduced to zero, flow will come to a complete stop despite finite arteriovenous pressure gradients and without coagulation ("collateral bloodviscidation").

4) The general biologic consequences of a loss of blood fluidity are highly dependent on local factors (driving pressures, angioarchitectonics and geometry, vasomotor effect and extravascular pressure, wall permeability, and O_2 requirements) and are likely to be quite different at different organs (e.g., brain, retina, skin, lung, muscle).

5) Despite these important organ-specific differences, it is justified to institute measures that maintain the hematocrit at normal levels (45–48% Hct). Such measures include prophylaxis (i.e., complete fluid and electrolyte balance), compensation, and therapy (by isovolemic or slightly hypervolemic hemodilution).

6) For present surveillance of high altitude mountaineers, hematocrit measurements and a rapid, simple test sensitive to high molecular weight plasma proteins (e.g., measurement of tendency to red cell aggregation) are recommended. For future research, a more detailed analysis of all factors influencing blood fluidity is required. However, this will yield biologically significant information *only* if accompanied by sophisticated measurements of cardiopulmonary and peripheral vascular performance. Experiments attempting to clarify the relative role of hemorheologic factors must include quantification of the recruitment of vasomotor reserve.

References

1. Braasch, D.: Red cell deformability and capillary blood flow. Physiol. Rev. 51:679, 1971.
2. Burton, A.C.: Physiology and Biophysics of the Circulation. Chicago, Year Book Medical, 1965.
3. Charm, S.E. and Kurland, G.S.: Blood Flow and Microcirculation. New York, Wiley, 1974.
4. Chien, S.: The present state of blood rheology. In Messmer, K. and Schmid-Schönbein, H. (eds.): Theoretical Basis and Clini-

cal Application. Basel, Karger, 1972, pp. 1–40.
5. Chien, S.: Biophysical behavior of red cells in suspension. In Surgenor, D. McN. (ed.): The Red Blood Cell, Vol. 2, 2nd ed. New York, Academic Press, 1975, pp. 1031–1121.
6. Chmiel, H., Effert, S., and Methey, D.: Rheologische Veränderungen des Blutes beim akuten Herzinfarkt und dessen Risikofaktoren. Dtsch. Med. Wschr. 98:1641, 1973.
7. Cokelet, G.V.: The rheology of human blood. In Fung, Y.C., Perrone, N., and Anliker, M (eds): Biomechanics, Its Foundation and Objectives. Englewood Cliffs, N.J., Prentice Hall, 1972, pp. 63–104.
8. Copley, A.L., Hung, C.R., and King, R.G.: Rheogoniometric studies of whole human blood at shear rates from 1000 to 0.0009 sec^{-1}. Part 1: Experimental Findings. Biorheology 10:17, 1973.
9. Dintenfass, L.: Rheology of Blood in Diagnostic and Preventive Medicine. London, Butterworths, 1976.
10. Driessen, G., Heidtmann, H., and Schmid-Schönbein, G.: Effect of hemodilution and hemoconcentration on red cell flow velocity in the capillaries of rat mesentery. Pflügers Arch. 380:1, 1979.
11. Dührssen, C., Gaehtgens, P., Kreutz, F., and Zierold, K.: Deformation of human red blood cells during capillary flow. Pflügers Arch. 377:R11, 1978.
12. Gaehtgens, P., Schmid-Schönbein, H., Schmidt, F., Will, G., and Stöhr-Liesen, M.: Comparative microrheology of nucleated avian (NRBC) and non-nucleated human (HRBC) erythrocytes during viscometric and small tube flow. La Jolla, 2nd World Congress for Microcirculation, 1979.
13. Heidtmann, H., Driessen, G., Haest, C.W.M., Kamp, D., and Schmid-Schönbein, H.: The influence of rheological factors on the recovery of the microcirculation following arterial hypotension. Microvasc. Res. (in press).
14. Kiesewetter, H., Kotitschke, G., and Schmid-Schönbein, H.: Yield stress measurement in red cell suspensions. La Jolla, Congress on Biorheology, 1978.

15. Kiesewetter, H., Schmid-Schönbein, H., Radtke, H. and Stolwerk, G.: In vitro demonstration of collateral blood viscidation: Flow measurement in a model of vascular networks. 2nd World Congress for Microcirculation, La Jolla. Microvasc. Res. 17(3,2):72, 1979.
16. Klitzman, B. and Duling, B.R.: Causes of low microvascular hematocrit in hamster cremaster capillaries. Microvasc. Res. 17:3S70, 1979.
17. Lessin, L.S. and Bessis, M.: Morphology of the Erythron. In Williams, W.J., Beutler, E., Erslev, A.J., and Rundles, R.W. (eds.): Hematology. New York, McGraw Hill, 1977, pp. 103–134.
18. Merrill, E.W.: Rheology of blood. Physiol. Rev. 49:863, 1969.
19. Schmid-Schönbein, H.: Microrheology of erythrocytes, blood viscosity, and the distribution of blood flow in the microcirculation. In Guyton, A.C. and Cowley, A.W. (eds.): Cardiovascular Physiology II, Vol. 9. Baltimore, University Park Press, 1976.
20. Schmid-Schönbein, H.: Valediction to "blood viscosity"; Salutation to "blood fluidity." The present state of hemorheology. Editorial. In: Basic Research in Cardiology (in press).
21. Schmid-Schönbein, H.: Blood fluidity as a consequence of red cell fluidity: Flow properties of blood and flow behavior of blood in vascular diseases. In: Angiology (in press).
22. Schmid-Schönbein, H., Gallasch, G., Volger, E., and Klose, H.J.: Microrheology and protein chemistry of pathological red cell aggregation (blood sludge) studied in vitro. Biorheology 10:213, 1973.
23. Sobin, S.S. and Tremer, H.M.: Three-dimensional organization of microvascular beds as related to function. In Kaley, G. and Altura, B.M. (eds.): Microcirculation, Vol. 1. Baltimore, University Park Press, 1977, pp. 43–67.
24. Stökenius, M.: Die Kapillarisierung verschiedner Vogelgehirne. Gegenbaurs Morphol. Jahrb. 105:343, 1964.
25. Zander, R. and Schmid-Schönbein, H.: Intracellular mechanisms of oxygen transport in flowing blood. Respir. Physiol. 19:279, 1973.

16
Oxygen Transport Capacity

K. MESSMER

Under physiological conditions oxygen content of the blood, blood flow, and oxygen delivery are balanced to meet the tissue requirements resulting from the metabolic activity of a given organ or the whole body. Changes in oxygen content, total flow rate, or flow distribution can seriously affect this equilibrium; blood fluidity, which is of limited significance in normal circulatory conditions, might become the crucial factor for tissue supply by virtue of governing the distribution and the rates of flow within the different segments of the microcirculation.

High altitude exposure results in acute significant changes of the oxygen content of the blood due to increased erythropoiesis and hemoconcentration from dehydration (28). It is generally agreed that the acute increase in red cell mass, whether from a real increase in red cell mass or from a shrunken plasma volume, e.g., hemoconcentration, can severely affect not only the supply of oxygen and nutrients but also the removal of waste materials from the tissues due to alterations of both the flow properties and the flow conditions of blood (23).

Hemoconcentration

High hematocrit implies both increased oxygen content and increased viscosity of blood. To evaluate whether the benefits from an augmented oxygen content can outweigh the potential disadvantages of an impaired blood fluidity due to hemoconcentration, several groups have conducted experiments in resting animals exposed to isovolemic blood exchange for concentrated red cells (6,11,14,17,22).

The findings shown in Fig. 16-1 are considered representative for acute hemoconcentration; the experiments were performed in splenectomized and heparinized dogs in pentobarbital anesthesia and with muscle relaxation.

Arterial blood pressure remains virtually unchanged; however, cardiac output declines in close relation to the increase in peripheral resistance. The direct relation between the peripheral vascular resistance and whole blood viscosity (as measured in vitro at a shear rate of 60.3 (sec^{-1}) is evident. The decline of cardiac output results

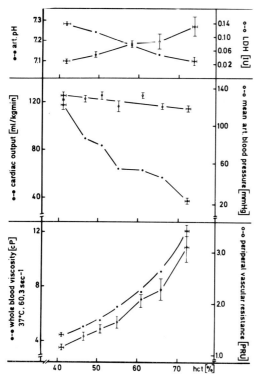

Fig. 16-1. Isovolemic hemoconcentration in splenectomized dogs. Changes in cardiac output, arterial blood pressure, peripheral resistance, whole blood viscosity, arterial pH, and LHD-serum concentration are shown in relation to the increase in hematocrit. From Sunder-Plassmann, L., et al.: In: Proc. 6th Europ. Conf. Microcirculation. Basel, Karger, 1971.

not astonishing that acute hemoconcentration is followed by (1) a reduction in venous return; (2) a fall in both stroke volume and cardiac output; and (3) a decrease in tissue perfusion to an extent that oxygen consumption diminishes. Consequently, the animal experiments with acute hemoconcentration have revealed tissue hypoxia as reflected by a decrease in local venous pH, and an increase in P_{CO_2}, lactate, and lactate dehydrogenase (11).

These findings are highly indicative that the decreased fluidity of blood has outweighed the advantage of an increased oxygen content under resting conditions. The clinical importance of high blood viscosity due to high hematocrit is particularly illustrated by the considerably higher risk of cerebrovascular disease, myocardial infarction, vascular occlusive episodes, and thromboembolic complications (19,26), phenomena well known also in individuals acutely exposed to high altitudes.

Although the increase in oxygen content of the blood is considered often as a physiologic mechanism of adaptation to high altitudes, there is no evidence that hematocrits above 45% provide benefits either in normoxia or in hypoxia (14).

Systemic Oxygen Transport Capacity

The determinants of systemic oxygen transport are cardiac output and the oxygen content of the blood. By calculating the product of both at different hematocrits the systemic oxygen transport capacity curve is obtained. Hint (5) proposed, on the basis of theoretical considerations, that systemic oxygen transport (SO_2T) as a function of the actual hematocrit would yield a bell-shaped curve with a peak at around 30% hematocrit (dotted curve in Fig. 16-2). Combining the results of acute hemoconcentration and acute normovolemic hemodilution studies, our group was able to define this relationship over a hematocrit range from 7 to 70% (17,24).

primarily from a fall in stroke volume with heart rate remaining constant (6,11); this fact emphasizes in particular the presence of an augmented afterload and the causal role of whole blood viscosity and venous return for the adjustment of the circulation.

Lipowsky and coworkers (13) have computed the apparent viscosity and wall shear stresses in the arterial and venous microvessels of the cat's mesentery; the lowest wall shear stresses and the highest viscosity values were found in the venules. With unchanged driving pressure (as seen in Fig. 16-1), hemoconcentration is therefore likely to impede the flow most in the vessels yielding the lowest shear stresses (post-capillary venules) (3). It is therefore

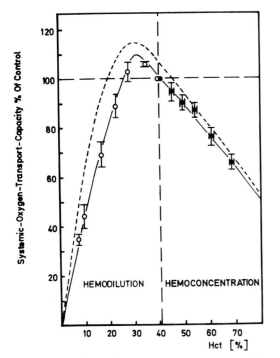

Fig. 16-2. Systemic oxygen transport capacity (*ordinate*) in relation to arterial hematocrit of splenectomized dogs undergoing isovolemic hemoconcentration by means of homologous packed red cells and isovolemic hemodilution with dextran 60, respectively. [Dotted curve: data from Hint (5).] From Sunder-Plassmann, L., et al.: In: Proc. 6th Europ. Conf. Microcirculation. Basel, Karger, 1971.

The experiments were performed on anesthetized dogs which had been splenectomized several weeks prior to the acute experiment. It was found that under the experimental conditions used the systemic oxygen transport curve presented a bell shape with a remarkable degree of conformity to the theoretical curve of Hint (5). Maximum oxygen transport was found at lower than the control levels of hematocrit.

Increments of hematocrit from the control value were associated with an immediate decline of systemic oxygen transport; this finding is in agreement with the results of Jan and Chien (6) as well as McGrath and Weil (14). The data of McGrath and Weil are especially pertinent when considering oxygen transport under high altitude conditions because these authors have combined acute polycythemia (hemoconcentration) with hypoxia (Pa_{O_2} 40.5 mmHg); they found that the combination of hypoxia and polycythemia decreased SO_2T more than either alone.

Jan and Chien (6) furthermore demonstrated that oxygen transport to the myocardium was significantly depressed beyond hematocrits of 60%; it is reasonable to assume that the decrease in total coronary flow underlying the reduced oxygen transport in the coronary circulation is not uniform in nature, but in favor of the epicardial layers with consequent underperfusion of the subendocardium. One is tempted to assume that the increased myocardial vascularity observed in animals chronically exposed to high altitude presents a protection from such a redistribution of coronary blood flow.

In contrast to acute hemoconcentration, acute normovolemic hemodilution might become favorable for oxygen transport. The absence of an instant fall of SO_2T with the linear reduction in hematocrit (see Fig. 16-2) is due to the rise in cardiac output, the most typical feature of normovolemic hemodilution (2,6,9,15,20–22).

Normovolemic dilution of blood with colloids not only increases cardiac output but improves the blood flow to most of the vascular beds with a profound rise in coronary blood flow (Fig. 16-3). This preferential increase in coronary flow was also observed by Jan and Chien (6) and was the most prominent finding in the study of von Restorff et al. (21) on awake dogs undergoing hemodilution and exercise.

The peak oxygen transport at the hematocrit around 30% was not confirmed by Jan and Chien (6) in their anesthetized dogs and could not be demonstrated by von Restorff et al. (21) in awake animals. Both groups of investigators report systemic oxygen transport values of about 90% of control for hematocrits of 30% (Fig. 16-4).

This apparent discrepancy with our own findings (Fig. 16-2) might be attributed to

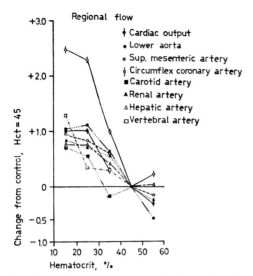

Fig. 16-3. Isovolemic hemodilution with dextran 70 in dogs. Changes of cardiac output and regional blood flow. From Race, C., et al.: J. Thorac. Cardiovasc. Surg. 53:578, 1967.

differences concerning the experimental techniques (splenectomized versus non-splenectomized dogs, control of normovolemia, time of measurement, dextran 60 versus homologous plasma). It seems, however, noteworthy to recall that Laks et al. (12) have found that oxygen transport in

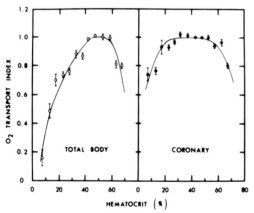

Fig. 16-4. Acute variations in hematocrit by isovolemic blood exchange for packed red cells and isovolemic hemodilution with homologous plasma. Changes in oxygen transport rates in systemic (*left*) and coronary (*right*) circulations. From Jan, K.M. and Chien, S.: Am. J. Physiol. 233:H107, 1977.

patients undergoing preoperative hemodilution had augmented to 106% at 27.5% hematocrit. The important aspect for the present discussion is that hemodilution is the only means to avoid and to reverse the sequelae of hemoconcentration.

Limited hemodilution—again in the resting animal—neither deleteriously affected the distribution of blood flow nor the oxygen supply to the tissues. Measurements with locally applied multiwire P_{O_2}-electrodes (7,8) have convincingly demonstrated that the adequacy of local tissue oxygenation is maintained even at hematocrits as low as 20% (16). The concept of intentional hemodilution as a means to improve the flow properties and flow conditions of blood has been validated by its successful clinical application (10,12, 15,25,27). Keeping in mind the importance of normovolemia as a prerequisite for an adequate increase in cardiac output as well as the necessity of a normal coronary vascular bed (1,2,9), side effects from limited normovolemic hemodilution are rarely encountered (15,25).

Enhancement of Oxygen Transport at High Altitude

High altitude hypoxia curtails transport and delivery of oxygen by diminishing the arterial oxygen content. From secondary polycythemia and/or hemoconcentration a further depression of SO_2T has to be taken into account. Therefore it seems advisable to prevent rises of hematocrit at high altitude. This recommendation seems appropriate even though it is based on the animal experiments mentioned above, which are distinctly different from the high altitude situation. The sympathoadrenal drive and the degree of exercise of climbers or inhabitants of high altitudes most certainly interfere with the circulatory effects of hemoconcentration per se. Nevertheless, several facts call for a careful control of the hematocrit value during high altitude acclimatization or exposure:

1) An additional stress is added to the heart working against an increased resistance which might be both of viscous and hypoxic origin, especially in the pulmonary circulation (pulmonary hypertension due to hypoxia).
2) Hemoconcentration causes redistribution of blood flow, particularly in the microcirculation, with stagnant or sluggish flow. This might further limit oxygen uptake in the lung by ventilation-perfusion disturbances. At the periphery injuries from cold might result from the severely disturbed microcirculation.
3) By virtue of decreasing the flow velocity hemoconcentration is likely to favor thromboembolic complications.
4) Tissue hypoxia with anaerobic metabolism, acidosis, and transmineralization are the final results of the impaired oxygen transport and delivery due to severe hemoconcentration.
5) Hemoconcentration with all its sequelae will finally interfere with the ability of the circulation to adjust for the metabolic demands during and after physical exercise.

The target for prophylactic and therapeutic interventions at high altitude can consequently be identified rather easily: the fluidity of the blood. From the experimental findings available one can conclude that hemoconcentration from dehydration should be prevented by all means; overt hemoconcentration should be treated immediately. Dehydration with a consequent rise in hematocrit can be partly suppressed by oral intake of *additional* loads of water and electrolytes. If oral fluid intake is not possible or does not prevent the hematocrit from rising above 60%, infusions of colloids and electrolytes are indicated to replenish the plasma and the extracellular volumes. Overt polycythemia is best treated by isovolemic or hypervolemic hemodilution; if performed with oncotically active solutions and strictly isovolemically (simultaneous blood-letting and infusion!), hemodilution can abolish the sequelae from hemoconcentration rather rapidly (11).

The aim of intentional hemodilution at high altitudes is to reduce the hematocrit below 60%, but never below its initial sea level value.

These suggestions arise from clinical experience with hemodilution in patients suffering from secondary polycythemia, which were found to respond to hemodilution by improved oxygen uptake and an increased peripheral energy metabolism as a result of the improved fluidity of the blood (4,18).

Zink and coworkers (28) have meanwhile proven the applicability of intentional hemodilution to the extreme conditions of high altitude expeditions.

References

1. Anderson, H.T., Kessinger, J.M., McFarland, W.J., Laks, H., and Geha, A.S.: Response of the hypertrophied heart to acute anemia and coronary stenosis. Surgery 84:8, 1978.
2. Buckberg, G.D. and Brazier, J.: Coronary blood flow and cardiac function during hemodilution. In Messmer, K. and Schmid-Schönbein, H. (eds.): Intentional Hemodilution. Bibl. Haematol. 41:173, 1975.
3. Chien, S.: Present state of blood rheology. In Messmer, K. and Schmid-Schönbein, H. (eds.): Hemodilution. Theoretical Basis and Clinical Application. Basel, Karger, 1972.
4. Gregory, R.J.: Technique for treatment of polycythemia by exchange transfusion. Lancet 1:858, 1971.
5. Hint, H.: The pharmacology of dextran and the physiological background of the clinical use of Rheomacrodex. Acta Anaesthesiol. Belg. 19:119, 1968.
6. Jan, K.M. and Chien, S.: Effect of hematocrit variations on coronary hemodynamics and oxygen utilization. Am. J. Physiol. 233:H107, 1977.
7. Kessler, M.: Oxygen supply to tissue in normoxia and oxygen deficiency. Microvasc. Res. 8:283, 1974.
8. Kessler, M. and Grunewald, W.: Possibilities of measuring oxygen pressure fields in

tissue by multiwire platinum electrodes. Progr. Resp. Res. 3:147, 1969.
9. Kettler, D., Hellberg, K., Klaess, G., Kontokollias, J.S., Loos, W., and de Vivie, R.: Hemodynamics, oxygen demand, and oxygen uptake of the heart during isovolaemic hemodilution. Anaesthesist 25:131, 1976.
10. Klövekorn, W.P., Pichlmaier, H., Ott, E., Bauer, H., Sunder-Plassmann, L., and Messmer, K.: Akute präoperative Hämodilution—eine Möglichkeit zur autologen Bluttransfusion. Chirurg 45:452, 1974.
11. Klövekorn, W.P., Sunder-Plassmann, L., Siegle, M., and Messmer, K.: Exchange transfusion with colloids in acute polycythemia. Anaesthesist 23:142, 1974.
12. Laks, H., O'Connor, N.J., Pilon, R.N., Anderson, W., MacCallum, J.R., Klövekorn, W.P., and Moore, F.D.: Acute normovolemic hemodilution: Effects on hemodynamics, oxygen transport, and lung water in anaesthetized man. Surg. Forum 24:201, 1973.
13. Lipowsky, H.H., Kovaldeck, S., and Zweifach, B.W.: The distribution of blood rheological parameters in the microvasculature of cat mesentery. Circ. Res. 43:738, 1978.
14. McGrath, R.L. and Weil, J.V.: Adverse effects of normovolemic polycythemia and hypoxia on hemodynamics in dogs. Circ. Res. 43:793, 1978.
15. Messmer, K.: Hemodilution. Surg. Clin. North Am. 55:659, 1975.
16. Messmer, K., Sunder-Plassmann, L., Jesch, F., Görnandt, L., Sinagowitz, E., and Kessler, M.: Oxygen supply to the tissues during limited normovolemic hemodilution. Res. Exp. Med. 159:152, 1973.
17. Messmer, K., Sunder-Plassmann, L., Klövekorn, W.P., and Holper, K.: Circulatory significance of hemodilution: rheological changes and limitations. Adv. Microcirc. 4:1, 1972.
18. Möller, P., Brandt, R., Fagrell, B., Hellström, K., and Strandell, P.: The effect of isovolemic hemodilution in patients with chronic respiratory insufficiency. Abstract No. 111. Prague, Int. Symp. Pulmonary Circulation III, 1979.
19. Pearson, T.C. and Wetherley-Mein, G.: Vascular occlusive episodes and venous hematocrit in primary proliferative polycythemia. Lancet 2:1219, 1978.
20. Race, C., Dedichen, H., and Schenk, W.G., Jr.: Regional blood flow during dextran induced normovolemic hemodilution in the dog. J. Thorac. Cardiovasc. Surg. 53:578, 1967.
21. Restorff, W.v., Höfling, B., Holtz, J., and Bassenge, E.: Effect of increased blood fluidity through hemodilution on general circulation at rest and during exercise in dogs. Pflügers Arch. 357:15, 1975.
22. Richardson, T.Q. and Guyton, A.C.: Effects of polycythemia and anemia on cardiac output and other circulatory factors. Am. J. Physiol. 197:1167, 1959.
23. Schmid-Schönbein, H.: Microrheology of erythrocytes, blood viscosity, and the distribution of blood flow in the microcirculation. In Guyton, A.A. and Cowley, A.W. (eds.): Cardiovascular Physiology, Vol. 9. Baltimore, University Park Press, 1976.
24. Sunder-Plassmann, L., Klövekorn, W.P., Holper, K., Hase, U., and Messmer, K.: The physiological significance of acutely induced hemodilution. Aalborg, Proc. 6th Europ. Conf. Microcirculation, 1970. Basel, Karger, 1971.
25. Sunder-Plassmann, L. and Messmer, K.: Akute präoperative Hämodilution. Chirurg 50:410, 1979.
26. Thomas, D.J., Marshall, J., Ross-Russell, R.W., Wetherley-Mein, G., DuBoulay, G.H., Pearson, T.C., Symon, L., and Zilkha, E.: Effect of hematocrit on cerebral blood flow in man. Lancet 1:941, 1977.
27. Watkins, G.M., Rabelo, A., Bevilacqua, R.G., Brennan, M.F., Dmochowski, J.R., Ball, M.R., and Moore, F.D.: Bodily changes in repeated hemorrhage. Surg. Gynecol. Obstet. 139:161, 1974.
28. Zink, R.A., Schaffert, W., Messmer, K., Brendel, W., and Hackett, P.: Hemodilution: practical experiences in high altitude expeditions. In: Topics in Environmental Physiology and Medicine. New York, Springer-Verlag. (This volume.)

17

Skeletal Muscle Perfusion, Exercise Capacity, and the Optimal Hematocrit

P. GAEHTGENS AND F. KREUTZ

The concentration of red cells in the blood is the most important determinant of apparent blood viscosity and therefore one of the determinants of tissue perfusion. Even if it is known that alterations of hematocrit due to their influence on blood viscosity lead to changes of blood flow through various organs, it is not clear to what extent the distribution of flow within the microcirculation is affected by such changes. Therefore it is not directly possible to estimate the optimal level of hematocrit at which O_2 delivery to the tissue cells is maximal. Experimentally, this optimal hematocrit (H_{opt}) was found to be lower than the physiologic hematocrit level under resting conditions (6); these observations have led to the therapeutic concept of hemodilution as a means to improve the nutritive supply of the tissues. However, the optimal hematocrit level is less well defined for conditions of physical exercise (1,11). In addition, the alteration of exercise capacity under conditions of a chronic elevation of hematocrit remains to be evaluated. The present experimental study was therefore performed in order to determine H_{opt} for skeletal muscle during sustained rhythmic exercise.

Materials and Methods

The experimental techniques used in this study have been described in detail elsewhere (3). The experiments were carried out on the surgically isolated and autoperfused gastrocnemius muscle of the dog. The muscles were alternately perfused with blood of varying hematocrit at constant perfusion pressure. Electric stimulation was used to induce rhythmic muscle contractions under isotonic conditions. Measurements were made of blood flow, O_2 uptake, and the amplitude of contraction of the muscle. The hematocrit of the perfused blood was varied by addition of autologous red cells or plasma, respectively, which had been obtained from another dog.

Results and Discussion

In a first series of experiments the hematocrit of the perfused blood was acutely changed while the muscle was working continuously under steady state conditions. Analysis of the changes of exercise hyperemia demonstrated an approximately linear decrease in blood flow with increasing arterial hematocrit in the range between 0.2 and 0.8 (3). These data were used to calculate the relationship between O_2 delivery and hematocrit, which is shown in Fig. 17-1. These results showed a distinct maximum of O_2 delivery at hematocrit levels of approximately 0.52 for the working muscle in comparison to a value of 0.3 for the resting muscle. The latter value is in good agreement with findings reported earlier for other tissues by various investigators (6–8). The observed shift of the calculated H_{opt} between resting and exercising state is the result of the alteration of the relationship between hematocrit and muscle blood flow, the slope of which is significantly lower under conditions of exercise. In resting muscle, the hematocrit-dependent alterations of O_2 transport capacity can be compensated for by changes of vascular conductivity.

In the working muscle, however, such compensatory changes are impossible due to the reduced vasodilatory reserve remaining after exercise dilatation. The occurrence of vascular dilatation in resting muscle can be inferred from results of earlier studies (2) in which blood flow changes in working and resting muscle were determined following dilution of the blood. Simultaneously, whole blood viscosity was measured in vitro at high shear rates (230 sec^{-1}). It is admittedly not easily possible to predict quantitatively the changes in skeletal muscle flow on the basis of these in vitro viscosity determinations, since owing to the Fahraeus-Lindqvist effect the apparent viscosity in the living vascular system is somewhat lower and varies in different vessels of the microcirculation (4,10). Nevertheless, it is certainly not without relevance that the product of blood flow and apparent viscosity, which is theoretically proportional to vascular conductivity, increased significantly during hemodilution in the resting, but not in the exercising muscle (Fig. 17-2). This can be regarded as an indication of compensatory vasodilatation in the resting tissue, which may explain the larger flow changes upon hematocrit alteration found in resting muscle compared to exercise conditions. It should be added that the mechanical effects of muscle contraction on tissue perfusion may play an additional role: previous studies (2,5) showed that the flow changes resulting from hematocrit alterations were smaller in the working compared to the nonworking but pharmacologically dilated muscle.

Under conditions of submaximal work levels the reduction of O_2 delivery at reduced and at substantially elevated hematocrit levels can at least partially be compensated for by an increased O_2 extraction from the perfused blood. In the present experiments an increase in O_2 extraction was indeed observed at hematocrits below 0.3 and above 0.6. It is therefore not surprising that the present results demonstrate a rather flat "maximum" of the relationship between O_2 uptake as well as muscle per-

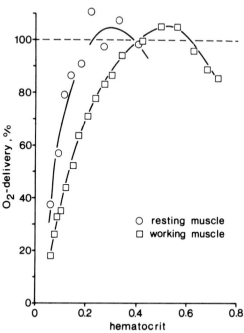

Fig. 17-1. Relationship between hematocrit and O_2 delivery to skeletal muscle during rest (*circles*) and during rhythmic exercise (*squares*).

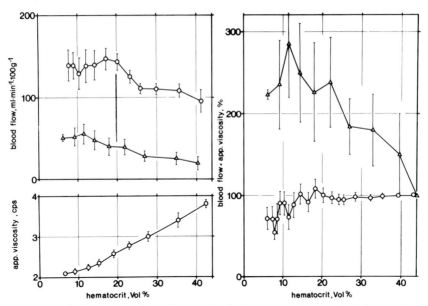

Fig. 17-2. Changes in blood flow in resting (*triangles*) and exercising (*circles*) muscle and in high-shear blood viscosity during dilution of the blood. The product of both parameters increases in the resting muscle, but not during exercise.

formance and the hematocrit of the blood, as shown in Fig. 17-3. A significant reduction of these parameters was seen only if the hematocrit was decreased below approximately 0.4 or increased beyond approximately 0.6. Thus, from the viewpoint of working capacity a distinct hematocrit optimum is not observed. It should be added that a compensatory increase in O_2 extraction is not possible at maximum working levels; therefore a clear-cut H_{opt} might be present under these conditions, and it appears on the basis of the present data that this H_{opt} is again higher than the physiologic hematocrit level. It must also be stressed that the increase in O_2 delivery to the working muscle, which occurs if the hematocrit in the blood is elevated from normal values to the H_{opt}, is relatively small, indeed. Therefore an increase in working capacity, if present, is likely to be small, too. Nevertheless, it can be concluded from these observations that increased hematocrit levels of the blood are

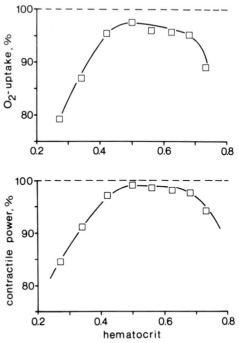

Fig. 17-3. Relationship between hematocrit and O_2 uptake (*top*) and contractile power (*bottom*) during steady state exercise.

in principle advantageous (or at least not detrimental) for steady state exercise capacity of skeletal muscle.

The above conclusion, however, is based on the results of acute experiments which cannot readily be compared to conditions of chronic hematocrit alterations such as during adaptation to high altitudes. In the experiments described thus far the hematocrit alteration was induced after exercise hyperemia had been fully established; during the resting periods, however, the muscles were perfused with blood of a physiologic hematocrit value. This experimental procedure was different in a second series of experiments in which muscle exercise was initiated only after the hematocrit of the perfused blood had been altered.

The results observed in these additional experiments showed that under resting conditions muscle blood flow became almost unmeasurably low at hematocrits exceeding 0.7. As a consequence of this significant flow reduction the calculation of O_2 delivery at elevated hematocrits gave extremely low values for the resting muscle. This observation may thus support earlier evidence for an accumulation of metabolites in the tissue which were seen in venous blood after hemodilution (9). Therefore it appears that hemoconcentration is associated with a significant reduction of supply of substrates and removal of waste products in the resting muscle. In addition, the present experiments showed that the development of exercise hyperemia upon initiation of muscle contractions was impaired (Fig. 17-4): the time needed to establish maximum blood flow increased significantly with increasing hematocrit of the perfused blood (Fig. 17-5). In one experiment, which is however not representative of the whole series, it was observed that the contractile power of the muscle was significantly reduced immediately after the onset of stimulation: in this case it took several minutes for the muscle to gradually increase both contraction amplitude and exercise blood flow (Fig. 17-6). Under these extreme conditions the me-

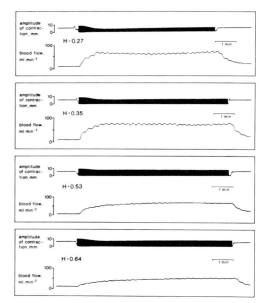

Fig. 17-4. Original recordings of exercise hyperemia and amplitude of muscle contractions at four hematocrit levels in the perfused blood. The hematocrit was altered during the resting periods.

chanical effect of muscle contraction may have been favorable to establish the final level of both flow and work capacity.

In contrast to this single observation, the steady state exercise capacity of the muscles seen after complete flow adaptation was not reduced in the whole second series of experiments, even at significantly elevated hematocrits. All data indicated a nor-

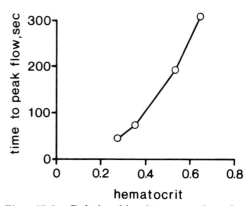

Fig. 17-5. Relationship between the time needed for development of full exercise hyperemia and the hematocrit of the perfused blood.

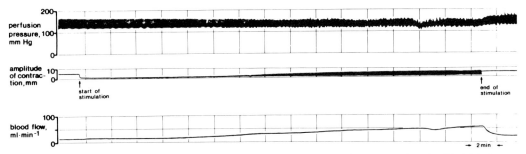

Fig. 17-6. Original recording of an experiment in which the muscle was perfused with high hematocrit ($H = 0.72$) before the onset of exercise. Note the extremely slow increase of blood flow and contraction amplitude.

mal or even slightly increased exercise capacity. Nevertheless, the observations made in this series of experiments appear to suggest that the value of H_{opt} found in the previous experiments may not be quite relevant, if conditions of chronically elevated hematocrit are considered.

Care must be taken if the present results are applied to conditions prevailing at high altitude. Although the optimal hematocrit (in terms of steady state O_2 delivery) for exercising skeletal muscle was found to be distinctly higher than the physiologic hematocrit level, the decreased fluidity of the blood at higher hematocrits is obviously detrimental for the rapid adaptation of blood flow (and thus O_2 delivery) to the varying metabolic needs of skeletal muscle. There is experimental evidence which indicates that under conditions of chronically elevated hematocrit the mechanism of flow adjustment may be severely hampered by the alteration of the flow properties of the blood. Rheologically, this could be interpreted in terms of the high shear forces needed to accelerate the sluggish flow of the concentrated red cell suspension passing the microcirculation. Under these conditions the higher than physiologic value of H_{opt} is possibly misleading if the exercise capacity of skeletal muscle is to be evaluated. On one hand exercise endurance may be limited by the reduced removal rate of metabolic waste products associated with the reduction in the absolute level of blood flow during exercise. More importantly, however, the impairment of muscle perfusion during the resting periods may provide a limiting factor for the exercise capacity under conditions of chronically elevated hematocrit levels.

References

1. Ekblom, B., Goldberg, A.N., and Gullbring, B.: Response to exercise after blood loss and reinfusion. J. Appl. Physiol. 33:175, 1972.
2. Gaehtgens, P., Benner, K.U., and Schickendantz, S.: Effect of hemodilution on blood flow, O_2 consumption, and performance of skeletal muscle during exercise. Bibl. Haematol. 41:54, 1975.
3. Gaehtgens, P., Kreutz, F., and Albrecht, K.H.: Optimal hematocrit for canine skeletal muscle during rhythmical isotonic exercise. Eur. J. Appl. Physiol. 41:27, 1979.
4. Gaehtgens, P. and Uekermann, U.: The apparent viscosity of blood in different vascular compartments of the autoperfused canine foreleg, and its variation with hematocrit. Bibl. Anat. 11:76, 1973.
5. Gustavsson, L., Appelgren, L., and Myrvold, H.E.: Polycythemia, viscosity and blood flow in working and non-working skeletal muscle in the dog. Cagliari, Italy, Xth World Conf., Europ. Soc. for Microcirc., 1978.
6. Messmer, K., Sunder-Plassmann, L., Klövekorn, W.P., and Holper, K.: Circulatory significance of hemodilution: rheological changes and limitations. Adv. Microcirc. 4:1, 1972.

7. Murray, J.F., Gold, P., and Johnson, B.L.: Systemic oxygen transport in induced normovolemic anemia and polycythemia. Am. J. Physiol. 203:720, 1962.
8. Richardson, T.Q. and Guyton, A.C.: Effects of polycythemia and anemia on cardiac output and other circulatory factors. Am. J. Physiol. 197:1167, 1959.
9. Sunder-Plassmann, L., Lewis, D.H., Klövekorn, W.P., Mendler, N., Holper, K., and Messmer, K.: Rheological and microcirculatory effects of acute hemodilution. Proc. Int. Union Physiol. Sci. 9:545, 1971.
10. Whittacker, S.R.F. and Winton, F.R.: Apparent viscosity of blood flowing in the isolated hindlimb of the dog, and its variation with corpuscular concentration. J. Physiol. (Lond.) 78:339, 1933.
11. Williams, M.H., Goodwin, A.R., Perkins, R., and Bocrie, J.: Effect of blood reinjection upon endurance capacity and heart rate. Med. Sci. Sports 5:181, 1973.

18

Cardiac Output and Regional Blood Flows in Altitude Residents

J. DURAND, P. VARENE, AND C. JACQUEMIN

It is now well documented that residence in altitude induces circulatory changes. The aim of this presentation is not to give an exhaustive picture of the altitude resident's circulatory status, but only to describe some of its aspects: (1) the regional distribution of cardiac output; (2) the functional consequences of the circulatory changes, taking the thermoregulatory disturbances induced by the reduction of cutaneous blood flow as an example; and (3) the altitude threshold of the hypoxic pulmonary vasoconstriction.

Regional Distribution of Cardiac Output

Cardiac Output

Cardiac output ($\dot{Q}$) and general arteriovenous oxygen difference $(a - \bar{v})O_2$, at rest, were claimed not to be significantly different in lowlanders (LL) and in highlanders (HL) (4,13,40,48,55,57,58). However, other data available in the literature and unpublished observations from I.B.B.A.[1] indicate the presence of a moder-

[1]Instituto Boliviano de Biologia de Altura, La Paz, Bolivia.

ate, but significant reduction of $\dot{Q}$; this has been reported by several authors (25,27,28). The decrease in $\dot{Q}$ parallels the increase in O_2 concentration in arterial blood (Ca_{O_2}, which itself is related to an overcompensation of the fall in Pa_{O_2}) by an increase in hematocrit (Hct) and hemoglobin concentration. This has been noticed or suggested by several authors working on polycythemic subjects at both sea level (SL) and altitude (HA) (2,28,37). These two changes in opposite direction result in a slight increase in arterial oxygen flow ($\dot{Q} \cdot Ca_{O_2}$) (Table 18-1).

During submaximal exercise $\dot{Q}$ increases as a linear function of the oxygen consumption ($\dot{V}_{O_2}$), as it does at sea level. It was first stated that the slope and the intercept of the relationship were the same at both SL and HA (1). However, it is possible now to gather more data on $\dot{Q}$ during exercise at HA and chiefly to have results in which $\dot{Q}$ and $\dot{V}_{O_2}$ are measured independently (and hence may be correlated directly). It appears, then, that the intercept with the ordinate is not significantly different, but that the slope decreases in HL. This observation agrees well with the fact that Ca_{O_2} increases with altitude despite the fall in Pa_{O_2} (Fig. 18-1) (18).

Table 18-1. Cardiac Output ($\dot{Q}$), Oxygen Concentration in Arterial Blood (Ca_{O_2}), and Oxygen Arterial Flow ($\dot{Q} \cdot Ca_{O_2}$) in Sea Level and Altitude Residents.

	Sea level		3750 m (P_B, 493 torr)	p	4375 m (P_B, 451 torr)
$\dot{Q}$					
liter · min^{-1}	6.16 ± 0.51	($p < 0.001$)	5.58 ± 0.68	(NS)	5.23 ± 0.79
liter · min^{-1} · m^{-2}	3.52 ± 0.23	(NS)	3.43 ± 0.40	(NS)	3.59 ± 0.41
ml · min^{-1} · 100 g^{-1}	8.6 ± 0.07	($p < 0.001$)	8.2 ± 0.02	($p < 0.001$)	8.0 ± 0.05
Ca_{O_2} (ml$_{O_2}$ · ml^{-1})	0.183 ± 0.011	($p < 0.001$)	0.201 ± 0.012	(NS)	0.213 ± 0.22
$\dot{Q} \cdot Ca_{O_2}$ (ml · min^{-1} · m^{-2})	644 ± 31	($p < 0.001$)	689 ± 53	(NS)	722 ± .58
Number of subjects	119		43		6
References	(18)		(29,35,62, and unpublished data from I.B.B.A.)		(35)

Measurements were made at rest in recumbent position. Mean ± 1 SD.
NS = not significant.

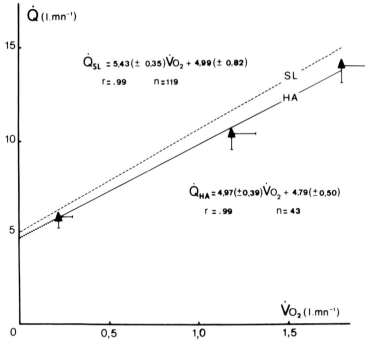

Fig. 18-1. Cardiac output ($\dot{Q}$) as a function of oxygen consumption ($\dot{V}_{O_2}$) at rest and during the steady state of submaximal exercise in recumbent posture, in lowlanders at sea level (*SL*), and in altitude residents at 3750 m (*HA*). Sea level values are taken from Durand and Mensch-Dechene (18). (HA values are mean ± SE.)

Distribution of Cardiac Output

Functional Circulations

Renal and splanchnic circulations. Renal blood flow, measured by paraaminohippurate (PAH) perfusion, was found to be lower in HL than in LL (6,30). However, in a more recent study, Zelter et al. (62) failed to find any significant difference. This is consistent with plasma flow values obtained after radiohippuran injection (12). This discrepancy may be explained by the fact that the PAH extraction coefficient is only 0.91 in HL, so that PAH concentration in the renal venous blood cannot be considered as negligible.

To our knowledge, the only measurements of hepatosplanchnic blood flow in HL were published by Capderou et al. (10). Determinations were made by calculating the hepatic clearance of indocyanine green perfused at constant rate. The values reported are slightly, but not significantly, higher than those obtained in SL dwellers by the same group or by others (9,49).

Cutaneous circulation. The decrease in skin blood flow ($\dot{Q}_{sk}$) at altitude is now well documented (1,16,20,21,51,60,61). Figure 18-2 shows blood flow in the right hand, measured by venous occlusion plethysmography, which is supposed to be representative of $\dot{Q}_{sk}$. One can notice that the difference is only significant for skin temperatures above 33°C. For that skin temperature which corresponds to an ambient air temperature of 22°C, a rough extrapolation yields 100 ml · min^{-1} reduction in total $\dot{Q}_{sk}$ (about 20%). In turn, this decrease would result in a 7 W · °C^{-1} decrease in the physiologic heat conductance. The thermoregulatory consequences of this diminution in active heat convection will be discussed later.

The cold-induced vasodilatation (shunting) is considerably less in altitude than at

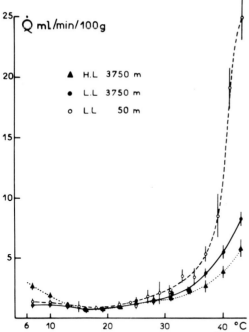

Fig. 18-2. Blood flow in the right hand ($\dot{Q}$) plotted against local temperature (°C) in high altitude residents (*HL*, ▲) and in lowlanders (*LL*) at sea level (○) and after 1 month of acclimatization to altitude (●). $\dot{Q}$ is expressed as milliliters per minute and per 100 ml of hand volume. Measurements were made in a water-filled plethysmograph using venous occlusion technique. Mean ± SE. Data from Durand and Martineaud (16).

sea level. So also is the postischemic reaction (53).

Probably of greater physiologic significance than $\dot{Q}_{sk}$ reduction is the decrease in cutaneous vascular compliance. Volume changes induced by a given increase or decrease in transmural pressure are smaller at high altitude than at sea level. Since this decrease in vascular compliance could be due either to the presence of a larger volume of blood in the capacitance vessels or to active constriction of the walls of those vessels, it was necessary to determine the absolute value of the total volume of blood which can be shifted away from the hand by increasing the pressure in the plethysmograph up to the point where there was no further decrease in the volume of the hand.

By this method the decrease in compliance appears to reflect an increase in capacitance vessel tone and is associated with a decrease in the volume of blood present in the skin (Fig. 18-3). This is observed both in residents and in sojourning subjects.

Here again a possibly hazardous extrapolation would give an approximate figure of 300 ml of blood mobilized from the skin to other vascular beds by such a decrease in the compliance of cutaneous capacitance vessels, at a skin temperature of 33°C (8).

Nutrient Circulations

Cerebral blood flow ($\dot{Q}_c$). In 1972, Milledge and Sorensen reported that the cerebral oxygen arteriovenous difference $(a - v)O_2$ was larger in HL than in LL (33). More recently, several studies (15, 19,31,56), except one (50), confirmed that $\dot{Q}_c$ was lower in HL than LL at sea level, which contrasts with the increase in brain perfusion in newcomers during the first few days after arrival at HA (50,54). The reduction of $\dot{Q}_c$ in HL is consistent with the broadening of the local $(a - v)O_2$ which results in an identical local oxygen consumption in HL studied at HA and in LL at SL.

The relationship between $\dot{Q}_c$ and Pa_{CO_2} is comparable in HL and in LL, but with a higher setting in the former. Thus, for an equivalent Pa_{CO_2}, $\dot{Q}_c$ would be higher in HL than in LL. In HL, $\dot{Q}_c$ is more sensitive to Pa_{O_2} changes than in LL: correction of hypoxia leads to a further decrease in $\dot{Q}_c$; correction of both hypoxia and hypocapnia finally results in a $\dot{Q}_c$ lower in HL than in LL (Fig. 18-4).

Cerebral mean transit time increases as a hyperbolic function of the hematocrit (Hct), so that red cell flow through the brain circulation is the same in HL and in LL (Fig. 18-5).

In conclusion, $\dot{Q}_c$ in HL is reduced conversely to the increase in Hct, the effect of hypocapnia being compensated by the increase in the hypoxic response.

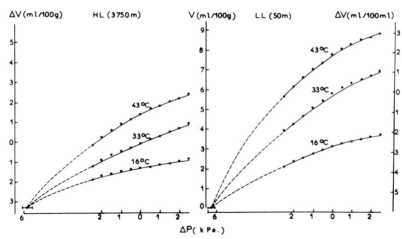

Fig. 18-3. Volume-pressure curves of capacitance vessels of the right hand in lowlanders (*right*; *LL*) and in highlanders (*left*; *HL*). Curves have been obtained for three different local temperatures: 16, 33, and 43°C. Changes in venous transmural pressure (ΔP) are shown on the abscissa and volume changes (ΔV; *right* and *left scales*) and blood volume (V; *middle scale*), all expressed as ml/100 ml of hand volume, are shown on the ordinates. Data from Durand and Martineaud (16) and Durand et al. (17).

Coronary blood flow ($\dot{Q}_{co}$). All authors agree that myocardial blood flow is reduced in HL as compared to LL (23,35,36). Contrary to what was found in the cerebral circulation, there is no increase in local $(a - v)O_2$ differences, so that local oxygen uptake is decreased and myocardial efficiency is improved.

This amazing conclusion induced some authors to wonder whether this was not a failure of the inert gases washout technique. Thus, Maseri et al. (32) utilized tritiated water (THO) instead of N_2O or ^{133}Xe. As expected, they found larger values than those obtained with liposoluble indicators, at both sea level and altitude, but again HL values were lower than LL values (Table 18-2). These results are comparable to those obtained in congenital heart disease (52) or in chronic obstructive lung diseases (34) with hypoxemia; however, neither change in blood flow repartition between right and left coronary arteries, morphometric data (3), nor enzymatic changes (24,47) can satisfactorily explain these findings. In a preliminary study Prioux-Guyonneau et al. (38,39,41) found what they describe as an increase in noradrenaline uptake by the heart muscle in HL as compared to LL. If this was confirmed, it could explain at least part of the increase in myocardial efficiency. Undoubtedly, coronary circulation in altitude residents needs further investigation.

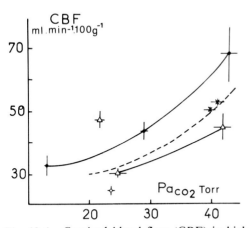

Fig. 18-4. Cerebral blood flow (CBF) in high altitude residents in supine posture ($\dot{Q}_C$) as a function of carbon dioxide partial pressure in arterial blood (Pa_{CO_2}) and for different values of inspired oxygen fraction: $Fi_{O_2} = 0.21$ (●); 0.33 (△); 0.75 (○); 0.16 (▲). Dotted line represents the curve obtained in lowlanders at sea level and (*) the values obtained by the authors of the present study. Mean ± SE. 10 torr ~ 1.33 kPa. From Durand et al. (15).

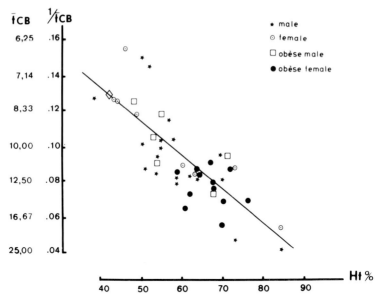

Fig. 18-5. Mean cerebral circulatory transit time (T_{CB}) and its inverse as a function of the venous hematocrit (Ht) in altitude residents in sitting posture; individual values. Mean values ± SE in lowlanders at sea level are given for comparison (◇). Data from Durand et al. (15).

Blood flow in skeletal muscle ($\dot{Q}_m$). To our knowledge, the only study of $\dot{Q}_m$ in HL was published by Bidart et al. (7). $\dot{Q}_m$ was measured by ^{133}Xe washout at rest and during local exercise and recovery in the tibialis anterior muscle. Results obtained show a considerable reduction of $\dot{Q}_m$ in HL sojourners as compared with LL at sea level (Fig. 18-6). No local $(a - v)O_2$ was measured; however, since the general $\dot{V}_{O_2}$ is the same in both HL and LL during exercise for the same work load (43), this would mean that exercising muscle $(a - v)O_2$ would be greater in HL than in LL. This again is unexpected since local $(a - v)O_2$ is maximal and steady in exercising muscle (59) and further increase in local $\dot{V}_{O_2}$ is obtained by increasing local flow rather than O_2 extraction.

Most likely the Kety and Schmidt technique does not allow comparison of $\dot{Q}_m$ in HL and LL. One possible explanation could be an uneven distribution of red blood cells and plasma in the muscular capillary network which would result in an apparent shunt, as suggested by Cerretelli

Table 18-2. Comparison of Myocardial Blood Flow ($\dot{Q}_{CO}$, ml · min^{-1} · 100 g^{-1}), Oxygen Arteriovenous Difference $(a-v)_{O_2}$, and Oxygen Consumption ($\dot{V}_{O_2}$, ml · min^{-1} · 100 g^{-1}) in Sea Level and Altitude Residents at Rest.

		Sea level	3750 m	p
THO n = 5 and 7	$\dot{Q}_{CO}$ $(a-v)_{O_2}$ $\dot{V}_{O_2}$	146 ± 56 0.116 ± 0.010 16.7 ± 6.0	84 ± 12 0.111 ± 0.017 9.4 ± 2.5	($p < 0.05$) (NS) ($p < 0.05$)
N$_2$O n = 7 and 10	$\dot{Q}_{CO}$ $(a-v)_{O_2}$ $\dot{V}_{O_2}$	72 ± 7 0.123 ± 0.076 8.7 ± 0.9	55 ± 10 129 ± 0.019 7.1 ± 1.4	($p < 0.001$) (NS) ($p < 0.05$)

Results were obtained using either tritiated water, or nitrous oxide. n, number of subjects at sea level and at altitude, respectively. Mean ± SD. (THO) THO data, from Maseri et al. (32); N$_2$O data from Moret et al. (35).

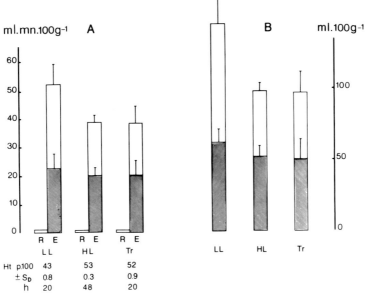

Fig. 18-6. Muscular blood flow measured by ^{133}Xe washout in the tibialis anterior muscle, at rest (R) and during exercise (E) in lowlanders at sea level (LL), in altitude residents (HL), and in translocated lowlanders (Tr) after sojourning 30 days at 3750 m (La Paz, Bolivia). **A** Columns represent plasma flow (*open*) and red cell flow (*shaded*). E corresponds to the maximal blood flow measured. Values of the hematocrit (Ht), mean ± SD, and number of subjects are indicated below the corresponding flow column. **B** Extra volume of blood perfusing the muscle during the exercise period (1 min) and during recovery (3 min). *Open column,* plasma; *shaded column,* red blood cells. Mean ± SD. After Bidart et al. (7).

(11). Another argument in favor of a biased measurement of muscular blood flow in HL is given by the comparison of the cardiac output with the sum of regional flows. In LL the part left for "miscellaneous" unmeasurable blood flows (e.g., in bone, joints, adipose tissue) is only 1.4 ml/100 g of tissue, whereas in HL one arrives at the unreasonable figure of 6.2 ml/100 g (Fig. 18-7).

Finally, regional blood flows when measured by the direct application of the Fick principle (mass conservation) are found not to be different in HL and LL. On the contrary, when measured by determining the time constant of a single tissue compartment, they are found to be lower in HL than in LL. In the case of cerebral blood flow this finding is consistent with the broadening of local $(a - v)O_2$. But with respect to blood flow in the myocardium and skeletal muscle, no such compensation exists and these unexpected results may reflect some methodologic error.

In conclusion, blood flow distribution is different in HL and in LL. Among the factors controlling blood repartition, hypoxia and hypocapnia are paramount; however, hematocrit also plays an important role. Progress in understanding the cardiocirculatory strategy, if any, of "altitude-adapted" men now depends mainly on research at the microcirculatory level.

Altitude and Thermoregulation

Reduction in skin blood flow and consequently in physiologic heat conductance modifies HL thermoregulatory mechanisms, especially when subjects are exposed to external and/or internal heat loads. To investigate these consequences, Raynaud et al. (44,45) compared the energy

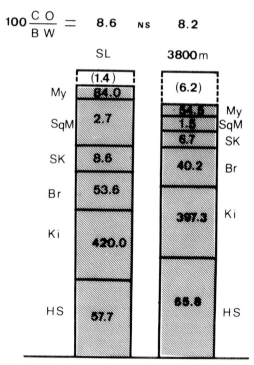

Fig. 18-7. Comparison of the distribution of cardiac output in lowlanders and highlanders. Cardiac output (CO; ml·min^{-1}) related to body weight (BW; g) is not significantly different at sea level (SL) and at 3800 m. The height of each rectangle gives the value of the corresponding local blood flow in ml·min^{-1}·100 g^{-1}. Figures in rectangles are absolute values (ml·min^{-1}). My, coronary; SqM, skeletal muscle; SK, skin; Br, cerebral; Ki, renal; HS, hepatosplanchnic. Open rectangles at the top of the columns represent the difference between cardiac output and the sum of measurable local blood flows.

equilibrated at $\bar{T}_{sk}$, this increase in the thermal arteriovenous difference would compensate the decrease in cutaneous blood flow.

At altitude T_{re} increases more during exercise than at sea level; on the contrary, $\bar{T}_{sk}$, compared to rest values, decreases in altitude during exercise. Consequently, heat loss by radiation and convection are reduced and the heat balance is reequilibrated by an increase in sweating rate (Fig. 18-8).

Figure 18-9 shows that the gain of the

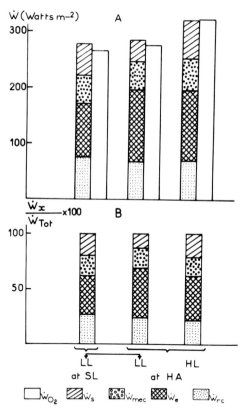

Fig. 18-8. Heat balance during submaximal exercise in lowlanders (LL) at sea level (SL) and at 3800 m (HA) and in highlanders (HL) at the same altitude. **A** Absolute values. **B** Percentage of the total energy expenditure. $\dot{W}_{O_2}$, energy production calculated with oxygen consumption; $\dot{W}_s$, heat storage; $\dot{W}_{mec}$, mechanical energy production; $\dot{W}_e$ evaporative heat loss; $\dot{W}_{rc}$, convection and radiation heat loss. From Raynaud et al. (44).

balance in HL and LL at SL and 3 weeks after translocation to altitude. Measurements were made at rest and during exercise and at two different room temperatures (20 and 33°C).

At rest mean skin temperature ($\bar{T}_{sk}$) is the same at both SL and HA, but central temperature (T_{re}) is higher in subjects at altitude than at sea level. There is no significant difference between residents and sojourners. Assuming that T_{re} is close to the arterial temperature, the precooling being negligible at T_{sk} above 30°C (5), and that the cutaneous venous blood is thermally

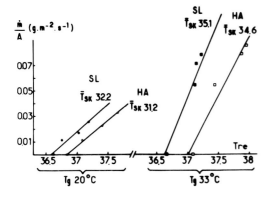

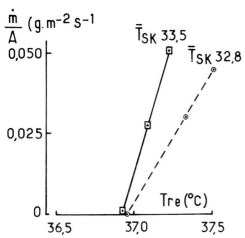

Fig. 18-9. **A** Sweating rate per square meter of body surface area ($\dot{m}/A$) as a function of rectal temperature (Tre) at rest and at two different levels of exercise, and for two different room temperatures (T_g 20°C and T_g 33° C), at sea level (SL) and at 3750 m (HA). Mean skin temperature ($\bar{T}_{sk}$) is calculated near the corresponding regression line. **B** Sweating rate per square meter of body surface area ($\dot{m}/A$) as a function of rectal temperature (Tre) at rest and at two different levels of exercise. Measurements made at 3750 m while breathing room air (⊙) and while breathing a hypercapnia mixture (⊡) in order to raise Pa_{CO_2} from 31 to 37 torr. The mean skin temperature ($\bar{T}_{sk}$) is calculated near the corresponding regression line. From Raynaud et al. (45).

thermoregulatory response, the sweating rate in that particular case, is not altered but is reset at a higher central temperature (22). When altitude hypocapnia is corrected by breathing an appropriate gas mixture, the slope of the sweating response is steeper (26).

When an organism is adapted to a given strain, such as altitude, it loses some compensatory options, and when it is exposed to another occasional stress, such as heat, it must find a compromise, sacrificing partially its homeostasis.

Physiologic Threshold of Altitude

Low barometric pressure is not the only physical characteristic of terrestrial elevation. However, from a physiologic viewpoint, it is the main and most specific component of altitude. Associated factors such as thermal stresses and solar radiation are undoubtedly important in altitude biology, but man is able to find adequate protection against them. On the contrary, a decrease in atmospheric pressure is a strain to which human technology cannot oppose any practical answer, except for brief exposure. Hence, it is reasonable to define altitude by its influence on the partial pressure of respiratory gases in arterial blood. Because of the shape of the hemoglobin dissociation curve, altitude hypoxia becomes significant when barometric pressure stands below 67 kPa (3300–3500 m) and "normal" life appears not to be possible below 50 kPa (5300–5500 m).

The hypoxic threshold close to 3000 m (i.e., $Pa_{O_2} \simeq 9$ kPa) is confirmed by the existence of pulmonary artery hypertension in subjects residing above this level.

Figure 18-10 shows the value of mean pulmonary artery pressure at different altitudes and also the position of the QRS axis of the electrocardiogram in the frontal plane. These field observations agree well with observations in human pathology (Fig. 18-11) and experimental data (Fig. 18-12).

Acknowledgment

This work was supported by C.N.R.S. (RCP 538).

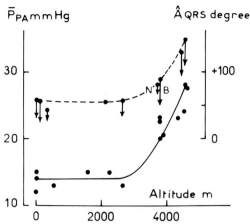

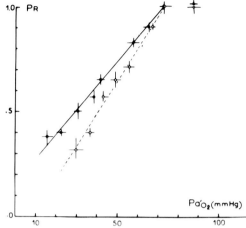

Fig. 18-10. Mean pulmonary artery pressure (*solid line*) and position of the QRS axis in the frontal plane of the electrocardiogram (*dashed line*) as a function of altitude. The arrows indicate the amplitude of the axis rotation to the left with aging. N and B correspond to the values obtained in residents at the same altitude in Nepal and Bolivia, respectively. Data from Reeves and Grover (46) and Raynaud et al. (42).

Fig. 18-12. Redistribution of pulmonary blood flow as a function of local oxygen tension. In this experiment under bronchospirometry one lung is made hypoxic while the other stays normoxic. Repartition of pulmonary blood flows is determined by injection into the right atrium of ^{85}Kr dissolved in an indocyanine green solution. Radioactivities of expired gas from the right and the left lung are measured separately. *Ordinate*, ratio of blood flows in hypoxic/normoxic lungs. *Abscissa*, the calculated oxygen tension in the capillary blood of the hypoxic lung. Measurements were done at two different values of carbon dioxide tension in arterial blood (mean ± SE). From Durand et al. (14).

References

1. Alarcon-Castillo, L.: Varaciones de la pression venosa peripherica. Thesis. Lima, Faculty of Medicine, 1956.
2. Anderson, C.B. and Gray, F.D.: The circulatory and ventilatory effects of normovolemic polycythemia. Yale J. Biol. Med. 35:233, 1962.
3. Arias-Stella, J. and Topilsky, M.: Anatomy of the coronary circulation at high altitude. In Porter, R. and Knight, J. (eds.): High Altitude Physiology: Cardiac and Respiratory Aspects. Ciba Foundation Symposium. Edinburgh, Churchill Livingstone, 1971, pp. 149–157.
4. Banchero, N., Sime, F., Penaloza, D., Cruz, J., Gamboa, R., and Marticorena, E.: Pulmonary pressure, cardiac output, and arterial oxygen saturation during exercise at high altitude and at sea level. Circulation 33:249, 1966.
5. Bargeton, D., Durand, J., Mensch-Dechene, J., and Decaud, J.: Echanges de chaleur de la main. Rôle des réactions circulatoires et des variations de la température locale du sang artériel. J. Physiol. (Paris) 51:111, 1959.

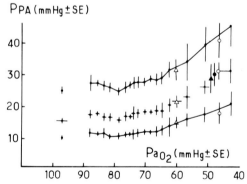

Fig. 18-11. Pulmonary artery pressure (systolic mean and diastolic) as a function of arterial blood oxygen tension in normal subjects and in patients with chronic obstructive lung disease. Values obtained in altitude residents are figured for comparison (△, 3750 m; ▲, 4375 m; ○, 4800 m), and during acute hypoxia, by breathing 12% oxygen in nitrogen (●). Mean values ± SE.

6. Becker, L., Schilling, A., and Harvey, R.: Renal function in man acclimatized to high altitude. J. Appl. Physiol. 10(1):79, 1957.
7. Bidart, Y., Drouet, L., and Durand, J.: Débit sanguin dans le muscle squelettique chez les sujets résidant et transplantés en altitude (3800 m). J. Physiol. (Paris) 70:333, 1975.
8. Bidart, Y., Mensch-Dechene, J., Bourdarias, J.P., and Seroussi, S.: Circulation cutanée du pied lors d'un exercice des membres supérieurs chez le sujet normal et chez le sujet atteint de section traumatique de la moëlle. J. Physiol. (Paris) 62(Suppl. 1):130, 1970.
9. Bradley, S.E., Ingelfinger, F.J., Bradley, G.P., and Curry, J.J.: The estimation of hepatic blood flow in man. J. Clin. Invest. 24:890, 1945.
10. Capderou, A., Polianski, J., Mensch-Dechene, J., Drouet, L., Antezana, G., Zelter, M., and Lockhart, A.: Splanchnic blood flow O_2 consumption, removal of lactate, and output of glucose in highlanders. J. Appl. Physiol. 43(2):204, 1977.
11. Cerretelli, P.: O_2 breathing at altitude: effects on maximal performance. Hypoxia Symposium (The Arctic Institute of North America). Banff, Canada, February 21–24, 1979.
12. Chavez, C., Barragan, M., and Mariaca, J.: Determinacion del flugo plasmatico renal en la Altura con Hippuran I-131. Inst. Boliviano Biol. Alture 4(2)25, 1972.
13. De Michelli, A., Villaces, E., Guzzy, P., and Rubio, V.: Observaciones sobre los valores hemodynamicos y respiratorios obtenidos en subjectos normales. Arch. Inst. Cardiol. Mexico 30:507, 1960.
14. Durand, J., Leroy Ladurie, M., and Ranson-Bitker, B.: Effects of hypoxia and hypercapnia on the repartition of pulmonary blood flow in supine subjects. Prog. Resp. Res. 5:156, 1970.
15. Durand, J., Marc-Vergnes, J.P., Coudert, J., Blayo, M.C., and Pocidalo, J.J.: Cerebral blood flow, brain metabolism, and CSF acid-base balance in highlanders. In Paintal, A.S. and Gill-Kumar, P. (eds.): Respiratory Adaptations, Capillary Exchange, and Reflex Mechanisms. New Delhi, Nauchetan Press, 1977, pp. 27–37.
16. Durand, J. and Martineaud, J.P.: Resistance and capacitance vessels of the skin in permanent and temporary residents at high altitude. In Porter, R. and Knight, J. (eds.): High Altitude Physiology: Cardiac and Respiratory Aspects. Ciba Foundation Symposium. Edinburgh, Churchill Livingstone, 1971, pp. 159–17a.
17. Durand, J., Martineaud, J.P., Pradel, M., and Massoum, M.: Influence de l'altitude sur les résistances et capacitances vasculaires cutanées chez l'homme. J. Physiol. (Paris) 59(4):400, 1967.
18. Durand, J. and Mensch-Dechene, J.: Relation théorique entre le débit cardiaque et la consommation d'oxygène au cours de l'exercice. Bull. Physiopath. Resp. 15:977, 1979.
19. Durand, J., Olesen, J., Coudert, J., David, P., and Marc-Vergnes, J.P.: High altitude resident's cerebral blood flow. IX Interamerican Congress of Cardiology. San Francisco, April 23–29, 1972.
20. Durand, J., Verpillat, J.M., Pradel, M., and Martineaud, J.P.: Influence of altitude on the cutaneous circulation of residents and newcomers. Fed. Proc. 28(3):1124, 1969.
21. Elsner, R.W., Bolstad, A., and Forno, E.: Maximum oxygen consumption of Peruvian Indian natives at high altitudes. In Weike, H. (ed.): The Physiological Effects of High Altitude. Oxford, Pergamon Press, 1964, p. 217.
22. Greenleaf, J.E., Greenleaf, C.J., Card, D.H., and Saltin, B.: Exercise temperature regulation in man during acute exposure to simulated altitude. J. Appl. Physiol. 26:290, 1969.
23. Grover, R.F., Lufschanowski, R., and Alexander, J.K.: Decreased coronary blood flow in man following ascent to high altitude. In Vogel, J. (ed.): Hypoxia, High Altitude, and Heart. Basel, Karger, 1970, pp. 72–79.
24. Harris, P.: Some observations on the biochemistry of the myocardium at high altitude. In Porter, R. and Knight, J. (eds.): High Altitude Physiology: Cardiac and Respiratory Aspects. Ciba Foundation Symposium. Edinburgh, Churchill Livingstone, 1971, pp. 125–129.
25. Hartley, L.H., Alexander, J.K., and Modelski, M.: Subnormal cardiac output at rest and during exercise in residents at 3100 m altitude. J. Appl. Physiol. 23:839, 1967.
26. Houdas, Y., Bonaventure, M., Sauvage, A.,

Houdas, C., and Ginestet, A.: Quantitative influence of CO_2 inhalation on thermal sweating in man. Aerospace Med. 44:265, 1973.
27. Hultgren, H.N., Kelly, J., and Miller, H.: Pulmonary circulation in acclimatized man at high altitude. J. Appl. Physiol. 20:233, 1965.
28. Klausen, K.: Cardiac output in man at rest and work during and after acclimatization to 3800 m. J. Appl. Physiol. 21:609, 1966.
29. Lockhart, A., Zelter, M., Mensch-Dechene, J., Antezana, G., Paz-Zamora, M., Vargas, E., and Coudert, J.: Pressure-flow-volume relationships in the pulmonary circulation of normal highlanders. J. Appl. Physiol. 41(4):449, 1976.
30. Lozano, R. and Monge, C.: Renal function in high altitude natives and in natives with chronic mountain sickness. J. Appl. Physiol. 20(5):1026, 1965.
31. Marc-Vergnes, J.P., Antezana, G., Coudert, J., Gourdin, D., and Durand, J.: Débit sanguin et métabolisme énergétique du cerveau et équilibre acido-basique du L.C.R. chez les résidents en altitude. J. Physiol. (Paris) 68:633, 1974.
32. Maseri, A., L'Abbate, A., Coudert, J., Biagini, A., Michelassi, C., and Distante, A.: Study of the adaptation of the coronary circulation in natives at high altitude. In: Anthropologie des Populations Andines. I.N.S.E.R.M. 63:363, 1976.
33. Milledge, J.S. and Sørensen, S.C.: Cerebral arteriovenous oxygen difference in man native to high altitude. J. Appl. Physiol. 32:687, 1972.
34. Moret, P.: Circulation coronaire et insuffisance du myocarde dans l'insuffisance respiratoire. Poumon Coeur 4:421, 1968.
35. Moret, P., Covarrubias, E., Coudert, J., and Duchosal, F.: Cardiocirculatory adaptation to chronic hypoxia: comparative study of coronary flow, myocardial oxygen consumption, and efficiency between sea level and high altitude residents. Acta Cardiol. 27:283, 1972.
36. Moret, P., Covarrubias, E., Coudert, J., and Duchosal, F.: Cardiocirculatory adaptation to chronic hypoxia: comparative study of myocardial metabolism of glucose, lactate, pyruvate, and free fatty acids between sea level and high altitude residents. Acta Cardiol. 27:483, 1972.
37. Murray, J.F., Gold, P., and Johnson, B.L.: Systemic oxygen transport in induced normovolemic anemia and polycythemia. Am. J. Physiol. 203:720, 1962.
38. Prioux-Guyonneau, M., Cretet, E., Jacquot, C., Rapin, J.R., and Cohen, Y.: The effect of various simulated altitudes on the turnover of norepinephrine and dopamine in the central nervous system of rats. Pflügers Arch. 380:127, 1979.
39. Prioux-Guyonneau, M., Jacquot, C., Cohen, Y., and Rapin, J.R.: Influence de l'hypoxie normobare et hypobare sur le taux de renouvellement de la noradrénaline cardiaque. J. Physiol. (Paris) 72:579, 1976.
40. Pugh, L.G.C.E.: Cardiac output in muscular exercise at 5800 m (19,000 ft). J. Appl. Physiol. 19:441, 1964.
41. Rapin, J.R., Coudert, J., Drouet, L., Durand, J., and Cohen, Y.: Uptake of noradrenaline in high altitude natives' heart. Experientia 33:739, 1977.
42. Raynaud, J., Corone, P., Escourrou, P., and Drouet, L.: Electrocardiographic observations in high altitude residents of Nepal and Bolivia. Int. J. Biometeorol. (in press).
43. Raynaud, J., Martineaud, J.P., Bordachar, J., Tillous, M.C., and Durand, J.: Oxygen deficit and debt in submaximal exercise at sea level in high altitude. J. Appl. Physiol. 37(1):43, 1974.
44. Raynaud, J., Varene, P., Jacquemin, C., and Durand, J.: Heat exchange and energetic balance during exercise in highlanders and lowlanders at sea level and high altitude. Intern. Symp. on Environmental Physiology: Bioenergetics and Temperature Regulation. Dublin, July 18–23, 1971. Environ. Physiol. 30A:161, 1972.
45. Raynaud, J., Varene, P., Vieillefond, H., and Durand, J.: Circulation cutanée et échanges thermiques en altitude (3800 m). Arch. Sci. Physiol. 27:A247, 1973.
46. Reeves, J.T. and Grover, R.F.: High altitude pulmonary hypertension and pulmonary edema. In: Progress in Cardiology, Vol. 4. Philadelphia, Lea & Febiger, 1975, pp. 99–117.
47. Reynafarge, B.: Effect of chronic hypoxia on the kinetics of energy transformation in heart mitochondria. In: Metabolism of the Hypoxic and Ischaemic heart. Cardiology 56:206, 1971–1972.
48. Rotta, A., Canepa, A., Hurtado, A., Velas-

quez, T., and Chavez, R.: Pulmonary circulation at sea level and at high altitude. J. Appl. Physiol. 9:328, 1956.
49. Rowell, L.B., Blackmon, J.R., and Bruce, R.A.: Indocyanine green clearance and estimated hepatic blood flow during mild to maximal exercise in upright man. J. Clin. Invest. 43:1677, 1964.
50. Roy, S.B.: Circulatory and ventilatory effects of high altitude-acclimatization and de-acclimatization of Indian soldiers. A prospective study, 1964-1972. New Delhi, ICMR. 1973.
51. Roy, S.B., Guleria, J.S., Khanna, P.K., Talwar, J.R., Manchanda, S.C., Pande, J.N., Kaushik, V.S., Subba, P.S., and Wood, J.E.: Immediate circulatory response to high altitude hypoxia in man. Nature 217:1177, 1968.
52. Rudolph, W.: Myocardial metabolism in cyanotic congenital heart disease. In: Metabolism of the Hypoxia and Ischaemic Heart. Cardiology 56:209, 1971-1972.
53. Seroussi, S., Durand, J., Verpillat, J.M., Pradel, M., and Martineaud, J.P.: Débit et distensibilité vasculaire cutanées post-ischémiques chez l'homme. J. Physiol. (Paris) 60(Suppl. 2):545, 1968.
54. Severinghaus, J.W., Chiodi, H., Eger, E.I., Brandstater, B., and Hornbein, T.F.: Cerebral blood flow in man at altitude. Role of cerebrospinal fluid pH in normalization of flow in chronic hypoxia. Circ. Res. 19:274, 1966.
55. Sime, F., Banchero, N., Penaloza, D., and Cruz, J.: Hemodynamic changes in the pulmonary circulation of high altitude natives after two years residence at sea level. Circulation 34(Suppl. 3):217, 1966.
56. Sørensen, S.C., Lassen, N.A., Severinghaus, J.W., Coudert, J., and Paz-Zamora, M.: Cerebral glucose metabolism and cerebral flow. J. Appl. Physiol. 37(3):305, 1974.
57. Thielen, E.O., Gregg, D.E., and Rotta, A.: Exercise and cardiac work response at high altitude. Circulation 12:383, 1955.
58. Vogel, J.H.K., Weaver, W.F., Rose, R.L., Blount, S.G., and Grover, R.F.: Pulmonary hypertension on exertion in normal man living at 10,150 feet (Leadville, Colorado). Med. Thoracalis 19:461, 1962.
59. Wahren, J.: Quantitative aspects of blood flow and oxygen uptake in the human forearm during rhythmic exercise. Acta. Physiol. Scand. 67(Suppl. 269):5, 1966.
60. Weil, J.V., Byrne-Quinn, E., Battock, D.J., Grover, R.F., and Chidsey, C.A.: Forearm circulation in man at high altitude. Clin. Sci. 40:235, 1971.
61. Wood, J.E. and Roy, S.B.: The relationship of peripheral venomotor responses to high altitude pulmonary edema in man. Am. J. Med. Sci. 259:56, 1970.
62. Zelter, M., Capderou, A., Polianski, J., and Mensch-Dechene, J.: Renal blood flow in normal highlanders. J. Appl. Physiol. (in press).

19
The Pulmonary Circulation of High Altitude Natives

G. Antezana, L. Barragán, J. Coudert, L. Coudkowicz, J. Durand, A. Lockhart,
J. Mensch-Dechene, M. Paz Zamora, H. Spielvogel, E. Vargas, and M. Zelter

La Paz, the capital of Bolivia, a city with about 700,000 inhabitants, extends itself from an altitude of 4000 m above sea level down to 3200 m. The altitude of the Bolivian Institute of Altitude Biology (I.B.B.A.) is 3600 m or 12,200 ft. At this altitude the atmospheric pressure averages about 499 mmHg, which provides an inspired oxygen tension (PI_{O_2}) of 95 mmHg.

Since Rotta et al. (8) first published their data, it has been known that the pulmonary circulation behaves quite differently at high altitude than at sea level with respect to hemodynamics. In particular, a state of pulmonary hypertension exists due to the low alveolar oxygen tension ($Pa_{O_2} = 60$ mmHg in La Paz) which increases pulmonary vascular resistance (12). Rotta et al., who made the first high altitude studies with right heart catheterization at Morococha, Peru (4540 m or 14,900 ft.), found an increase in mean pulmonary artery pressure to 24 mmHg in natives of that altitude and to 18 mmHg in newcomers. Peñaloza et al. (7), who studied larger series also at Morococha, indicated that the mean pulmonary artery pressure of such people is in fact higher. In a study of 38 native residents of Morococha they found mean pulmonary artery pressure to be 28 mmHg. Blount and Vogel (3) recorded an average value of 24 mmHg in newcomers to Leadville, Colorado (altitude 10,150 ft).

The differences in magnitude of the mean pulmonary artery pressure in natives and newcomers have been attributed to an increased bulk of medial muscle in the peripheral pulmonary arteries of those permanently living at high altitude (1) and to pulmonary vasoconstriction stemming from the critical reduction in alveolar oxygen tension in newcomers (11). Above the altitude of 10,000 ft the mean pulmonary artery pressure can be predicted from an empiric relationship, which states that the mean pulmonary artery pressure varies directly with the inspired oxygen tension (PI_{O_2}) and inversely with the product of the existing barometric pressure multiplied by the alveolar oxygen tension. It is possible to set up a linear regression equation relating the mean pulmonary artery pressure at altitude to the alveolar oxygen tension (6). In this way the mean pulmonary artery pressure can be estimated and predicted; for instance, for the altitude of La Paz (12,200 ft), the predicted pulmonary artery pressure was approximately 23 mmHg.

The first cardiac catheterization data derived from nine male and two female vol-

Table 19-1. *Right Heart and Pulmonary Vascular Pressures of 11 Normal High Altitude Natives at La Paz (3600 m).*

	RAP (mmHg)			RVP (mmHg)			MPAP (mmHg)			PCP (mmHg)		
	S	D	$\bar{X}$	S	D	$\bar{X}$	S	D	$\bar{X}$	S	D	$\bar{X}$
$\bar{X}$	6.2	0.90	3.81	36.7	0	18.9	38.0	14.5	22.9	10.1	4.2	6.8
SE	0.73	0.54	0.32	2.01	0	2.05	2.47	0.76	1.34	1.07	0.46	0.59

RAP, right atrial pressure; RVP, right ventricular pressure; MPAP, main stem pulmonary artery pressure; PCP, pulmonary wedge pressure; S = systolic; D = diastolic; $\bar{X}$ = mean.

unteers at La Paz, whose average age was 22.4 years, showed a mean pulmonary artery pressure of 22.9 mmHg, which is very close to that predicted from the regression equation (9). On exercise this pressure rose to a mean of 49 mmHg which constitutes an increase of 109%. Pulmonary wedge or capillary pressure in these subjects was 6.8 mmHg and increased only slightly during exercise, i.e., to 9.8 mmHg. The results of this first study are shown in Tables 19-1–19-3. In these normal individuals, oxygen breathing at that altitude for 4 min resulted in a mean pulmonary artery pressure drop from 22.9 to 16 mmHg, which is about 30%. No change was demonstrated with regard to the mean capillary pressure, and cardiac output was not affected. An infusion of acetylcholine caused mean pulmonary artery pressure to fall, but not as much as that noticed with oxygen. The mean fall was from 22.9 to 18 mmHg and there was no change in the pulmonary wedge pressure.

Coudert and the French-Bolivian team (4) studied 67 normal male high altitude natives with a mean age of 24 years and compared the results with those obtained by Banchero et al. from a comparable group in Lima (2). The results, which are similar to those of the first study, are shown in Tables 19-4–19-7. During exercise (75 W) cardiac output increased from 7.0 ± 1.5 to 11.9 ± 1.8 liter/min and mean pulmonary artery pressure increased from 20.0 ± 3 to 32 ± 8 mmHg. Pulmonary wedge pressure remained unchanged (7 ± 4 at rest; 7 ± 3 mmHg during exercise).

It could be supposed that the increase in pulmonary artery pressure would favor the perfusion of the upper lung zones and thus improve the ventilation-perfusion rate, especially in the sitting and standing positions. In order to establish whether or not there exists a relationship between pulmonary artery pressure and regional distribution of pulmonary blood flow, two studies were undertaken (5,10).

The first study compared 15 normal male high altitude natives to five normal lowlanders. The observations were confined to the right upper lung (RUZ) and right lower lung zones (RLZ). Simultaneous isotope dilution curves derived from these lung zones by surface scanning in the sitting and recumbent positions showed in

Table 19-2. *Pulmonary and Systemic Vascular Resistances of 11 Normal High Altitude Natives at La Paz (3600 m).*

	$\dot{Q}$ FICK (liter/min)	TPR (dyn/s/cm^{-5})	PAR (dyn/s/cm^{-5})	SAP			SVR (dyn/s/cm^{-5})
				S	D	$\bar{X}$	
$\bar{X}$	5.71	312.0	219.7	127	80		2207
SE	0.32	21.4	21.6	3.5	2.5	2.7	82.7

TPR, total pulmonary vascular resistance; PAR, pulmonary arteriolar resistance; SAP, systemic artery pressure; SVR, systemic vascular resistance; Q Fick = cardiac output by direct Fick method; S = systolic; D = diastolic; $\bar{X}$ = mean.

Table 19-3. Blood Gas Tensions, pH, and Oxygen Saturations of 11 Normal High Altitude Natives at La Paz (3600 m).

	PA				SA				a-v
	P_{O_2} (mmHg)	pH	Saturation (%)	O_2 content (vol/100 ml)	P_{O_2} (mmHg)	P_{CO_2} (mmHg)	pH	Saturation (%)	O_2 content (vol/100 ml)	O_2 content difference
$\bar{X}$	37.6	7.3584	72.04	15.24	60.2	34.73	7.4172	89.56	18.95	3.71
SE	0.02	0.004	2.54	0.293	1.61	0.65	0.002	0.84	0.37	0.20

PA, pulmonary artery blood; SA, systemic artery blood.

Table 19-4. Biometric Data of 67 Normal Male High Altitude Natives Studied at La Paz (3600 m) and Data from a Similar Group of 22 Normal Controls Studied at Lima (150 m).

		Age (years)	Weight (kg)	Height (cm)	BSA (m^2)	Hct (%)	Hb (g%)
La Paz	$\bar{X}$	23.6	60.2	163	1.65	48.8	16.5
	SD	4.8	6.7	6	0.11	4.1	2.6
	SE	0.6	0.8	1	0.01	0.67	0.5
Lima	$\bar{X}$	20.7	–	–	1.63	44.1	14.8
	SD	1.28	–	–	0.073	2.59	0.88
	SE	0.27	–	–	0.015	0.55	0.18

BSA, body surface area.
Lima data from Banchero et al. (2).

male high altitude natives a distribution of about 17% of cardiac output (Q) to RUZ sitting contrasted with 26% in recumbency. This is almost identical with the flow distribution to RUZ in sitting lowlanders and appears to be uninfluenced by the elevation in mean pulmonary artery pressure of 7.7 mmHg at La Paz above that found at sea level.

By contrast, observations in a sitting young high altitude woman showed a more even distribution of Q to the RUZ than in vertical male high altitude natives, i.e., a Q_{RUZ} of 21.6%. This observation prompted us to extend the investigation of the regional distribution of pulmonary blood flow to both lungs, by specifically recording additional isotope dilution curves from the left upper (LUZ) and the left lower lung zones (LLZ) from the two body postures. In this study 15 normal male and eight normal female high altitude natives who had lived all their lives at or above the altitude of 3600 m and were physically fit and normal were observed. The males had a mean age of 20.9 years; the females, 27.3 years. As expected the females had lower mean weight, height, BSA, Hb, and Hct (Table 19-8).

Inspection of the isotope indicator dilution time concentration curves recorded simultaneously from LUZ (continuous lines) and LLZ (interrupted lines) in the vertical male high altitude natives shows markedly smaller areas for Q_{LUZ} (Fig. 19-1) than in female high altitude natives, who in the vertical position yielded LUZ and LLZ areas of similar magnitude (Fig. 19-2); whereas in recumbency the areas of LUZ and LLZ in both male and female natives are approximately equal (Figs. 19-3 and 19-4).

Table 19-9 summarizes the percentage distribution of total Q to the four lung zones in normal high altitude natives. RUZ and RLZ values are derived from the first study; left lung zone values are derived from the second study. In male high altitude natives in the vertical posture the means of Q_{RUZ} and Q_{LUZ} were 17.5 and 16.9% of Q, respectively; whereas the means in recumbency for Q_{RUZ} and Q_{LUZ} were 25.9 and 22.8%, respectively. In female high altitude natives in the vertical position the means for Q_{RUZ} and Q_{LUZ} were 21.6 and 22.5%, respectively; whereas in recumbency they measured 29.9 and 22.1%, respectively. The change in Q_{LUZ} in female high altitude natives with posture thus seems insignificant. Regardless of similar mean pulmonary artery pressures in both sexes at La Paz, the upper lung distribution zones in the vertical position appear to be at least 15% larger in female than in male high altitude natives.

The results shown represent a summary of the studies carried out at the Bolivian Institute of Altitude Biology in order to establish normal pulmonary circulation values for the population born and living at the altitude of 3600 m. They demonstrate the difference that exists between the pulmonary circulation of high altitude natives and

Table 19-5. Blood Gas Data from 67 Normal Male High Altitude Natives Studied at La Paz (3600 m) and from a Similar Control Group Studied at Lima (150 m).

		Pa_{O_2} (mmHg)	Pa_{CO_2} (mmHg)	pHa	Sa_{O_2} (%)	Ca_{O_2} (vol%)	Cv_{O_2} (vol%)	$a\text{-}v_{O_2}$ (vol%)	Pv_{O_2} (mmHg)	Sv_{O_2} (%)
La Paz	$\bar{X}$	58.5	28.0	7.51	90.1	4.3	36	72	3.1	5.5
	SD	92.2	4.6	0.06	4.38	2.12	2.09	0.9	0.98	1.6
	SE	1.3	0.7	0.01	0.65	0.31	0.32	0.1	—	—
Lima	$\bar{X}$	—	—	—	95.7	19.04	15.04	4.00	—	—
	SD	—	—	—	2.06	0.998	1.262	0.905	—	—
	SE	—	—	—	0.43	0.223	0.282	0.202	—	—

Pa_{O_2}, Pa_{CO_2}, Sa_{O_2}, and Ca_{O_2}: partial pressure of oxygen and carbon dioxide, and saturation and content of oxygen in arterial blood, respectively. Pv_{O_2}, Sv_{O_2}, and Cv_{O_2}: partial pressure, saturation, and content of oxygen in mixed venous blood, respectively.
$a\text{-}v_{O_2}$: arteriovenous oxygen content difference.
Lima data from Banchero et al. (2).

Table 19-6. *Right Heart and Pulmonary Vascular Pressures of 67 Normal High Altitude Natives Studied at La Paz (3600 m) and from a Similar Control Group Studied at Lima (150 m).*

		RAP (mmHg)	RVP (mmHg)		MPAP (mmHg)			PCP (mmHg)	SAP (mmHg)		
		$\bar{X}$	S	D_{II}	S	D	$\bar{X}$	$\bar{X}$	S	D	$\bar{X}$
La Paz	$\bar{X}$	5	34	6	29	13	21	9	120	90	
	SD	1.9	9.5	2.3	6.6	3.8	4.2	2.7	15.0	9.9	11.2
	SE	3.7	1.7	0.4	0.9	0.5		0.4	2.4	1.6	1.7
Lima	$\bar{X}$	2.6	—	—	22	6	12	6.5	128	70	95
	SD	1.36			3.5	2.1	2.2	1.68	8.5	6.7	8.1
	SE	0.28	—	—	0.7	0.4	0.4	0.35	1.8	1.4	7

RAP, right atrial pressure; RVP, right ventricular pressure; MPAP, main stem pulmonary artery pressure; PCP, pulmonary wedge pressure; SAP systemic artery pressure; D_{II} = end diastolic.
Lima data from Banchero et al. (2).

Table 19-7. *Cardiac Output, Cardiac Index, Heart Rate, Stroke Volume, and Pulmonary and Systemic Vascular Resistances of 67 Normal Male High Altitude Natives Studied at La Paz (3600 m) and of a Similar Control Group Studied at Lima (150 m).*

		$\dot{Q}$ FICK (liter/min)	CI (liter/min/m²)	HR (beats/min)	SV (ml)	TPR	PAR	SVR
						(dyn/s/cm^{-5})		
La Paz	$\bar{X}$	6.43	3.91	72	92	265	148	1191
	SD	1.69	1.09	9.9	25.3	80	43	376
	SE	0.25	0.16	1.6	4.2	12	7	66
Lima	$\bar{X}$	—	3.97	68	—	160	73	1248
	SD	—	0.977	7.9	—	46.7	24.4	304.4
	SE	—	0.218	1.6	—	10.4	5.4	69.8

CI, cardiac index; HR, heart rate; SV, stroke volume.
Lima data from Banchero et al. (2).

Table 19-8. *Biometric Data of 15 Male and 8 Female High Altitude Dwellers Whose Distribution of Pulmonary Blood Flow to the Left Lung Was Studied.*

	Age (years)	Height (cm)	Weight (kg)	BSA (m²)	Hb (g)	Hct (%)
Males (n = 15)						
Mean ± SD	20.9 ± 3.0	163 ± 4.4	61.2 ± 3.8	1.61 ± 0.51	16.4 ± 0.8	50.8 ± 1.5
Females (n = 8)						
Mean ± SD	27.3 ± 9.6	156 ± 6.1	54.7 ± 4.1	1.51 ± 0.18	15.3 ± 1.9	49.1 ± 0.9

BSA, body surface area in m².

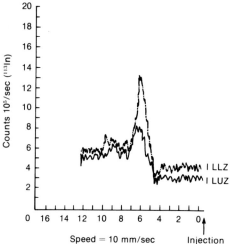

Fig. 19-1. ¹¹³In indicator dilution curves simultaneously recorded from the left upper (———) and left lower (–·–·–·) lung zones in a male high altitude native in vertical position. The total activity in (1 LUZ) is less than in (1 LLZ).

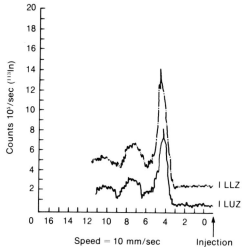

Fig. 19-2. ¹¹³In indicator dilution curves simultaneously recorded from the left upper (———) and left lower (–·–·–·) lung zones in a female high altitude native in the vertical position. The total activity of (1 LUZ) is almost identical with that of (1 LLZ).

that of lowlanders, inasmuch as high altitude natives have a higher mean pulmonary artery pressure, higher mean right ventricular pressure, and higher pulmonary vascular resistances; in addition, their mean pulmonary artery pressure rises during exercise by a much higher percentage than does that of lowlanders. Highlanders have a lower oxygen saturation of the blood, lower arterial oxygen tension, and

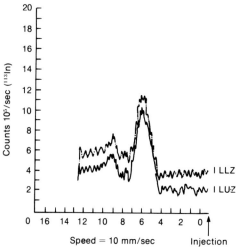

Fig. 19-3. ¹¹³In indicator dilution curves recorded simultaneously from left upper (———) and left lower (–·–·–·) lung zones in a male high altitude native in recumbency. The total activity of (I LUZ) is identical with that of (I LLZ).

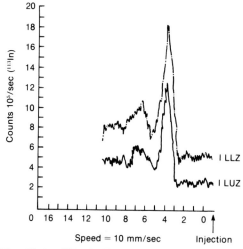

Fig. 19-4. ¹¹³In indicator dilution curves recorded simultaneously from left upper (———) and left lower (–·–·–·) lung zones in a female high altitude native in recumbency. The total activity of (I LUZ) is identical with that of (I LLZ).

Table 19-9. Distribution (%) of Total Pulmonary Blood Flow ($\dot{Q}$) to Four Lung Zones in Normal High Altitude Natives (3650 m).

Sex	$\dot{Q}_{RUZ}$	$\dot{Q}_{RLZ}$	$\dot{Q}_{LUZ}$	$\dot{Q}_{LLZ}$
Males (n = 15)				
Vertical ($\bar{X}$)	17.5	29.4	16.9	31.8
Recumbency ($\bar{X}$)	25.9	30.0	22.8	24.6
Females (n = 8)				
Vertical ($\bar{X}$)	21.6	31.2	22.5	26.1
Recumbency ($\bar{X}$)	29.0	30.6	22.1	23.1

RUZ, right upper lung zone; RLZ, right lower lung zone; LUZ, left upper lung zone; LLZ, left lower lung zone.

also lower arterial CO_2 tension, whereas the pH remains unchanged, in comparison to lowlanders. No difference between highlanders and lowlanders was demonstrable with respect to percentage distribution of pulmonary blood flow to the different lung zones despite the higher pulmonary artery pressure in highlanders.

References

1. Arias-Stella, J. and Saldaña, M.: The terminal portion of the pulmonary arterial tree in people native to high altitudes. Circulation 28:915, 1963.
2. Banchero, N., Sime, F., Peñaloza, D., Cruz, J., Gamboa, R., and Marticorena, E.: Pulmonary pressure, cardiac output, and arterial oxygen saturation during exercise at high altitude and at sea level. Circulation 33:249, 1966.
3. Blount, S.G., Jr. and Vogel, J.H.K.: Altitude and the pulmonary circulation. Adv. Intern. Med. 13:11, 1967.
4. Coudert, J.: La circulation pulmonaire du natif de la haute altitude a La Paz (3700m). In: Anthropologie des Populations Andines. I.N.S.E.R.M. 63:305, 1976.
5. Coudert, J., Paz-Zamora, M., Barragán, L., Briancon, L., Spielvogel, H., and Cudkowicz, L.: Regional distribution of pulmonary bloodflow in normal high altitude dwellers at 3650 m (12,200 ft). Respiration 32:189, 1975.
6. Cudkowicz, L.: Mean pulmonary artery pressure and alveolar oxygen tension in man at different altitudes. Respiration 27:417, 1970.
7. Peñaloza, D., Sime, F., Banchero, N., Gamboa, R., Cruz, J., and Marticorena, E.: Pulmonary hypertension in healthy men born and living at high altitudes. Am. J. Cardiol. 11:150, 1963.
8. Rotta, A., Canepa, A., Hurtado, A., Velasquez, T., and Chavez, R.: Pulmonary circulation at sea level and at high altitude. J. Appl. Physiol. 9:328, 1956.
9. Spielvogel, H., Otero-Calderon, L., Calderon, G., Hartmann, R., and Cudkowicz, L.: The effects of high altitude on pulmonary hypertension of cardiopathies at La Paz, Bolivia. Respiration 26:369, 1969.
10. Spielvogel, H., Vargas, E., Antezana, G., Barragán, L., and Cudkowicz, L.: Effects of posture on pulmonary diffusing capacity and regional distribution of pulmonary blood flow in normal male and female high altitude dwellers at 3650 m (12,200 ft). Respiration 35:125, 1978.
11. Vogel, J.H.K., Weaver, W.F., Rose, R.L., Blount, S.G., Jr., and Grover, R.F.: Pulmonary hypertension on exertion in normal man living at 10,150 ft (Leadville, CO). Med. Thorac. 19:461, 1962.
12. Von Euler, U.S. and Liljestrand, G.: Observations on the pulmonary arterial blood pressure in the cat. Acta Physiol. Scand. 12:301, 1946.

20

Comparison Between Newcomer Rats and First Generation of Rats Born at High Altitude, Particularly Concerning the Oxygen Supply to the Heart

F. KREUZER AND Z. TUREK

Considerable information is available concerning the high altitude adaptation of newcomers, but much less is known with respect to the behavior of the first generation born at high altitude. Monge (10) discusses this as follows:

> The well-known historian, Antonio de la Calancha [1], tells, as does the writer, Martínez Vela, in his *Annals of the Imperial City of Potosí* (4002 meters above sea level) [9], that when it was first founded there were 100,000 natives and 20,000 Spaniards taken with the fever for riches which made the city so famous. While the former went on reproducing with customary Indian fertility, the latter either did not succeed in having children or they did not survive. The birth of the first Spaniard did not take place until 53 years after the founding of the city. His birth was attributed to a miracle of Saint Nicholas of Tolentino.

Quoting de la Calancha, Monge continues:

> "It was born on Christmas Eve in the year 1598, he gave it the name of Nicholas and raised it there, curing it miraculously of many sick spells caused not by the cold but by other deadly diseases. He who was this child is now become a Doctor in this University and an Alderman on the Municipal Council known as Doctor Don Nicolás Flores; he was the first child of Spanish parents to survive in Potosí from among all those that were born there in fifty-three years."

Monge concludes:

> It is unquestionable that here may be seen a slow process of adaptation to life in the high altitude for which more than a generation was necessary.

For the past few years we have been paying particular attention to a study of the conditions prevailing in the first generation born at high altitude. Since, for obvious reasons, information concerning human beings is difficult, if not impossible, to obtain, we studied this problem in rats. We compared the responses of newcomers and natives at a simulated altitude of 3500 m (equivalent to the elevation of Jungfraujoch) in the low pressure chamber with control animals at sea level. At the time of the experiments the rats were about 70 days old and still growing. The newcomers had spent 4 to 5 weeks at high altitude. All measurements were performed in hypoxia induced by breathing 12% oxygen in nitrogen. The classic Krogh model and formula (7) were applied for the calculations in cardiac muscle, with the only modification that the hexagonal pattern suggested by Thews (14) was used rather than the cylindric arrangement. The present paper is a comprehensive comparative survey of our previous work which we hope to extend to the second generation in the future.

Not all the methods can be described here in detail; they may be found in the original papers referred to below. The conventional methods will not be mentioned; we will summarize here only the methods used for the measurements in the heart.

After killing the animal the beating heart was removed, washed in distilled water, and prepared according to Fulton et al. (2). The right and left ventricles and the ventricular part of the septum were prepared and weighed. Furthermore the dry weight was estimated after drying at 95°C to constant weight.

For the estimation of capillary and muscle fiber densities the hearts were removed as described, washed, fixed in 10% formalin solution, and embedded in paraffin. Histologic sections were prepared perpendicularly to the septum in the region about 4 mm below the atrioventricular border. Staining was performed according to Hort (PAS reaction; ref. 5). Using a microscope, the capillaries and muscle fibers were counted in the lateral part of the walls of both ventricles in the neighborhood of the cavity but not close to the subendocardial region. The data were not corrected for shrinkage of the tissue caused by embedding in paraffin. According to Hort (5) surface shrinkage amounts to about 35%. This means that the values of one-dimensional parameters may be about 20% too low. However, the shrinkage does not affect quantitative comparison between the control and altitude groups of animals.

The muscle fiber diameter was assessed as follows. In 20 fields from each ventricle we estimated, by the surface integration method (4) using the special grid of Carl Zeiss, the ratio of the surface area occupied by muscle fibers and capillaries to the total area of the field. By assuming the muscle fibers and capillaries to be cylindric, the diameter of the fibers and the external radius of the capillaries were calculated from the surface ratio and the number of muscle fibers and capillaries per mm^2.

From the obtained values of fiber and capillary density the ratio of fiber and capillary density was calculated (F/C ratio), and the diffusion distance ($D/2$) was calculated according to Rakušan and Poupa (13) as half of the mean distance between two capillaries. $D/2$ corresponds to the mean radius of Krogh's tissue cylinder (8).

Table 20-1 presents the factors generally relevant for oxygen transport in the organism. As expected, hematocrit and oxygen capacity are increased at altitude, particularly in the natives. Arterial oxygen content and oxygen saturation are increased only in the newcomers; in the natives arterial oxygen content is similar to that in the controls and arterial oxygen saturation is even lower than in the controls. Mixed-venous oxygen content and oxygen saturation are increased in the newcomers, but decreased in the natives as compared with the controls. The increase in arterial and mixed-venous oxygen content in the newcomers is mainly due to the increased blood oxygen capacity; therefore, mean capillary oxygen content and saturation also must be increased. The arteriovenous oxygen difference is increased in both altitude groups whereas cardiac output is decreased in both groups, which is different from the response in man where there is a temporary increase for a few days in newcomers and no change in natives. Both altitude groups show a decreased oxygen affinity of the blood as seen from the values of P_{50}, the oxygen pressure for 50% oxygen saturation. These results suggest that the natives are less fit than the newcomers and probably even the controls.

The cardiac data are listed in Table 20-2. Total cardiac weight is increased slightly in the newcomers and markedly in the natives. With respect to the partial weights of

Table 20-1. Comparison of Values Relevant to Oxygen Transport between Control, Newcoming, and Native Rates (Simulated Altitude, 3500 m) as Measured at Hypoxia (Breathing 12% Oxygen in Nitrogen).

Parameter	Controls	Newcomers	Natives
O_2 consumption (ml/min)	4.65	4.52	4.43
Hematocrit	42.7	*52.5*	*56.6*
O_2 capacity (ml/100 ml)	20.15	*24.90*	*26.30*
Arterial O_2 content (ml/100 ml)	12.9	16.9	13.3
Mixed-venous O_2 content (ml/100 ml)	5.9	8.2	4.8
Arterial O_2 saturation (%)	63.7	67.3	*50.1*
Mixed-venous O_2 saturation (%)	28.3	32.4	*17.6*
Arteriovenous O_2 difference (ml/100 ml)	7.0	8.7	8.5
Cardiac output (ml/min)	68.8	*54.3*	*55.2*
P_{50} (torr)	35.6	*38.3*	*39.7*
(kPa)	4.75	*5.11*	*5.29*

Newcomer data from Turek et al. (20), native data from Turek et al. (17).
Age of rats about 70 days, body weight about 224 g. Important differences italicized.

Table 20-2. Comparison of Cardiac Tissue Values between Control, Newcoming, and Native Rats (Simulated Altitude, 3500 m).

Parameter	Controls	Newcomers	Natives
Heart weight total (mg)	734	*771*	*1239*
Weight, LV (mg)	460	429	*634*
Weight, RV (mg)	136	*211*	*434*
Weight, septum (mg)	138	132	*171*
Capillaries per mm²			
LV	2321; 2583	2318 n.s.	2555 n.s.
RV	2533; 2599	*3306*	*2918*
Muscle fibers per mm²			
LV	2314; 2574	2376 n.s.	*2188*
RV	2389, 2583	*1181*	*1692*
Muscle fiber diameter (μm)			
LV	18.6; 17.0	19.7 n.s.	18.0
RV	18.7; 16.6	*27.2*	*21.1*
Fiber-capillary ratio			
LV	1.00; 1.00	1.03 n.s.	0.86
RV	0.94; 0.99	*0.36*	*0.58*
Diffusion distance (μm)			
LV	10.4; 9.8	10.4 n.s.	9.9 n.s.
RV	10.0; 9.8	*8.8*	*9.3*
Coronary blood flow (ml/min/g)			
LV	2.06	*2.85*	2.05
RV	1.86	*2.57*	*1.94*

Data from Turek et al. (15), Turek et al. (19), and Grandtner et al. (3). LV left ventricle; RV, right ventricle.
Age of rats about 70 days, body weight about 230 g. First value of controls belongs to newcomers, Second value to natives. n.s.—not significant as compared with respective control. Important differences italicized.

left ventricle, right ventricle, and septum, the newcomers show an exclusive right hypertrophy, whereas all three partial weights are increased in the natives, although this increase is most marked in the right ventricle, the weight of which is three times that of the controls. These differences are clearly seen in Fig. 20-1.

The capillary density in the heart muscle is increased only in the right ventricle of both altitude groups. The cardiac muscle fiber density is decreased in the right ventricle of both groups and in the left ventricle of the natives only. The muscle fiber diameter is increased particularly in the right ventricle of both altitude groups, predominantly so in the newcomers; this implies a right cardiac hypertrophy (not hyperplasia) probably due to the well-known increase in pulmonary arterial pressure (which, however, could not be measured here). The fiber-capillary ratio is markedly decreased in the right ventricle of both altitude groups which points to an augmented capillarization possibly due to an increased oxygen consumption concomitant with an increase in pulmonary arterial pressure. This increased capillarization implies a shortening of the diffusion distance in the right ventricle of both altitude groups. It may be calculated that as a consequence of this shortened diffusion distance the drop of the oxygen pressure across the tissue per unit oxygen consumption is mitigated and therefore the peripheral tissue oxygen pressure is increased, particularly in the presence of an increased oxygen consumption, although this compensatory shortening of the diffusion distance is unable to fully compensate for an increased oxygen consumption. An overall comparison between newcomers and natives again shows that the natives cannot measure up to the conditions prevailing in the newcomers and in the controls.

Coronary blood flow per gram of tissue is increased in the newcomers as compared with the controls; but in the natives, in the entire heart as well as in the three component parts of the heart, values are similar to control values (22,23). This pattern is quite different from that found in man: Human newcomers have a left ventricular coronary blood flow lower than that of sea level residents, and only patients suffering from chronic mountain sickness show an increased value (11). Coronary blood flow in the control rats at normoxia certainly is not higher than in acute hypoxia as is generally accepted. We have no ready explanation for this difference between man and rat.

Calculations based on the Krogh model for right and left ventricle showed that the peripheral tissue oxygen pressure over a range of five values of oxygen consumption

Fig. 20-1. Histologic sections from the hearts of a control animal (*left*), of a rat born at sea level and exposed to simulated altitude (3500 m) later in life (*middle*), and of a rat born and staying in the low pressure chamber (*right*). All rats had similar body weights. Note the dimensions of the right ventricles (on the left in each section). × 6.5. From Grandtner et al. (3).

decreased in the following sequence: newcomers > natives > controls (18). The right shift of the oxyhemoglobin dissociation curve (Table 20-1, P_{50}) could be shown to be slightly unfavorable in the newcomers and markedly unfavorable in the natives. An evaluation of the importance of the determinants of tissue oxygen pressure (12) shows that in the newcomers an increase in coronary blood flow is more important than an increase in blood oxygen capacity and arterial oxygen pressure, a shortening of the diffusion distance occurring only in the right ventricle; in the natives, however, the predominant factor is an increase in blood oxygen capacity, without which myocardial hypoxia would ensue. This mechanism, however, is rather risky because an increased hematocrit must involve an augmented circulatory peripheral resistance and therefore an increased load on the heart. In summary, it may be concluded that the newcomers have available a wider scale of adaptational mechanisms than do the natives. We were able to show (21) that a high altitude adaptation to 3500 m for 5 weeks increases the resistance of the heart muscle to isoprenaline-induced myocardial necrosis.

It was mentioned above that the diffusion distance was shortened only in the right ventricle of both altitude groups. What may be the reason for this finding? Might this increased capillary proliferation in the right ventricle be due to the heavier load on the right ventricle (increased oxygen consumption concomitant with increased pulmonary arterial pressure) and/or to the arterial hypoxemia, or might the right ventricle be more prone to develop capillary proliferation for some unknown reason? In order to approach an answer to this question we studied rats in which a myocardial infarction induced by ligation of the left coronary artery had led to a right cardiac hypertrophy in the absence of any hypoxemia after 4 to 5 weeks (16). Now the same changes occurred in the right ventricle as in the remaining undamaged tissue of the left ventricle: Muscle fiber density was decreased and muscle fiber diameter was increased, pointing to cardiac muscular hypertrophy, but a decreased capillary density and an unchanged fiber-capillary ratio implied an increased diffusion distance, i.e., there was no capillary proliferation. Accordingly, calculations based on the Krogh model showed that in the case of a stress situation (increased cardiac oxygen consumption or decreased coronary blood flow with all capillaries open) the decline of the oxygen pressure in the tissue per unit oxygen consumption was increased in both ventricles (although more so in the right ventricle). Thus tissue oxygen transport was impaired, particularly in situations of stress, contrary to what has been described above for altitude newcomers and natives. This suggests that arterial hypoxemia must play a role in the genesis of capillary proliferation in the heart muscle.

References

1. Calancha, A. de la: Crónica Moralizada de la Orden de San Agustín, Vol. 1. Barcelona, 1639.
2. Fulton, B.M., Hutchinson, E.C., and Jones, A.M.: Ventricular weight in cardiac hypertrophy. Br. Heart J. 14:413, 1952.
3. Grandtner, M., Turek, Z., and Kreuzer, F.: Cardiac hypertrophy in the first generation of rats native to simulated high altitude. Muscle fiber diameter and diffusion distance in the right and left ventricle. Pflügers Arch. 350:241, 1974.
4. Henning, A.: Das Problem der Kernmessung. Eine Zusammenfassung und Erweiterung der mikroskopischen Messtechnik. Mikroskopie 12:174, 1957.
5. Hort, W.: Quantitative Untersuchungen über die Capillarisierung des Herzmuskels im Erwachsenen- und Greisenalter, bei Hypertrophie und Hyperplasie. Virchows Arch. Pathol. Anat. 327:560, 1955.
6. Kreuzer, F. and Turek, Z.: Oxygen supply to the heart. In ten Hoor, F., Bernards, J.A., de Jong, J.W., and Kreuzer, F. (eds.): Oxygen Supply of Heart and Brain. Physiology,

Pathophysiology, Clinical Implications. The Hague, Dutch Heart Foundation, 1979, pp. 48–62.
7. Krogh, A.: The number and distribution of capillaries in muscles with calculations of the oxygen pressure head necessary for supplying the tissue. J. Physiol. 52:409, 1919.
8. Krogh, A.: The Anatomy and Physiology of Capillaries. New Haven, Yale University Press, 1922.
9. Martínez Vela, B.: Annales de la Villa Imperial de Potosí. La Paz, 1939.
10. Monge, C.: Acclimatization in the Andes. Historical Confirmations of "Climatic Aggression" in the Development of Andean Man. Baltimore, Johns Hopkins Press, 1948.
11. Moret, P., Covarrubias, E., Coudert, J., and Duchosal, F.: Cardiocirculatory adaptation to chronic hypoxia. I. Comparative study of coronary flow, myocardial oxygen consumption and efficiency between sea level and high altitude residents. Acta Cardiol. 27:283, 1972.
12. Rakušan, K.: Oxygen in the Heart Muscle. Springfield, Ill., Thomas, 1971.
13. Rakušan, K., and Poupa, O.: Changes in the diffusion distance in the rat heart muscle during development. Physiol. Bohemoslov. 12:220, 1963.
14. Thews, G.: Die Sauerstoffdiffusion im Gehirn; ein Beitrag zur Frage der Sauerstoffversorgung der Organe. Pflügers Arch. ges. Physiol. 271:197, 1960.
15. Turek, Z., Grandtner, M., and Kreuzer, F.: Cardiac hypertrophy, capillary and muscle fiber density, muscle fiber diameter, capillary radius and diffusion distance in the myocardium of growing rats adapted to a simulated altitude of 3500 m. Pflügers Arch. 335:19, 1972.
16. Turek, Z., Grandtner, M., Kubát, K., Ringnalda, B.E.M., and Kreuzer, F.: Arterial blood gases, muscle fiber diameter and intercapillary distance in cardiac hypertrophy of rats with an old myocardial infarction. Pflügers Arch. 376:209, 1978.
17. Turek, Z., Grandtner, M., Ringnalda, B.E.M., and Kreuzer, F.: Hypoxic pulmonary steady-state diffusing capacity for CO and cardiac output in rats born at a simulated altitude of 3500 m. Pflügers Arch. 340:11, 1973.
18. Turek, Z. and Kreuzer, F.: Changes in oxygen transport at high altitude, particularly concerning the heart muscle. In Paintal, A.S., Gill-Kumar, P. (eds): Respiratory Adaptations, Capillary Exchange and Reflex Mechanisms. New Delhi, India, Navchetan Press, 1977, pp. 75–88.
19. Turek, Z., Ringnalda, B.E.M., Grandtner, M., and Kreuzer, F.: Myoglobin distribution in the heart of growing rats exposed to a simulated altitude of 3500 m in their youth or born in the low pressure chamber. Pflügers Arch. 340:1, 1973.
20. Turek, Z., Ringnalda, B.E.M., Hoofd, L.J.C., Frans, A., and Kreuzer, F.: Cardiac output, arterial and mixed-venous O_2 saturation, and blood O_2 dissociation curve in growing rats adapted to a simulated altitude of 3500 m. Pflügers Arch. 335:10, 1972.
21. Turek, Z., Kubát, K., Ringnalda, B.E.M., and Kreuzer, F.: Experimental myocardial infarction in rats acclimated to simulated high altitude. Basic Res. Cardiol. 75:544, 1980.
22. Turek, Z., Turek-Maischeider, M., Claessens, R.A., Ringnalda, B.E.M., and Kreuzer, F.: Coronary blood flow in rats native to simulated altitude and in rats exposed to it later in life. Pflügers Arch. 355:49, 1975.
23. Turek, Z., Turek-Maischeider, M., Claessens, R.A., Ringnalda, B.E.M., and Kreuzer, F.: Coronary blood flow and its distribution under hypoxia in rats born and staying at a simulated altitude of 3500 m—comparison with sea level controls and those exposed to simulated altitude later in life. In Bhatia, B., Chhina, G.S., and Singh, B. (eds.): Selected Topics in Environmental Biology. New Delhi, India, Interprint Publ. 1976, pp. 283–290

21

Circulatory Flow of Oxygen Returning to the Lung During Submaximal Exercise in Altitude Residents

J. DURAND AND J. MENSCH-DECHENE

During muscular exercise the product of the cardiac output ($\dot{Q}$) and the mixed-venous concentration of oxygen ($C\bar{v}_{O_2}$) is constant and independent of the oxygen uptake ($\dot{V}_{O_2}$) (2).

This flow of oxygen returning from the tissues to the lung ($\dot{Q} \cdot C\bar{v}_{O_2}$) is modified by exposure to natural or simulated altitude (Fig. 21-1). In subjects acutely exposed, $\dot{Q} \cdot C\bar{v}_{O_2}$ remains independent of $\dot{V}_{O_2}$, but at a lower level than at sea level. In altitude residents, however, $\dot{Q} \cdot C\bar{v}_{O_2}$ is larger than in sea level residents since polycythemia overcompensates the decrease in arterial oxygen saturation. When a hyperoxic mixture is given, in order to reach an inspired oxygen partial pressure equal to that of sea level, $\dot{Q} \cdot C\bar{v}_{O_2}$ rises above its control value, since cardiac output does not decrease or decreases only slightly. This observation favors the assumption that metabolic demand or oxygen content in mixed-venous blood plays little or no role in cardiac control during exercise.

Acknowledgment

This work was realized at the Instituto Boliviano de Biología de Altura, La Paz (Bolivia), and was supported in part by the Ministère des Affaires Etrangères (Coopération Technique) and by the C.N.R.S. (RCP 538).

References

1. Banchero, N., Sime, F., Penaloza, D., Cruz, J., Gamboa, R., and Marticorena, E.: Pulmonary pressure, cardiac output, and arterial oxygen saturation during exercise at high altitude and at sea level. Circulation 33:249, 1966.
2. Durand, J. and Mensch-Dechene, J.: Relation théorique entre le débit cardiaque et la consommation d'oxygène au cours de l'exercice. Bull. Physiopath. Respir. 15:977, 1979.
3. Stenberg, J., Ekblom, B., and Messin, R.: Hemodynamic response to work at simulated altitude (4000 m). J. Appl. Physiol. 21:1589, 1966.

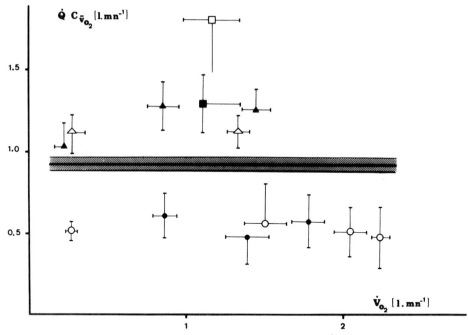

Fig. 21-1. Flow of oxygen returning from the tissues to the lungs ($\dot{Q} \cdot C\bar{v}_{O_2}$) as a function of oxygen consumption ($\dot{V}_{O_2}$) at rest and during submaximal muscular exercise: in sea level dwellers translocated to 3800 m (○) or at simulated altitude of 4000 m (●; data from Stenberg et al. (3)); in altitude residents at 3800 m (▲) and 4350 m (△; data from Banchero et al. (1)); and in altitude residents at 3800 m breathing room air (■) and a hyperoxic mixture: P_{IO_2}, 160 torr; kPa, 21.3 (□). Mean values and standard deviations. The horizontal line and shadowed area indicate mean value and standard error obtained in sea level dwellers at sea level.

22
Effect of the α-Adrenergic Blocking Agent Phentolamine (Regitine) on Acute Hypoxic Pulmonary Hypertension in Awake Dogs

SHU-TSU HU, HSUEH-HAN NING, CHAO-NIEN CHOU, AND HUA-YU HUANG

There has long been uncertainty regarding the involvement of the α-adrenergic mechanism in the genesis of acute hypoxic pulmonary hypertension (2). After our repeated failure to obtain any conclusive results in studying the effect of the α-adrenergic blocking agent phentolamine (Regitine) on the hypoxic pressor effect in pentobarbital anesthetized dogs, we thought it would be worthwhile to carry out the experiment in conscious animals. The present study shows that α-adrenergic blockade does inhibit to a great extent the pulmonary hypertension in awake dogs exposed to acute isobaric hypoxia.

Methods

Six mongrel dogs of 10–15 kg body weight were trained to stand, sit, or lie prone quietly in one posture for at least 1 h. After the training period, each animal was operated upon under general anesthesia for implantation of silicone rubber tubing (i.d., 1.0–1.5 mm) in the pulmonary artery trunk 3 cm distal to the pulmonary artery valves, with the tip pointing toward the right ventricle. The tubing was properly fixed to adjacent soft tissues by ties and sutures. The other end was exteriorized by piercing through a dorsal intercostal space and tunneling underneath the skin. Generally the operated animals were ready for experimentation on the 4th postoperative day.

Pulmonary arterial pressure was observed directly by connecting the saline-filled rubber tubing to a water manometer which was leveled with the right ventricle.

The animals inhaled a low oxygen mixture (P_{O_2}, 80 mmHg) through a face mask, the difference in P_{O_2} of the daily preparations being no more than 5 mmHg as determined by gas analysis. For α-adrenergic blockade, phentolamine methane sulfonate (Regitine, Ciba-Geigy Ltd., Switzerland) was infused at an even speed over 60 s (1.5 mg/kg) into the pulmonary circulation through the indwelling tubing. The adequacy of this dose was examined by infusing L-norepinephrine (5 μg/kg) into the pulmonary circulation; it was found that there was no or only a 1–2 cm H_2O rise in pulmonary arterial pressure.

The experiments conducted had two goals: (1) to test the effect of α-blockade on the pulmonary arterial pressure which had been already raised to a high basal level by continuous low oxygen inhalation; and (2) to give the α-blocking agent as a pretreatment before low oxygen inhalation, and to compare the size of the pulmonary pressor response with that of the control. In either case, control experiments in which normal saline was given instead of a blocking agent were done on the same animals.

A parallel study of the effect of β-adrenergic blockade with propranolol (1 mg/kg) on the hypoxic pulmonary hypertension was carried out in the same manner as in the case of α-blockade. The adequacy of β-adrenergic blockade was confirmed by the abolishment of the pulmonary pressor effect of a subsequent infusion of isoproterenol (1 μg/kg).

Results

α-Adrenergic Blockade in the Course of Acute Hypoxic Pulmonary Hypertension

In all 22 experiments on the six dogs the infusion of phentolamine into the pulmonary artery of conscious dogs invariably depressed the hypoxia-induced pulmonary hypertensive level down to the level present during normoxia (Fig. 22-1). The action seems to be immediate and may be long-lasting. In another series of 11 observations the action was followed for a longer time. At 20 min after phentolamine infusion there was still marked depression without any obvious sign of return to the previous hypertensive level (Fig. 22-2).

β-Adrenergic Blockade in the Course of Acute Hypoxic Pulmonary Hypertension

None of the six dogs (ten observations) showed any marked effect of β-adrenergic blockade with propranolol on the acute hypoxic pulmonary hypertensive level (Fig. 22-3).

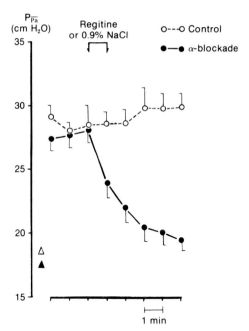

Fig. 22-1. Depressive effect of α-adrenergic blockade on steady state pulmonary hypertension produced by continuous inhalation of low oxygen mixture. Open and filled circles represent prehypoxia pulmonary arterial pressure levels in the control and α-blocking experiments, respectively. During the period between the two arrows, 0.9% NaCl (for control) and phentolamine (for α-blocking) were infused into the pulmonary artery. After α-blocking, decline of the pressure toward the prehypoxia level is evident. Vertical bars represent 1 SE of the mean. Number of observations: for control, $n = 15$; for blocking, $n = 22$.

α-Adrenergic Blockade Before Hypoxia

Ten min inhalation of the low oxygen mixture induced a very marked pulmonary pressor response which was quite reproducible in repetitions done on the same day. This served as a control.

In a series of seven experiments on two dogs, 10 min inhalation of low oxygen mixture, after a pretreatment of α-adrenergic blockade with phentolamine administered 8–15 min previously, led to a pressor response very much smaller than that seen in the control.

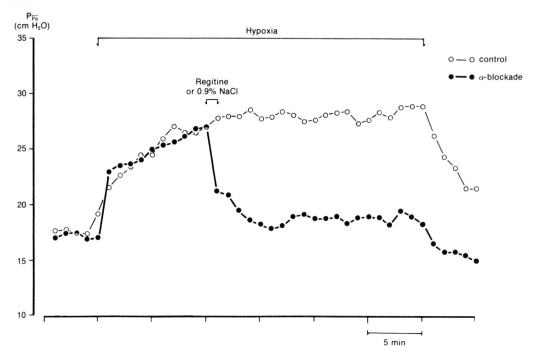

Fig. 22-2. The immediate and long-lasting depressive action of phentolamine α-blockade on the sustained pulmonary hypertension produced by continuous inhalation of low oxygen mixture. Note the tendency of the pulmonary arterial pressure to drop below the previous normoxic level in α-blocking experiments. Control curve represents mean of 8 observations; α-blocking curve, 7 observations.

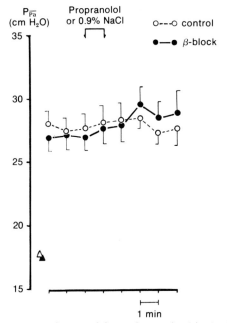

Fig. 22-3. Successful β-adrenergic blockade apparently has no effect on hypoxic pulmonary hypertension. Symbols as in Fig. 22-1, except that β-blocking with propranolol replaces α-blocking with phentolamine. Number of observations: for control, $n=10$; for β-blocking, $n=10$.

β-Adrenergic Blockade Before Hypoxia

An infusion of propranolol before inhalation of low oxygen mixture did not much affect the pulmonary pressor response (three experiments on two dogs) (Fig. 22-4).

The Influence of Pentobarbital Anesthesia

Five dogs were placed under pentobarbital anesthesia (35 mg/kg). Three of them (four experiments) failed to show any pressor response at all (Fig. 22-5). The other two dogs (five experiments) gave inconsistent results under anesthesia.

In contrast, all the five same dogs in a conscious state showed (73 out of 75 experiments) marked pulmonary pressor response to low oxygen inhalation. Only on two occasions in one dog the pressor response was not evident, but this dog did show marked pressor response on eight other occasions.

There may be many reasons for the discrepancies existing in the literature, but the present study has directed attention to an important point, i.e., anesthesia has some masking effect on the manifestation of the hypoxic pressor response (3,4). Recently, it has been reported that hypoxic pulmonary vasoconstriction is inhibited by two inhalation anesthetics in man (1). Evidently, the use of anesthetics in a study would bring about undesirable complications. Our study shows that the difficulties possibly due to the presence of anesthesia can be circumvented by eliminating the anesthetics.

References

1. Bjertnaes, L. J., Hauge, A., Nakken, K.F., Bredesen, J.E.: Hypoxic pulmonary vasoconstriction: Inhibition due to anesthesia. Acta Physiol. Scand. 96:283, 1976.

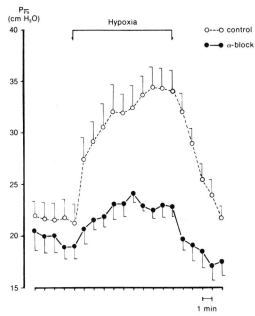

Fig. 22-4. Marked diminution of the hypoxia-induced pulmonary pressor response after α-blockade. Hypoxia is instituted by inhalation of low oxygen mixture (P_{IO_2}, 80 mmHg). Phentolamine and 0.9% NaCl are given 8–15 min prior to low oxygen inhalation in α-blocking and control experiments, respectively. Vertical bars represent 1 SE of the mean. Number of observations: for control, $n=7$; for α-blocking, $n=7$.

One of these dogs showed the following sequence of events: under anesthesia it twice did not respond to low oxygen inhalation; 3 days later, in its conscious state, it presented the typical marked pressor response which was confirmed by 17 subsequent repetitions.

Discussion

Judging from the positive evidence provided by the present study, the depressive action of the α-adrenergic blocking agent phentolamine on the hypoxic pulmonary pressor response seems to be clear-cut. Currently much doubt has been cast upon the involvement of the α-adrenergic mechanism in hypoxia-induced pulmonary vasoconstriction (2). The present study again raises the problem for reconsideration.

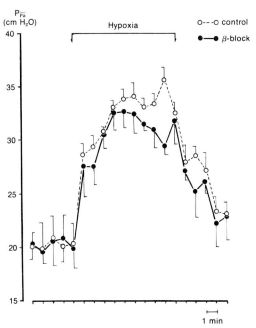

Fig. 22-5. Successful β-blockade produces no marked effect on the hypoxia-induced pulmonary pressor response. Symbols and experimental conditions as in Fig. 22-4, except that β-blocking with propranolol replaces α-blocking with phentolamine. Number of observations: for control, $n=3$; for β-blocking, $n=3$.

2. Fishman, A.P.: Hypoxia on the pulmonary circulation. How and where it acts. Circ. Res. 38:221, 1976.
3. Susmano, A., Passovoy, M., and Carleton, R.A.: Comparison of the effects of two anesthetic agents on the production of hypoxic pulmonary hypertension in dogs. Am. Heart J. 84:203, 1972.
4. Sykes, M.K., Loh, L., Seed, R.F., Kafer, E.R., and Chakrabarti, M.K.: The effect of inhalation anaesthetics on hypoxic pulmonary vasoconstriction and pulmonary vascular resistance in the perfused lungs of the dog and cat. Br. J. Anaesth. 44:776, 1972.

V. Hormonal, Hematologic, and Electrolyte Changes

23
Hormonal Responses to Exercise at Altitude in Sea Level and Mountain Man

JOHN R. SUTTON AND FAUSTO GARMENDIA

Physical exercise results in an increased plasma concentration of catecholamines (3,21), glucagon (1), cortisol (17), and growth hormone (18,23). With submaximal exercise, there is often no increase in growth hormone during exercise (13), but when the same subjects were studied under conditions of acute hypoxic exercise, a marked elevation in growth hormone secretion was found (14).

At a similar intensity of work, high altitude dwellers (HAD), who were born and lived all their lives at Morococha, Peru (4540 m), were found to have elevated basal serum levels of growth hormone, but these levels did not increase with submaximal exercise (Fig. 23-1) (14). The purpose of the present study was to examine the effect of maximal exercise on the hormonal responses in acclimatized mountaineers (AM) and HAD.

Methods

Eight active mountaineers, aged 23-28 years, who were born and lived at sea level, were studied during maximal exercise at sea level and later at Morococha, following 3 months acclimatization to altitudes between 4000 and 6000 m. The eight HAD, aged 23-31 years, were born and living at Morococha (4540 m). All studies were conducted after an overnight fast. Exercise consisted of running on a treadmill at an increasing speed and elevation, which was determined by a previous test to take approximately 20 min to reach exhaustion. Venous blood was taken from an antecubital vein at rest and following maximal exercise and the serum or plasma was separated and frozen until analyzed at the investigators' laboratory. Glucose was measured by autoanalyzer, using the ferricyanide method (5), and free fatty acids (FFA) were measured by the Dole titration (2). Growth hormone (4) and insulin (7) were measured by radioimmunoassay and cortisol (9) was measured by competitive protein binding.

In a second study, an oral glucose tolerance test (0.85 g/kg glucose) was performed on a separate group of 11 normal male HAD at Morococha (age 27.0 ± 1.8 years).

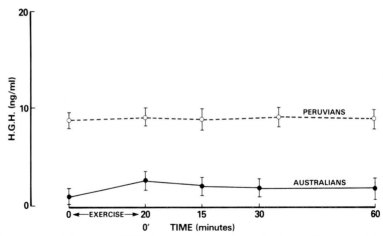

Fig. 23-1. Serum growth hormone (HGH) response to submaximal exercise at an altitude of 4540 m in acclimatized mountaineers and high altitude dwellers. From Sutton, J.: Med. J. Aust. 2:357, 1971.

Results

Maximal Exercise

The high altitude dwellers and the mountaineers were of similar age, but the latter were taller and heavier (Table 23-1). Time taken to reach exhaustion on the treadmill ranged from 17 to 26 min, but was similar in each study.

The HAD had a lower fasting plasma glucose than did the sea level subjects tested at sea level or at altitude. The increase with exercise was similar to that found in the sea level subjects (Fig. 23-2). Serum insulin was not significantly different from that of sea level subjects at rest and showed a minor, but nonsignificant, decrease with exercise (Fig. 23-3). Serum cortisol was similar at rest in all subjects and increased with exercise; the response was greatest in the acclimatized mountaineers tested at altitude and remained elevated 90 min postexercise (Fig. 23-4). Serum growth hormone, while high at rest in the HAD, showed a slightly smaller increase with exercise, although the absolute concentrations were high throughout and following exercise. Even 90 min postexercise, the serum growth hormone remained elevated in both the HAD and the acclimatized mountaineers compared with findings from studies performed on the mountaineers at sea level (Fig. 23-5).

Glucose Tolerance Test

Fasting levels of glucose were again lower in the HAD ($p < 0.001$). Plasma free fatty acids (FFA) were significantly higher in the basal state in the HAD ($p < 0.001$). Basal

Table 23-1. Anthropometric Data on the Mountaineers and the High Altitude Dwellers Who Performed Maximal Exercise.

Subjects	Location	Age (years)	Height (cm)	Weight (kg)
Mountainers	Sea level	25.4 ± 0.8	179.6 ± 1.5	76.2 ± 3.1
Mountaineers	Morococha	25.7 ± 0.8	179.6 ± 1.5	72.2 ± 3.5
High altitude dwellers	Morococha	27.7 ± 1.2	161.9 ± 2.3	60.9 ± 2.3

Mean ± 1 SEM.

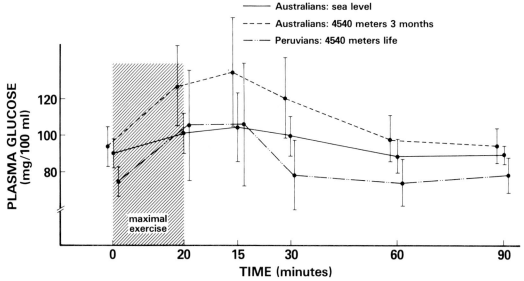

Fig. 23-2. Plasma glucose response to maximal exercise: (1) at sea level in sea level dwellers; (2) at 4540 m in sea level dwellers after 3 months acclimatization to altitude; and (3) in high altitude dwellers at 4540 m.

serum growth hormone levels were slightly higher in the HAD compared with the AM ($p < 0.005$) and there was a significant suppression of serum growth hormone following glucose administration and the subsequent hyperglycemia (Fig. 23-6).

Discussion

Maximal exercise at altitude is a major stimulus to the endocrine system in that elevations in glucose, cortisol, and serum growth hormone were found in both the

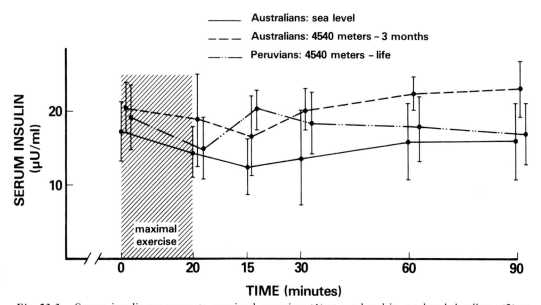

Fig. 23-3. Serum insulin response to maximal exercise: (1) at sea level in sea level dwellers; (2) at 4540 m in sea level dwellers after 3 months acclimatization to altitude; and (3) in high altitude dwellers at 4540 m.

168   J.R. Sutton and F. Garmendia

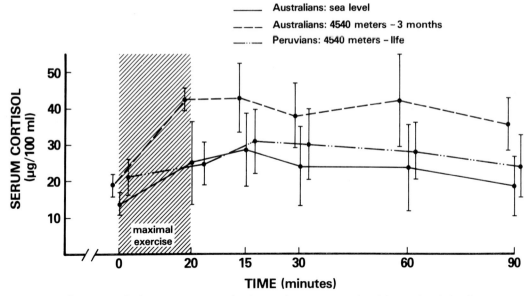

Fig. 23-4. Serum cortisol response to maximal exercise: (1) at sea level in sea level dwellers; (2) at 4540 m in sea level dwellers after 3 months acclimatization to altitude; and (3) in high altitude dwellers at 4540 m.

acclimatized mountaineers and in the HAD. After 3 months acclimatization to altitudes in excess of 4000 m, the responses to exercise in the mountaineers were still exaggerated when compared to maximal exercise at sea level and also when compared to responses of men born and living permanently at high altitudes.

The low fasting plasma glucose found in the HAD has been previously reported

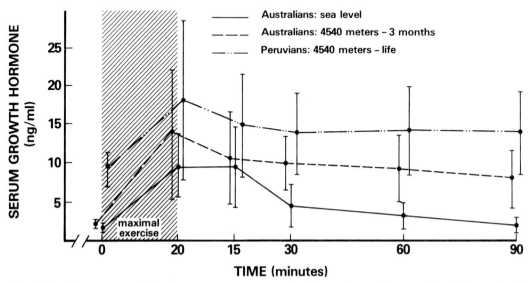

Fig. 23-5. Serum growth hormone response to maximal exercise: (1) at sea level in sea level dwellers; (2) at 4540 m in sea level dwellers after 3 months acclimatization to altitude; and (3) in high altitude dwellers at 4540 m.

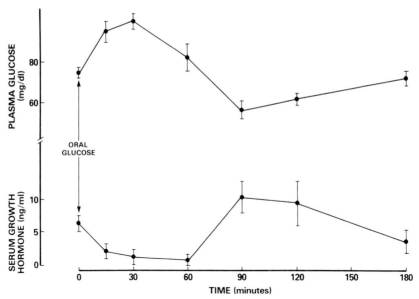

Fig. 23-6. Serum growth hormone and plasma glucose levels following an oral glucose load of 0.85 g/kg in high altitude dwellers, performed at 4540 m.

(10,11,24) and the reciprocal elevation in plasma FFA is consistent with fat being a major energy source in the HAD. In response to maximal exercise, plasma glucose increased in all studies, but was most marked in the AM. Submaximal exercise with acute hypoxia will also result in hyperglycemia (13), which may reflect an increased hepatic glycogenolysis in response to adrenaline and/or glucagon. The fasting serum insulin concentrations were similar in all studies and the previously noted tendency for insulin to decrease with exercise and increase following exercise (12,15) was again seen in this study. The unremarkable changes in serum insulin should be contrasted with the effects of acute hypoxia in which significant suppression of serum insulin was found (13) and may represent some adjustment of the pancreatic beta cell to chronic hypoxia. Basal serum cortisol levels were similar in all subjects, which contrasts with that found following acute exposure to high altitude (8,22). The changes in serum cortisol with exercise at sea level have been well described and the HAD have an identical response (17,18,23). Although the AM have similar basal serum cortisol levels with a normal diurnal rhythm (16) suggesting a perfect adaptation to high altitude, the augmented cortisol response to maximal exercise may reflect less than complete acclimatization which is uncovered only with a severe stress such as exhaustive exercise.

Elevated serum growth hormone in the HAD appears to be under normal physiologic regulation, as evidenced by the further increase when stimulated by maximal exercise and the suppression by oral glucose. It therefore would seem that the hypothalamic control of growth hormone secretion may well have a higher set point in the HAD. The prolonged elevation of serum growth hormone following maximal exercise is also intriguing and similar to previous findings at sea level in exhausted unfit or obese individuals (23). Such a finding could result from an increased pituitary release of the hormone, a delay in its metabolic clearance, or some combination of factors. The prolonged elevation following exercise could be due to a decreased metabolic clearance. For example, the elevation of testosterone (20) with exercise is due to a decrease in the metabolic clearance of the hormone, and not an increased production (19).

An elevation in serum growth hormone may have some importance in a somatotrophic role, i.e., in maintaining organ hypertrophy of heart and lungs, which is found in high altitude Peruvians. Alternatively, the elevated growth hormone may be important in a selective metabolic role. For instance, the elevated plasma FFA in the HAD could be the result of the growth hormone which will mobilize FFA from adipose tissue to be used as the principal energy substrate under hypoxic conditions (6).

Acknowledgments

This study was supported by a grant from the Instituto Hipolito Unanue and by the G.D. Searle Travelling Fellowship in Endocrinology from the Australian Postgraduate Federation in Medicine.

The authors wish to thank Dr. L. Lazarus, Dr. T. Velasquez, Dr. R. Tamayo, Dr. J. Torres, and the late Mr. John Young.

References

1. Böttger, I., Schlein, E.M., Faloona, G.L., Knochel, J.P., and Unger, R.H.: The effect of exercise on glucagon secretion. J. Clin. Endocrinol. 35:117, 1972.
2. Dole, V.P.: A relation between nonesterified fatty acids and the metabolism of glucose. J. Clin. Invest. 35:150, 1956.
3. Haggendal, J., Hartley, L.H., and Saltin, B.: Arterial noradrenaline concentration during exercise in relation to the relative work levels. Scand. J. Clin. Lab. Invest. 26:337, 1970.
4. Hales, C.N. and Randle, P.J.: Immunoassay of insulin with insulin-antibody precipitate. Biochem. J. 88:137, 1963.
5. Hoffman, W.S.: A rapid photoelectric method for the determination of blood glucose in blood and urine. J. Biol. Chem. 120:51, 1937.
6. Jones, N.L., Robertson, D.G., Kane, J.W., and Hart, R.A.: Effect of hypoxia on free fatty acid metabolism during exercise. J. Appl. Physiol. 33:733, 1972.
7. Lazarus, L. and Young, J.D.: Radioimmunoassay of human growth hormone in serum using ion-exchange resin. J. Clin. Endocrinol. Metab. 26:213, 1966.
8. Moncloa, F., Velasco, I., Betata, L.: Plasma cortisol concentration and disappearance rate of 4-14 C cortisol in newcomers to high altitude. J. Clin. Endocrinol. Metab. 28:379, 1968.
9. Murphy, B.E.P.: Some studies of the protein-binding of steroids and their application to the routine micro and ultramicro measurement of various steroids in body fluids by competitive protein-binding radioassay. J. Clin. Endocrinol. Metab. 27:973, 1967.
10. Picon-Reategui, E.: Studies on the metabolism of carbohydrates at sea level and at high altitudes. Metabolism 11:1148, 1962.
11. Picon-Reategui, E.: Intravenous glucose tolerance test at sea level and at high altitudes. J. Clin. Endocrinol. Metab. 23:1256, 1963.
12. Pruett, E.D.R.: Plasma insulin concentrations during prolonged work at near maximal oxygen uptake. J. Appl. Physiol. 29:155, 1970.
13. Sutton, J.R.: Effect of acute hypoxia on the hormone response to exercise. J. Appl. Physiol. 42:587, 1977.
14. Sutton, J.: Scientific and medical aspects of the Australian Andean Expedition. Med. J. Aust. 2:355, 1971.
15. Sutton, J.R.: Hormonal and metabolic responses to exercise in subjects of high and low work capacities. Med. Sci. Sports 10:1, 1978.
16. Sutton, J.R.: Acute mountain sickness. An historical review with some experiences from the Peruvian Andes. Med. J. Aust. 2:243, 1971.
17. Sutton, J.R. and Casey, J.H.: Adrenocortical response to competitive athletics in veteran athletes. J. Clin. Endocrinol. Metab. 40:135, 1978.
18. Sutton, J.R., Coleman, M.J., and Casey, J.H.: The adrenal cortical contribution to serum androgens in physical exercise. Med. Sci. Sports 6:72, 1974.
19. Sutton, J.R., Coleman, M.J., and Casey, J.H.: Testosterone production rate during

exercise. In Landry, F. and Orban, W.A.R. (eds.): 3rd International Symposium on Biochemistry of Exercise. Miami, Fla., Symposia Specialists, 1978, pp. 227–234.
20. Sutton, J., Coleman, M.J., Casey, J.H., and Lazarus, L.: Androgen responses during physical exercise. Br. Med. J. 1:520, 1973.
21. Sutton, J.R., Jurkowski, J.E., and Keane, P.M.: The effect of the menstrual cycle on the plasma catecholamine response to exercise in normal females. Clin. Res. 26:847A, 1978.
22. Sutton, J.R., Viol, G.W., Gray, G.W., McFadden, M., and Keane, P.M.: Renin, aldosterone, electrolyte, and cortisol responses to hypoxic decompression. J. Appl. Physiol. 43:421, 1977.
23. Sutton, J.R., Young, J.D., Lazarus, L., Hickie, J.B., and Maksvytis, J.: The hormonal response to exercise. Australas. Ann. Med. 18:84, 1969.
24. Urdanivia, E., Garmendia, F., Torres, J., Zubiate, M., and Tamayo, R.: Adrenal responses to tolbutamide-induced hypoglycemia in high altitude dwellers. J. Clin. Endocrinol. Metab. 40:717, 1975.

24
Time Course of Plasma Growth Hormone During Exercise in Man at Altitude

J. RAYNAUD, L. DROUET, J. COUDERT, AND J. DURAND

Several studies (5,6) tend to demonstrate that hypoxic conditions enhance the release of growth hormone (GH). However, results obtained in highlanders in Peru suggest that stimulation of GH release during exercise is not increased by altitude hypoxia (4). The present study was designed to define more accurately the time course of plasma GH concentration and that of other metabolites related to muscular exercise and to compare them in lowlanders at sea level, in subjects fully adapted to altitude hypoxia, and in lowlanders acutely exposed to hypoxia either by breathing a low pressure O_2 gas mixture or by a process of acclimatization after translocation to high altitude.

Subjects and Methods

Three groups of subjects were studied: (1) nine highlanders (HL) at high altitude (HA) (La Paz, 3850 m); (2) 20 lowlanders (LL) at sea level (SL) (Paris); and (3) five of the lowlanders, (a) on the third to sixth day after arrival at Pic du Midi (2850 m) and (b) 15 days after returning to Paris, in acute hypoxia, breathing 15% O_2 gas mixture which corresponds to the partial pressure of O_2 at 2850 m.

The following energetic and hormonal parameters were measured at rest, during a 1-h submaximal exercise performed on a bicycle ergometer, and during a period of 1-h recovery: blood glucose (BG), free fatty acids (FFA), lactic acid (LA), and growth hormone (IRHGH) plasma concentration as a function of time at the intervals indicated in Fig. 24-1 and $\dot{V}_{O_2}$ at steady state of exercise. Growth hormone was measured by radioimmunoassay. The time course of [IRHGH] is described in terms of peak concentration, delay of rise, and half-life. The area delimited by the curve of [IRHGH] above baseline during exercise

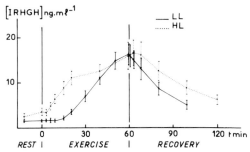

Fig. 24-1. Time course of [IRHGH] (mean ± SE) in 15 LL (*solid line*) and 8 HL (*dotted line*).

was determined by planimetry, thus providing a measurement which takes into account both delay and shape of the ascending curve. Two values of this area were calculated: one at 30, the other at 60 min of exercise.

Results

The results of the experiments are shown in Table 24-1. The mean maximal values of [LA] do not differ in LL and HL. However, LA_{max} is slightly higher in LL when they are placed in acute hypoxic conditions, compared to values under normoxia. The resting value of [BG] is significantly lower in HL compared to LL. The fall which occurs at the beginning of exercise is of the same order of magnitude in HL and LL. The time course of [FFA] is similar in the two groups in that [FFA] increases only at the end of the 1-h exercise period and the maximal value occurs during recovery. This maximal value is higher in HL than in LL.

The mean resting value of [IRHGH] is significantly higher in HL than in LL. During exercise, the time sequence of [IRHGH] shows differences when the conditions of oxygenation change (Fig. 24-1): in LL at SL, resting concentrations are maintained for some time during exercise, then [IRHGH] rises and reaches maximal value at the end of exercise. In HL, the delay is significantly reduced and half-life is much longer. The mean maximal value is not different from that of LL. Figure 24-2 confirms that the initial rise in [IRHGH] is definitely earlier and faster in HL than in LL despite comparable peak values: the mean area at 30 min is significantly ($p > 0.01$) larger for HL than for LL, and the mean area at 60 min in HL is also larger but the difference is not confirmed statistically. The five LL studied in acute hypoxic conditions exhibit similar variations in the time course of IRHGH to those of HL: faster rise in concentration is confirmed by the comparison of surface integrals of [IRHGH] as a function of time (Fig. 24-2).

Table 24-1. Experimental Results in Highlanders and Lowlanders.

	LL (SL)	HL (3850 m)	LL		
			SL	Acute hypoxia	2850 m
$\dot{V}_{O_2 s. st.}$ (liter · min^{-1} · m^{-2})	0.870± 0.05	0.820± 0.08	0.830± 0.05	0.829± 0.07	0.825± 0.05
Relative work load (percent $\dot{V}_{O_2 max}$)	54.3± 3.3	51.9± 4.0	59.8± 5.1		
[LA]$_{max}$ at exercise (mM · liter^{-1})	4.1± 0.3	4.3± 0.2	4.2± 0.4	5.2± 0.7	5.4± 0.6
[BG] at rest (g · liter^{-1})	0.95± 0.02[b]	0.77± 0.04[b]			
[FFA]$_{max}$ at exercise (μEq · liter^{-1})	760± 65[a]	1276± 254[a]			
[IRHGH] at rest (ng · ml^{-1})	1.58± 0.65[b]	3.58± 0.88[b]	1.50± 0.40	1.80± 0.65	1.75± 0.35
[IRHGH]$_{max}$ at exercise (ng · ml^{-1})	16± 2.3	16.5± 2.9	16.6± 5.3	19.8± 4.8	22.6± 6.1
Delay of rise (min)	15± 2.2[a]	5.5± 1.2[a]	20± 4.3	16± 4.1	10± 3.5
Half-life (min)	16± 2[a]	34± 6[a]	19± 2	16± 2	15± 5

Mean ± SE.
[a,b] $p < 0.05$ and $p < 0.01$, respectively, for LL at SL and HL at 3850 m.

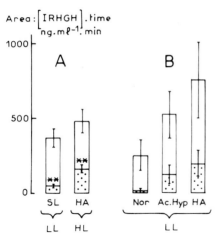

Fig. 24-2. The top of the columns correspond to the mean value of the areas determined by time and [IRHGH] above resting baseline (Fig. 24-1). Dotted columns correspond to the surface areas determined by 30 min of exercise, and open columns to the surface areas determined by 60 min of exercise.

Discussion

The following observations emerge from these results:

1) The similar time course and maximal values of [LA] which are observed in HL and LL favor the argument that anaerobic metabolism is the same for all subjects of similar physical fitness provided they are fully acclimatized to their environment, whatever the altitude.

2) At rest, the lowering of [BG] in HL as compared to LL was also noticed by Picon-Reategui et al. (2) in natives of 4000 m, but the mechanism is not clear.

3) The fact that the sudden rise in [FFA] during the early stages of recovery is more marked in HL suggests that FFA output is more important at HA than at SL. A similar finding is reported by others (3,5).

4) Two differences appear in [IRHGH] in LL and HL (Fig. 24-1): (a) the resting value is significantly higher in HL than in LL; and (b) the time sequences during muscular activity differ in that HL have a shorter delay in the onset of rise in concentration and a longer time constant of disappearance. However, the mean peak value reached after 1 h of exercise is similar to that of LL. This pattern of the [IRHGH] curve is also observed at exercise during the early stages of acclimatization to HA or during hypoxic mixture breathing. Thus, a shorter delay and a slower disappearance seem to be constant responses to exercise in hypoxia. Which hypotheses could be advanced to explain these results? At rest, the relatively but chronically low glycemia in HL might be responsible for the higher resting level of [IRHGH]. At exercise, no correlation is found between $[LA]_{max}$ and [IRHGH] peak values (1). Moreover, the time courses of [LA] are superimposable in LL and HL, precisely in the same time when the time courses of [IRHGH] differ. It would be hazardous also to assume that the slight drop in [BG] observed at the beginning of exercise, which is of the same magnitude in LL and HL, is responsible for the different time courses of [IRHGH]. In contrast, the present experiments seem to support the action of growth hormone as a stimulus in releasing FFA during exercise, since an earlier rise in [IRHGH] precedes a faster and greater output of FFA in HL.

In all the cases, stopping exercise plays a major role in the time course of [IRHGH]: its rise stops, followed by a decrease. This sudden change suggests that a balance between hormone release and catabolism, which is responsible for the observed variations in blood concentration, is immediately altered. This might be due to the sudden increase in hepatic blood flow as exercise stops, which would augment hormone clearance. We propose that the differences observed during exercise in the time course of [IRHGH] between LL and HL, which consist in a faster rise and a slower disappearance, are likely due to a more pronounced reduction in hepatic blood flow in HL as compared to LL.

In general, this study points out the importance of (1) characterizing the release pattern of IRHGH by variables other than maximal concentration, such as delay of rise, elevation rate, and half-life; and (2) dealing with long periods of exercise. If, in the present case, exercise duration had been reduced to 20 min, for instance, the actual difference observed at this time between the time sequence in HL and LL would have led to the conclusion that hormonal release was more important at HA that at SL, whereas the conclusion is different 40 min later.

References

1. Lassarre, C., Girard, F., Durand, J., and Raynaud, J.: Kinetics of human growth hormone during submaximal exercise. J. Appl. Physiol. 37:826, 1974.
2. Picon-Reategui, E., Buskirk, E.R., and Baker, P.T.: Blood glucose in high altitude natives and during acclimatization to altitude. J. Appl. Physiol. 29:560, 1970.
3. Stock, M.J., Chapman, C., Stirling, J.L., and Campbell, I.T.: Effects of exercise, altitude, and food on blood hormone and metabolite levels. J. Appl. Physiol. 45:350, 1978.
4. Sutton, J.R.: Scientific and medical aspects of the Australian Andean expedition. Med. J. Aust. 2:355, 1971.
5. Sutton, J.R.: Effect of acute hypoxia on the hormonal response to exercise. J. Appl. Physiol. 42:587, 1977.
6. Sutton, J.R., Jones, N.L., and Toews, C.J.: Growth hormone secretion in acid-base alterations at rest and during exercise. Clin. Sci. Mol. Med. 50:241, 1976.

25

Transcapillary Escape Rate of Albumin After Exposure to 4300 m

G. COATES, G.W. GRAY, C. NAHMIAS, A.C. POWLES, AND J.R. SUTTON

Pulmonary edema, retinal hemorrhage and exudates, cerebral edema, and peripheral edema are all known complications of exposure to high altitude. Thus, it seems that there must be some factor or factors in the high altitude environment which cause an increased leakage of fluid from the capillaries of these organs. The time of onset and mechanism of this increased capillary leakage is not known.

In an attempt to understand the nature of this increased leakage of fluid, we have measured the transcapillary flux of protein in eight normal volunteers at sea level and at a simulated altitude of 4268 m in a hypobaric chamber. To do this we used a technique developed by Parving and Rossing (4) which measures the rate at which ^{125}I-albumin leaves the intravascular space. This transcapillary escape rate (TER Alb) of albumin is increased in such diseases as hypertension (3), diabetes (4), and congestive heart failure (2). All these conditions produce an increased filtration of and/or permeability to fluid and proteins across capillaries.

Methods

To perform each study 10 μCi of ^{125}I-albumin was injected into a vein and nine blood samples were obtained from an indwelling catheter over the next 70 min. The subjects were supine for at least 30 min before the study started. We studied each subject at sea level and then after 4–8 h at a pressure of 446 torr (equivalent to 4268 m) in a hypobaric chamber. Decompression to this pressure took 28 min.

On each blood sample we measured the concentration of albumin, hematocrit, and counts from ^{125}I per min per ml. Plasma volume was calculated in four of the subjects from the dilution principles, and estimated in the other four from height and weight. To calculate the transcapillary escape rate we used the method of least squares to plot the line of best fit to the data of specific activity of ^{125}I-albumin against time. The $t_{1/2}$ of that line was obtained and the fraction of albumin leaving the vascular space per unit time (TER Alb) was calculated from the following relation:

$$\text{TER Alb} = \frac{0.693}{t_{1/2}} \% \cdot h^{-1}$$

Albumin flux (g · h⁻¹) was calculated as the product of total intravascular albumin (concentration × plasma volume) and TER Alb.

Results

TER Alb results of each subject at sea level and altitude are illustrated in Fig. 25-1. At sea level the mean TER Alb was $6.2\% \cdot h^{-1}$ and this increased to $7.6\% \cdot h^{-1}$ at altitude. This corresponds to an increase in albumin flux from 7.7 to 9.3 g · h⁻¹. There was no significant change in plasma volume or hematocrit.

Discussion

Although this change in albumin flux was not significant by the t test ($t = 1.14$), there was a clear tendency toward an increase.

This apparent increased disappearance of albumin from the intravascular space could be caused by a true increase in protein flux across capillaries by either an increased capillary permeability to protein or an increased filtration of fluid with a resultant increase in protein transport by convection. It could also be caused by an increased rate of protein metabolism or an increased rate of return of fluid and protein via the lymph producing a decrease in the specific activity of ^{125}I-albumin and thus an apparent increased loss from the vascular space. Surks (6) did show an increase in the metabolism of albumin in human volunteers during 8 days at Pikes Peak (14,100 ft). However, this was not significant until day 2 of their study and even then was only of the order of 2 g of albumin per day, whereas we found an increased loss from the vascular space of 1.6 g per hour.

Any increase in the rate of return of lymph to the vascular space would produce an increased plasma volume. We found no

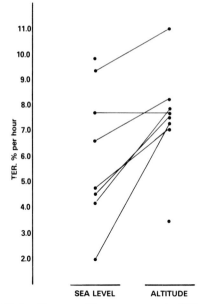

Fig. 25-1. Transcapillary escape rate (TER) of albumin in percent per h at sea level and at 4300 m in 8 subjects.

changes in plasma volume in the four subjects in whom it was measured and other investigators have reported similar findings up to 12 h at altitude (5). After 24 h at altitude there is a measurable decrease in plasma volume (1,5,6).

We conclude then that an exposure of 4–8 h at a simulated altitude of 4300 m produces an increase in albumin flux across capillaries. These findings are similar to those of Siggaard-Andersen et al. (5) who found an increase in protein flux in volunteers made acutely hypoxic by breathing carbon monoxide. However, these same investigators failed to show an increase in protein flux in response to acute exposure to a simulated altitude of 3454 m. This altitude is substantially lower than that to which our subjects were exposed and may explain the different results. There is a clear need to repeat these experiments at an even higher altitude and to relate any increase in protein flux to symptoms of mountain sickness or signs of pulmonary or cerebral edema.

References

1. Frayser, R., Rennie, I.D., Gray, G.W., and Houston, C.S.: Hormonal and electrolyte response to exposure to 17,500 ft. J. Appl. Physiol. 38:636, 1975.
2. Hesse, B., Parving, H.H., Lund-Jacobson, H., and Noer, I.: Transcapillary escape rate of albumin and right atrial pressure in chronic congestive heart failure before and after treatment. Circ. Res. 39:358, 1976.
3. Parving, H.H. and Gyntelberg, F.: Transcapillary escape rate of albumin and plasma volume in essential hypertension. Circ. Res. 32:643, 1973.
4. Parving, H.H. and Rossing, N.: Simultaneous determination of the transcapillary escape rate of albumin and IgG in normal and long-term juvenile diabetic subjects. Scand. J. Clin. Lab. Invest. 32:239, 1973.
5. Siggaard-Andersen, J., Petersen, F.B., Hansen, T.I., and Mellemgaard, K.: Plasma volume and vascular permeability during hypoxia and carbon monoxide exposure. Scand. J. Clin. Lab. Invest. (Suppl.) 103:39, 1968.
6. Surks, M.I.: Metabolism of human serum albumin in man during acute exposure to high altitude (14,100 ft.). J. Clin. Invest. 45:1442, 1966.

26
Platelet Survival and Sequestration in the Lung at Altitude

G. COATES, G.W. GRAY, C. NAHMIAS, A.C. POWLES, AND J.R. SUTTON

It is clear from the diving literature that platelet function and survival are altered by decompression from depth (8). These changes are partially preventable with platelet suppressant drugs (9). Gray et al. (2) suggested that similar platelet changes may occur on decompression from sea level to altitude and that these may be implicated in the development of high altitude pulmonary edema.

There are conflicting data in the literature about the effect of exposure to high altitude on the platelet. In 1970 Genton et al. (3) reported a significant fall in platelet count and a reduction in platelet survival in calves exposed to an altitude of 14,253 ft. Gray et al. (2) showed similar changes in platelet count in human volunteers both in a mountain environment at 17,000 ft and at a simulated altitude of 20,000 ft in a hypobaric chamber. This same group also demonstrated sequestration of labeled platelets in the lungs of rabbits decompressed to 18,000 ft. On the other hand, Maher et al. (7) found no change in platelet function or platelet count in eight volunteers exposed to a simulated altitude of 4400 m. More recently, Hyers et al. (5) measured platelet function at 10,000 ft in eight subjects who had previously had high altitude pulmonary edema. They found no change in platelet count.

We have taken a slightly different approach to the problem. The hypothesis is that decompression to altitude causes platelets to aggregate and sequester in the lungs. This was clearly demonstrated by Gray et al. (2) in the rabbit. In our experiments we observed the effect of rapid decompression to altitude on platelet survival in human volunteers. We also sought evidence of sequestration of platelets in their lungs.

Methods

The studies were performed on two groups of four subjects, all healthy males aged 22 to 39 years. Their platelets were labeled with ^{51}Cr by the method of Aster and Jandl (1) 48–72 h before decompression. Multiple blood samples were taken during 24 h before and 24 h after decompression to 446 torr (equivalent to 4268 m) in a hypobaric chamber. Decompression to this pressure took 28 min. Group I stayed at this pressure for 24 h, group II for only 8 h. In order to detect an accumulation of platelets in the lungs, we used a sensitive sodium iodide detector system to measure changes in ^{51}Cr counts from the right upper lobes. We also measured ^{51}Cr over the liver and thigh.

In order to be certain that any change in the amount of ^{51}Cr-platelets in the lungs was not simply due to a change in the amount of blood in the lungs, we monitored changes in regional lung blood volume with a modification of the technique of Lindsey and Guyton (6). Ten days before decompression each subject in group I received an intravenous injection of $10\mu Ci$ ^{59}Fe. By the time of the study 85–95% of the ^{59}Fe would be incorporated into red cells and thus any change of ^{59}Fe count in an organ would reflect a change in the amount of blood in that organ. We used ^{99m}Tc to label the red cells in group II (7a). We monitored changes in ^{99m}Tc or ^{59}Fe at the same time as we measured ^{51}Cr in the lungs, i.e., in group I 26, 20, 12, and 3 h before decompression and after 1, 9, and 14 h at altitude; and in group II 1 h before decompression and after 1, 4, and 7 h at altitude. The subjects were supine for all of these measurements.

Analysis of Data and Results

Platelet Survival

The data of ^{51}Cr counts per min per ml of whole blood against time were analyzed with a least squares fit program on a digital computer. Two fits were performed, one to the data at sea level and one to the data taken during and after exposure to altitude. The slopes of these lines were compared by the F-test (10) and were not significantly different.

Figure 26-1 shows the platelet survival data for the group I subjects. The data from the group II subjects are the same.

Platelet Accumulation in Lungs

Figure 26-2 illustrates the changes in ^{51}Cr and ^{59}Fe counts from the right upper lobes of the subjects in group I. There was no significant change in ^{59}Fe counts from the lung. There was also no change in the ratio of ^{59}Fe lung/^{59}Fe liver or ^{59}Fe lung/^{59}Fe thigh, i.e., we were unable to detect changes in lung blood volume. There was a fall in counts from ^{51}Cr in the lung, which reflected the fall in ^{51}Cr in the blood due to platelet senescence. There was no change in the ratio of ^{51}Cr lung/^{51}Cr liver or ^{51}Cr lung/^{51}Cr thigh, i.e., we were unable to detect an accumulation of platelets in the lung. The data from the group II subjects are identical.

Discussion

In these experiments we were unable to show decreased platelet survival or sequestration in the lung at a barometric pressure of 446 torr. Both Genton et al. (3) in the calf and Gray et al. (2) in humans and rabbits have shown changes in platelet survival or platelet count at altitude. In a preliminary communication Hyers et al. (4) also reported a significant reduction in platelet count at 4350 m in seven human volunteers. The discrepancy between these data and ours can be explained by different experimental conditions such as rate of decompression, or by biologic differences between humans and the experimental animals used. It is also possible that there is a threshold altitude below which the relatively crude tests of platelet function are unable to detect changes which are present or below which there are in fact no changes to detect.

We are testing this hypothesis in our laboratory with the rabbit as the constant biologic model and the barometric pressure as the only variable.

References

1. Aster, R.H. and Jandl, J.H.: Platelet sequestration in man. I. Methods. J. Clin. Invest. 43:843, 1964.

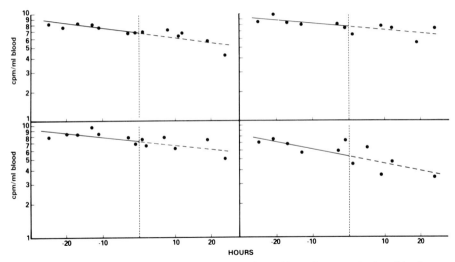

Fig. 26-1. Platelet survival data of group I subjects. Line of best fit was calculated by least squares. The vertical dotted line at time 0 represents the time of decompression to a simulated altitude of 4268 m.

2. Gray, G.W., Bryan, A.C., Freedman, M.H., Houston, C.S., et al.: Effect of altitude exposure on platelets. J. Appl. Physiol. 39:648, 1975.
3. Genton, E., Ross, A.M., Takeda, Y.A., and Vogel, J.H.K.: Alterations in blood coagulation at high altitude. Hypoxia, high altitude, and the heart. First Conference on Cardiovascular Disease. Aspen, Colo. Adv. Cardiol. 5:32, 1970.
4. Hyers, T.M., Reeves, J.J., and Grover, R.F.: Effect of altitude on platelets, fibrinol-

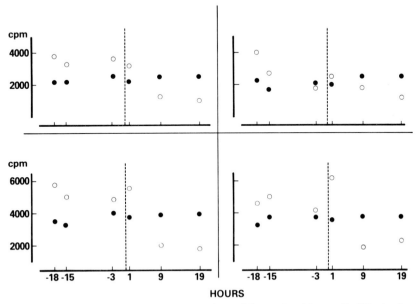

Fig. 26-2. Counts per min from the right upper lobes of Group I subjects. ●, ^{59}Fe (red cells); ○, ^{51}Cr (platelets). The vertical dotted line at time 0 represents the time of decompression to a simulated altitude of 4268 m.

ysis, and spirometry in man: influence of a platelet inhibitor. Clin. Res. 25:133A, 1977.
5. Hyers, T.M., Scoggin, C.H., Will, D.H., Grover, R.F., and Reeves, J.T.: Accentuated hypoxemia with high altitude exposure in subjects susceptible to high altitude pulmonary edema. J. Appl. Physiol. 46:41, 1979.
6. Lindsey, A.W. and Guyton, A.C.: Continuous recording of pulmonary blood volume: pulmonary pressure and volume changes. Am. J. Physiol. 197:959, 1959.
7. Maher, J.T., Levine, P.H., and Cymerman, A.: Human coagulation abnormalities during acute exposure to hypobaric hypoxia. J. Appl. Physiol. 41:702, 1976.
7a. Pavel, D.G., Zimmer, A.M., and Patterson, V.N.: In vivo labelling of red blood cells with ^{99m}Tc: a new approach to blood pool visualization. J. Nucl. Med. 18:305, 1977.
8. Philp, R.B.: A review of blood changes associated with compression-decompression: relationship to decompression sickness. Undersea Biomed Res. 1:117, 1974.
9. Philp, R.B., Inwood, M.J., Ackles, K.N., and Radon, M.W.: Effects of decompression on platelets and hemostasis in man and the influence of antiplatelet drugs (RA 233 and VK 744). Aerospace Med. 45:231, 1974.
10. Snedecor, G.W. and Cochrane, W.G.: Statistical Methods, 6th ed. Ames, Iowa, Iowa State Univ. Press, 1967.

27
Electrolyte Changes in the Blood and Urine of High Altitude Climbers

C. Rupp, R.A. Zink, and W. Brendel

During the Kanchenjunga expedition in 1975, the plasma concentrations of K^+, Na^+, Ca^{++}, Mg^{++}, Zn^{++}, and Cl^-; the urinary output of K^+, Na^+, Mg^{++}, and Zn^{++}; the osmolality; and the 24-h urine output had been measured in 11 climbers. These measurements have been evaluated by computer and the mean values for this group are compared in Figs. 27-1 and 27-2. All relative values are compared to initial values, which were taken in Munich before the expedition. The lower abscissa in both figures shows the altitude exposure (square units, cf. Chapter 48). The upper abscissa represents the days, day zero being the day of arrival in the base camp (5550 m). The negative numbers indicate the approach days before reaching the base camp. The ordinates show the dependent variables.

24-h Urine Output

During the approach to the Kanchenjunga base camp, which was reached after 26 days and an altitude exposure index of 35,000 SU, the 24-h urine output (initial value, 1130 ml/day) first decreased to less than 70% of initial value (740 ml/day) and then increased shortly before arrival at the base camp to more than 150% (1670 ml/day) (Fig. 27-1, top). These high values were found after 2 days of rest. Thereafter the 24-h urine output dropped below the original level (950 ml/day) during the ascent, increased again during the stay in camps III (7200 m) and IV (7800 m), and rose after resting periods of various length (> 24 h; < 5 days), when the climbers came back from the summit (1440 ml/day). Both decreases in 24-h urine volume were seen during periods of exercise and both increases after physical inactivity.

Osmolality

The urine osmolality (Fig. 27-1, top) showed a fairly contrary course: first a small rise from 980 to 1010 mosmol and then a decline to less than 85% (840 mosmol), although the rise in 24-h urine output was constant. After the return from the summit the osmolality ranged between 68 and 83% (670–820 mosmol).

Potassium

The concentration of plasma potassium (Fig. 27-1, *middle, circles*) remained between 3.88 mmol/liter and 1.54 mmol/liter (117%) during the whole expedition. The urinary output of potassium (*triangles*) fell at the beginning of the approach from 45.4 mmol/day to less than 50% (21.0 mmol/day), returned to about 90% (41.0 mmol/day) after 2 days rest, and was about 66% (30.0 mmol/day) after coming back from the summit. Finally, we found a rise to 81% (36.8 mmol/day) (1,5,6,7).

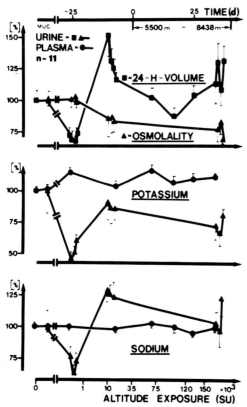

Fig. 27-1. Urine osmolality, 24-h urine volume, relative plasma sodium and potassium concentrations, and urinary sodium and potassium output per 24 h. *Upper abscissa:* time (day 0 represents arrival at the base camp at 5500 m; negative days represent the approach). *Lower abscissa*: altitude exposure (square units).

Sodium

The plasma sodium level (Fig. 27-1, *bottom, circles*), which was 144.2 mmol/liter initially, showed no clear deviations during the expedition. It varied maximally ± 5% (135.6–146.4 mmol/liter). The urinary sodium output (*triangles*) fell from 203.6 mmol/day at the beginning of the approach to about 62% (127.1 mmol/day) during exercise. After a recreation period of 2 days the urinary sodium output showed a definite increase, as did the urinary potassium output and 24-h urine volume, to 129% (263.6 mmol/day). The sodium excretion decreased to 96% after the summit ascent and finally increased to 122% (248.1 mmol/day) after a rest of 1 to 5 days (2,4,7).

Chloride

The plasma chloride level (Fig. 27-2, *bottom, squares*) rose slowly during the approach from 105.5 mmol/liter to 102% (107.8 mmol/liter), and reached a peak during the ascent to camp II (6650 m) of 112% (118.2 mmol/liter). The further ascent to higher camps brought a decrease to 104% (110.6 mmol/liter) and 100% (106.0 mmol/liter). Finally the plasma chloride concentration again reached 103% (108.7 mmol/liter) (3,4).

Calcium

The changes in the plasma concentration of calcium (Fig. 27-2, *bottom, circles*), similar to the plasma chloride changes, were not very distinct. A small rise at the beginning from 2.21 mmol/liter to 101% (2.24 mmol/liter), and a later increase to about 114% (2.5 mmol/liter) could be shown. During the ascent to the higher camps the plasma calcium concentration fell to 92% (2.0 mmol/liter) and rose again to 101% after a recreation period of 1 to 5 days (4).

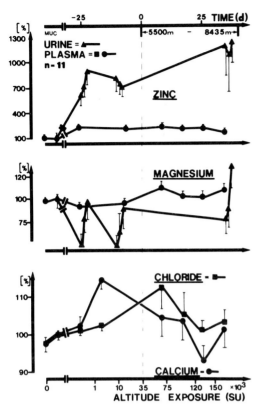

Fig. 27-2. Relative plasma concentrations and urinary output per 24 h of zinc, magnesium, chloride, and calcium with respect to the time course and altitude exposure. Abscissas as in Fig. 27-1.

from 77% (3.10 mmol/day) to 133% (5.32 mmol/day) (4).

Zinc

Both the plasma zinc level (Fig. 27-2, *top, circles*) and urinary zinc output (*triangles*) showed an astonishing increase. Plasma zinc concentration rose immediately from 10.9 mmol/liter to about 230% (24.0 mmol/liter) and remained there until the end of the expedition. Urinary zinc output increased steadily from 14.8 to 183.4 mmol/day, a 12-fold increase.

Conclusion

The plasma concentrations of sodium, potassium, magnesium, and chloride, as well as the accompanying 24-h urinary output, were not markedly impaired in comparison to the normal range, although the climbers were exposed to extreme physical, thermal, and altitude stresses. It may be that this was due to the daily intake of electrolyte beverages.[1]

We cannot explain the excessively high zinc levels in the urine and plasma. However, we might speculate that they were caused by either the local food (rice, potatoes, water) or the tinned food, which might have contained zinc released from the cans themselves.

We also found a steeply increased 24-h urine volume in expedition climbers immediately at the onset of a resting period after a longer period of exercise. An overshooting adrenal stimulation might be the reason for this (1), but we rather think that this amount of water was taken up by the muscles during the working stage and released from the muscles at rest.

Magnesium

The plasma magnesium concentration (Fig. 27-2, *middle, circles*) decreased during the approach from 0.78 mmol/liter to 91% (0.72 mmol/liter), and increased while climbing to the higher camps to 112% (0.87 mmol/liter). This level was maintained until the end of the expedition. Urinary magnesium output (*triangles*) showed two decreases, from 3.99 mmol/day to 50% (2.00 mmol/day) and from 94% (3.78 mmol/day) to 48% (1.94 mmol/day), and returned twice to its original level. After returning from the peak and a few days rest the magnesium excretion increased significantly

[1]Champ, Fresenius, Bad Homburg, West Germany; Elektrolyt-Drink-Pfrimmer, Pfrimmer, Erlangen, West Germany.

Acknowledgment

This study was supported by the Deutsche Forschungsgemeinschaft Zi 125/5. The electrolytes were measured in the laboratories of the German Heart Center, Munich, Federal Republic of Germany.

References

1. Ayres, P.J., Hurter, R.C., and Williams, E.S.: Aldosterone excretion and potassium retention in subjects living at high altitude. Nature 19:78, 1961.
2. Epstein, M., and Saruta, T.: Effects of simulated high altitude on renin-aldosterone and Na homeostasis in normal men. J. Appl. Physiol. 33:204, 1972.
3. Frayser, R., Rennie, D.I., Gray, G.W., and Houston, C.S.: Hormonal and electrolyte response to exposure to 17,500 feet. J. Appl. Physiol. 38:635, 1975.
4. Hannon, J.F., Chinn, K.S.K., and Shields, J.L.: Alteration in serum and extracellular electrolytes during high altitude exposure. J. Appl. Physiol. 31:266, 1971.
5. Hogan, R.P., Kotchen, T.A., Boyd, A.E., and Hartley, L.M.: Effect of altitude renin-aldosterone system and metabolism of water and electrolytes. J. Appl. Physiol. 35:385, 1973.
6. Slater, J.D.H., Williams, E.S., Edwards, R.H.T., Ekins, R.P., Sonksen, P.H., Beresford, C.A., and McLaughlin, M.: Potassium retention during the respiratory alkalosis of mild hypoxia in man: Its relationship to aldosterone secretion and other metabolic changes. Clin. Sci. 37:311, 1969.
7. Sutton, J.R., Viol, G.W., Gray, G.W., McFadden M., and Keane, P.M.: Renin, aldosterone, electrolyte, and cortisol responses to hypoxic decompression. J. Appl. Physiol. 43:421, 1977.
8. Zink, R.A., et al.: Proposals for International Standardization. In Brendel, W. and Zink, R.A. (eds.): High Altitude Physiology and Medicine, Chapter 48, *this volume*.

28
The Influence of Trekking on Some Hematologic Parameters and Urine Production

R.A. ZINK, H.P. LOBENHOFFER, B. HEIMHUBER, C. RUPP, AND R. SCHNEIDER

It is well known that physical exercise and exposure to altitude may cause hematologic changes (5,6,10,14,15) as well as alterations of body hydration (2,4,11). It is very difficult to differentiate if such changes are merely the effects of altitude or whether they are due to additional workloads (1,6,7,10,11,12,13).

In a preliminary investigation we studied 53 participants (aged 43 ± 7.2 years) in short-time trekking tours (STT). They hiked leisurely at a slight exercise level, for not more than 10 days, between 2500 and 4500 m. This group was compared to 21 long-time trekkers (LTT; aged 39 ± 8.7 years), who spent up to 3 weeks walking and climbing in the same area as the STT group. However, their walking distances per day were greater and they reached a higher altitude exposure during the first 10 days, due to short acclimatization excursions to the surrounding mountain ranges (up to 5100 m). The main interest of the LTT group was to climb some high peaks (up to 6200 m) during the second part of their sojourn.

The hematocrit, hemoglobin concentration, and red blood cell count (RBC) were measured using battery-operated microcentrifuges and microphotometers.[1] The 24-h urine volume was measured by simply collecting the urinary output.

The investigations were performed in the Khumbu area in Nepal during the premonsoon season of 1978. The initial values after arrival from the lowlands are compared with those obtained on the tenth day of altitude exposure and in the LTT group, with those from day 40, at the end of the trek.

Short-Time Trekking

The STT group (Fig. 28-1), showed no major alterations in the hematocrit, hemoglobin concentration, and red blood cell count between the beginning and the end of the trekking period. However, the 24-h urine volume was slightly decreased to 93% at day 10.

---

[1] N1000 and M1100, Compur Corp., D-8000 München 70, Federal Republic of Germany.

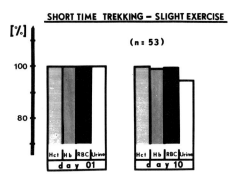

Fig. 28-1. Hematocrit (Hct), hemoglobin concentration (Hb), red blood cell count (RBC), and 24-h urine volume (Urine) at different stages of short-time trekking (slight exercise).

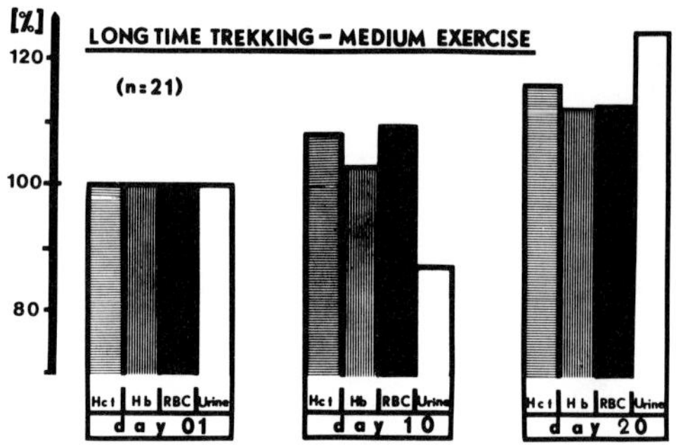

Fig. 28-2. Hematocrit (Hct), hemoglobin concentration (Hb), red blood cell count (RBC), and 24-h urine volume (Urine) at different stages of long-time trekking (medium exercise).

Long-Time Trekking

All hematologic parameters were markedly increased (Fig. 28-2; Hct, 108%; Hb, 103%, RBC, 120%) in the LTT group after 10 days during medium exercise at altitude. Urine production per 24 h dropped significantly to 88% of the initial amount ($p \leq 0.05$).

Ten more days later, at the end of the exercise stage and after one resting day, the hematocrit obtained was 116%, the hemoglobin concentration was 112%, and the RBC 114% of initial values ($p \leq 0.05$). The urinary output was significantly increased to 124% ($p \leq 0.01$).

Conclusion

Leisurely hiking between 2500 and 4200 m at slight exercise levels seems neither an adequate stimulus for an increase in hematocrit, hemoglobin concentration, and blood cell count, nor does it cause a marked hemoconcentration due to increased water losses. Tougher hiking with a greater workload and a somewhat higher altitude exposure does, however, cause such changes. These hematologic alterations become more pronounced when the altitude exposure and the exercise continues (3,8,9). After a resting period of more than 15 h, a large amount of water is released which might have been bound by the muscles at work.

Acknowledgments

This study was supported by the Deutsche Forschungsgemeinschaft (Zi 125/5).

References

1. Cohen, J. and Zimmerman, A.L.: Changes in serum electrolyte levels during marathon running. S. Afr. Med. J. 53:449, 1978.
2. Frayser, R., Rennie, I.D., Gray, G.W., and Houston, C.S.: Hormonal and electrolyte response to exposure to 17,500 feet. J. Appl. Physiol. 38:636, 1975.
3. Hannon, J.P., Chinn, K.S.K., and Shields, J.L.: Effects of acute high-altitude exposure on body fluids. Fed. Proc. 28:1178, 1969.
4. Hannon, J.P., Chinn, K.S.K., and Shields, J.L.: Alterations in serum and extracellular electrolytes during high altitude exposure. J. Appl. Physiol. 31:266, 1971.
5. Hultgren, H.N., Bilisoly, J., Fails, H., Stone, M.E., Pfeifer, J., and Marticorina, E.: Plasma volume changes during acute exposure to high altitude. In: Hypoxia Symposium, Banff, February, 1979. The Arctic Institute of North America, Calgary.
6. Jung, R.C., Dill, D.B., Horton, R., and Horvath, S.M.: Effects of age on plasma aldosteron levels and hemoconcentration at altitude. J. Appl. Physiol. 593, 1971.

7. Krzywicki, H.J., Consolazio, F.C., Johnson, H.J., Nielson, W.C., and Barnhart, R.A.: Water metabolism in humans during acute high altitude exposure. J. Appl. Physiol. 30:806, 1971.
8. Pugh, L.G.C.E., Gill, M.B., Lahivi, S., Milledge, J.S., Ward, M. P., and West, J.B.: Muscular exercise at great altitude. J. Appl. Physiol. 19:431, 1964.
9. Reynafarje, C.: In Weike, W.H. (ed.): The Physiological Effects of High Altitude, Macmillan, New York, 1964, p. 73.
10. Schmidt-Schönbein, H.: Micro-rheology of erythrocytes, blood viscosity and the distribution of flow in the microcirculation. Internat. Rev. Physiol. 9:1, 1976.
11. Surks, M.J., Chinn, K.S.K., and Matoush, L.O.: Alterations in body composition in man after acute exposure to high altitude. J. Appl. Physiol. 21:1741, 1966.
12. Wilkerson, J.E., Raven, P.B., Bolduan, N.W., and Horvath, S.M.: Adaptations in man's adrenal function in response to acute cold stress. J. Appl. Physiol. 36:183, 1974.
13. Williams, F.S.: Hormone response to hypoxia in humans. In: Hypoxia Sympopsium, Banff, 1979. The Arctic Institute of North America, Calgary.
14. Zink, R.A.: Ärztlicher Rat für Bergsteiger, Georg Thieme Verlag, Stuttgart, 1978, p. 24.
15. Zink, R.A., Schaffert, W., Brendel, W., Messmer, K., and Schmiedt, E.: Hemodilution in high altitude mountain climbing: A method to prevent, or treat frostbite, high altitude pulmonary edema and retinal hemorrhage. In: Abstracts of scientific papers. Amer. Soc. Anesthesiologists, Chicago, October 1978, p. 93.
16. Zink, R.A., et al.: Proposals for international standardization. In Brendel, W. and Zink, R.A. (eds.): High Altitude Physiology and Medicine. Chapter 48, *this volume*.

Part II
Disturbances Due to High Altitude and Therapy of High Altitude Complaints

I. Cerebral and Ophthalmologic Changes

29

High Altitude Complaints, Diseases, and Accidents in Himalayan High Altitude Expeditions (1946 – 1978)

H.R. WEINGART, R.A. ZINK, AND W. BRENDEL

Introduction

There has been a sharp increase in the number of mountaineers participating in high altitude expeditions during the last 30 years. Many reports and books have been written on those expeditions but no review made of the risks of mountain climbing and the frequency of high altitude-induced complaints. The following study tries to fill this gap by providing for climbers and medical advisors data on the dangers which may be encountered in high altitude expeditions. We are well aware of the fact that such a review does not meet all the requirements of statistical accuracy for the simple reason that the comparable parameters are continuously changing (e.g., climbing experience, duration of expedition, quality of equipment, etc.), and also because only a tiny fraction of climbers who participate in expeditions can be questioned. Moreover, it is absolutely impossible to estimate and quantify the degree of alpine difficulties of a single expedition. Nevertheless, we can give a rough estimate of the frequency of medical complications on high altitude expeditions. For the participants of this symposium it is of interest to know how often these events, which are discussed with such intensity, are likely to occur.

The present study was conducted (1) to investigate the frequencies of alpine accidents, general diseases, and high altitude complaints; (2) to analyze relationships between the altitude exposure and the number of high altitude complaints; and (3) to analyze the developments and changes of the last 30 years which influence the risks of high altitude mountaineering.

Methods and Procedures

It is estimated that 1600 high altitude expeditions were conducted in the Hindu Kush, the Karakoram, and the Himalayas between 1946 and 1978. Our study includes 402 parties. We defined the parties as expeditions if they lasted at least 14 days, reached at least 5000 m, used a base camp with or without additional advanced camps, and intended to climb a chosen mountain. These characteristics distinguished them from, for instance, trekking tours.

We found detailed reports for 80% of the expeditions, while for 20% there were brief notes. For about 30 expeditions it was possible to verify the reports by comparing different sources to make sure that the severe accidents and sicknesses were correctly reported. The selected expeditions comprised approximately 3200 participants. The number of the porters (Sherpas) involved could not be used in the study, since the exact number was mentioned only in some cases. Not more than 30 expeditions were considered for any one year in order to avoid overrepresentation of certain years. All data were collected from expedition reports, which were checked for the following: (1) the year of the expedition; (2) the proposed destination and the highest point reached; (3) the number of participants and summit climbers; (4) the frequency, kind, grading, and altitude at which alpine accidents took place (if reported); (5) general diseases and high altitude complaints.

Alpine accidents were classified as follows: accidents caused by avalanches, falls, crevasses, rockfalls, and other causes. General diseases were defined as diseases which can occur anywhere and do not depend on altitude.

High altitude complaints were distinguished according to pathophysiological criteria and divided by severity into high altitude discomfort (headache, insomnia, nausea and dizziness) and high altitude complications, which are severe or potentially severe disturbances. High altitude complications were subdivided (29) into retinal hemorrhage (HARH), thromboembolism (HATE), cold injuries (HACI), pulmonary edema (HAPE), and cerebral edema (HACE).

Three altitude intervals were selected for the computation of the frequencies: 2500–4000 m, 4000–6000 m, and above 6000 m.

The high altitude interference factors and their pathophysiological effects will not be discussed in detail in this study (1–4,6,18,20,22,26).

Results

Population

There were 3200 participants. Since the percentage of females was too small (less than 2%), data referring to the female sex were not specially evaluated in the analysis. Figure 29-1 and Table 29-1 sum up all events.

Alpine Accidents

Altogether 264 alpine accidents occurred, and 80 participants were killed. As seen in Fig. 29-2, the different causes of alpine accidents had different frequencies and different mortality rates.

Falls had the highest relative mortality rate (54%, 43 subjects), followed by cre-

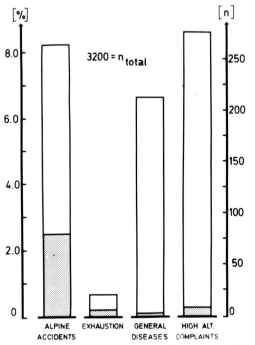

Fig. 29-1. Incidence and mortality rates of alpine accidents, exhaustion, general diseases, and high altitude complaints. *Open bar,* incidence; *stippled bar,* mortality rate.

Table 29-1. Accident Incidence and Mortality Rate on Himalayan Expeditions 1946–1978.

	Incidence, %	Mortality rate, %
Alpine accidents	8.2	2.5
General diseases	6.6	0.1
Exhaustion	0.7	0.2
High altitude complaints	8.6	0.3

n = 3200.

vasse accidents (30%, 6 subjects), avalanches (22%, 23 subjects), rockfalls (14%, 3 subjects), and others (14%, 5 subjects). In 116 reported accidents the mountaineers were uninjured.

We differentiated between the alpine accidents due to subjective and objective causes. Objective causes include natural dangers such as avalanches, crevasses, and rockfalls, whereas the subjective causes are determined by individual factors as, for instance, lack of experience, poor judgement of risks, carelessness, insufficient equipment, and others. We found 41 deaths were due to subjective causes and 39 to objective causes. The total incidence of accidents was 8.2%, the total mortality rate 2.5%, and the relative mortality rate 30.3%.

General Diseases

General diseases were reported in 213 cases. About 25 different diseases were described, which could be separated into four categories: respiratory (n = 45), intestinal (n = 49), infections (n = 40), and other diseases (n = 79). Of this total, 27 cases (12.7%) occurred at 2500–4000 m, 102 (47.9%) at 4000–6000 m, and 45 (21.2%) above 6000 m. In 39 cases (18.2%) the altitude levels were not reported.

The respiratory diseases included 21 cases of pneumonia, three of which were fatal. The intestinal diseases included 16 cases with stomach complications and 33 cases with diarrhea. Unspecified infections were reported by 40 climbers. "Other diseases" is a collective term for diseases of low frequency or unspecified sickness. The total incidence of general diseases was 6.6%, the mortality rate 0.1%.

Exhaustion

The term exhaustion denotes a general breakdown without further specification. Twenty-two mountaineers suffered from exhaustion; 6 died of it. Of these 19 (including 5 fatalities) occurred at 6000 m. The incidence was 0.7%, the mortality rate 0.2%, and the relative mortality rate 27.3%.

High Altitude Complaints

The normal high altitude adaptation is accompanied by no or minor symptoms of high altitude complaints. High altitude discomfort with the symptom complex of headache, insomnia, nausea, and dizziness, known as "mountain sickness," was reported in 128 cases, which is a morbidity of approximately 4%. Approximately 56% (71 subjects) of the high altitude discomfort cases occurred from 4000 to 6000 m, 36% above 6000 m (45 subjects), and 8% (10 subjects) below 4000 m.

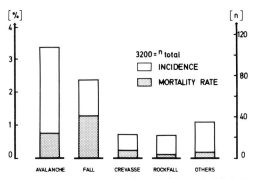

Fig. 29-2. Causes of alpine accidents in high altitude expeditions. Key as in Fig. 29-1.

The incidence of high altitude complications was 148 or 4.6% of the total. The following frequencies were found for the different complications:

	Incidence	Mortality
high altitude retinal hemorrhage (HARH)	6	0
high altitude thromboembolisms (HATE)	7	2
high altitude cold injuries (HACI)	106	0
high altitude pulmonary edema (HAPE)	21	5
high altitude cerebral edema (HACE)	8	2

Cold injuries were reported in 106 cases (Fig. 29-3). Of these, 102 (96%) occurred at altitudes above 6000 m. Distribution according to grading was: 19 cases in grade I, 63 cases in grade II, and 24 cases in grade III.

In addition to pulmonary edema and cerebral edema, there is a third life-endangering high altitude complication, mostly overlooked: the thromboembolic events. They have often been described as cerebral edema.

Discussion

We know about the problems of carrying out a retrospective analysis of expedition reports. The authors of the expedition reports were normally interested just in reporting about mountaineering and not about medical facts. This is also reflected in the results of our study. Furthermore, we realize that often causal relationships cannot be checked retrospectively, for example, the course of events of alpine accidents, and the reasons for a few cases of exhaustion. Nevertheless our data can indicate some possible relations.

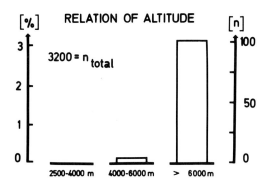
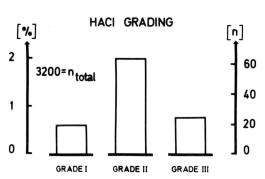

Fig. 29-3. High altitude cold injuries (HACI) in high altitude expeditions, relation of altitude and grading.

Since it was impossible to send questionnaires to the participants, we had to analyze expedition reports carefully. In some cases (about 7.5%) a crosscheck was possible using different sources referring to the same expedition. The use of reports of course precludes a random sample, since only expeditions furnishing adequate reports could be taken into account.

The incidence of general diseases was found to be 6.6%. Several types of diseases were not adequately reported, e.g., helminthic infections, hemorrhoids, coughs, mild infections, and mild intestinal complaints. It seems, however, that the frequency of 6.6% is a reliable value for severe diseases such as pneumonia, heart attacks, severe snowblindness, poliomyelitis, hepatitis, and others. Personal communications with expedition leaders corroborate this statement.

In comparison with other studies, we found a low incidence of high altitude discomfort (13). The main reason might be the different conception of these studies. Finally it has to be kept in mind that sometimes only high altitude discomfort symptoms were reported when the mountaineers actually suffered from severe disturbances.

The incidence of high altitude retinal hemorrhages (HARH) reported in this study seems to be far less than indicated by other authors (8,11,21,24), perhaps because HARH usually is asymptomatic. It was and is still very rare for an expedition to include a doctor who is not only interested in the retinal changes of the climbers but is able to use a funduscope. It is possible that quite a few alpine accidents have been caused by impaired vision due to HARH, since central retinal hemorrhages can diminish or abolish stereoscopic vision. The cases of severe snowblindness should also be seen in this light.

High altitude mountaineers seem to accept cold injuries up to grade II as more or less normal, judging from their reports, but we can be fairly sure that more severe frostbites were properly reported. Since in the more than 3 decades our report covers there was an enormous improvement in mountain gear, the protection against cold was also significantly improved. Within this period we saw a corresponding decrease in the relative frequency of cold injuries.

Period	Number of mountaineers	Number of cold injuries	%
1946–1958	540	39	7.2
1959–1968	800	31	3.8
1969–1978	1860	34	1.8

Pooling our results with those of other studies, we believe that our reported incidences of general diseases, high altitude discomfort, mild frostbite and retinal hemorrhages are somewhat underestimated, since in many cases minor complaints may not have been reported. But we are sure that our data are accurate for the more severe cases of general diseases, high altitude discomfort, and high altitude complications. The incidences of high altitude complications correspond quite well with the results of other studies (13).

Summary

This study is based on reports of 402 high altitude mountaineering expeditions to the Hindu Kush, the Karakoram, and the Himalayas in the years 1946 to 1978. These expeditions comprise approximately 3200 members.

There were 264 alpine accidents, 80 of them fatal. General disease affected 213, of whom three died; 22 members suffered from exhaustion, and six of them died; and 277 mountaineers had high altitude complaints, including nine fatalities.

References

1. Adolph, E.F.: Perspectives of adaptation: Some general properties. In Dill, D.B., Adolph, E.F., Wilber, C.G.: Handbook of Physiology, Vol. IV. American Physiological Society, Washington D.C., 1964, pp. 27–35.
2. Aoki, V.S., and Robinson, S.M.: Body hydration and the incidence and severity of acute mountain sickness. J. Appl. Physiol. 31:363, 1971.
3. Baethmann, A. and Schmiedek, P.: Pathophysiology of cerebral edema: chemical aspects. In Schuermann, K., Brock, M., Reulen, H.J., and Voth, D.: Brain Edema Pathophysiology and Therapy. Berlin, Springer, 1973, pp. 5–18.
4. Brendel, W.: Höhenakklimatisation und Erschöpfung. Ärztliche Praxis 26:2837, 1974.
5. Brendel, W.: Der Mechanismus der Kältegegenregulation. Arch. physikal. Ther. 12:137, 1960.
6. Brendel, W.: Anpassung von Atmung und Kreislauf im Verlaufe einer Himalaya-Expedition, Sonderdruck aus 18. Deutscher Sportärzte-Kongress in Hamburg 1957, Frankfurt, Wilhelm Limpert, pp. 196–204.

7. Ceretelli, P.: Limiting factors to oxygen transport on Mount Everest. J. Appl. Physiol. 40:658, 1976.
8. Clarke, C.R.A., and Duff, J.: Mountain sickness, retinal hemorrhages, and acclimatisation on Mount Everest in 1975. Brit. Med. J. 6034:495, 1976.
9. Dejours, P., Kelogg, R.H., and Pace, N.: Regulation of respiration and heart rate response in exercise during altitude acclimatisation. J. Appl. Physiol. 18:10, 1963.
10. Duane, T.D.: Altitude, diarrhea and retinal hemorrhages. JAMA 240:214, 1978.
11. Frayser, R., Houston, C.S., Bryan, A.C., Rennie, I.C., and Gray, G.: Retinal hemorrhage at high altitude. N. Engl. J. Med. 282:1183, 1970.
12. Frisancho, A.R.: Functional adaptation to high altitude hypoxia. Science 187:313, 1975.
13. Hackett, P.H., Rennie, I.D., and Levine, H.D.: The incidence, importance and prophylaxis of acute mountain sickness, Lancet 7996:1149, 1976.
14. Haft, J.I. and Fani, K.: Intravascular platelet aggregation in the heart induced by stress. Circulation 47:353, 1973.
15. Houston, C.S. and Dickinson, J.: Cerebral form of high altitude illness. Lancet 7938:758, 1975.
16. Hultgren, H.N., Grover, R.F., and Hartley, L.H.: Abnormal circulatory responses to high altitude in subjects with a previous history of high altitude pulmonary edema. Circulation 43:759, 1971
17. Jenny, E.: Erschöpfung und Bergungstod im Hochgebirge. Internist. Prax. 15:109, 1975.
18. Lahiri, S.: Physiological responses and adaptations to high altitude. In Robertshaw, D. (ed.): International Review of Physiology, 15:217, 1977.
19. Lobenhoffer, H.P., Zink, R.A., and Brendel, W.: High Altitude Pulmonary Edema (HAPE): analysis of 166 cases. Chapter 34, *this volume*.
20. Rennie, D.I.: What causes altitude illness? Pathophysiology. In Summary of Abstracts of the Hypoxia Symposium, Banff, 1979, pp. 40–42. The Arctic Institute of North America, Calgary.
21. Rennie, D.I.: Retinal changes in Himalayan climbers. Arch. Ophthalmol. 93:359, 1975.
22. Ryn, Z.: Psychopathology in alpinism. H.J. 32:113, 1972–1973.
23. Staub, N.C.: Pathogenesis of pulmonary edema. Am. Rev. Resp. Dis. 109:358, 1974.
24. Vogt, C., Greite, J.-H., and Hess, J.: Netzhautblutungen bei Teilnehmern an Hochgebirgs-Expeditionen. Klin. Mbl. Augenheilk. 172:770, 1978.
25. Ward, M.: Frostbite. In Clarke, C., Ward, M., and Williams, E. (eds.): Mountain Medicine and Physiology. London, Moore, 1975, pp. 19–27.
26. Weingart, H.R.: Erkrankungen und Unfälle bei Hochgebirgs-expeditionen. Dissertation. Munich, 1981.
27. Wilson, R.: Acute high altitude illness in mountaineers and problems of rescue. Ann. Intern. Med. 78:421, 1973.
28. Zink, R.A.: Ärztlicher Rat für Bergsteiger. Stuttgart, Thieme, 1978.
29. Zink, R.A., Schaffert, W., and Lobenhoffer, H.P.: Proposals for international standardizations in the research and documentation of high altitude medicine. Chapter 48, *this volume*.

30

Cerebral Edema: The Influence of Hypoxia and Impaired Microcirculation

A. Baethmann

It is well known and has again been stressed at this symposium that the brain, together with the respiratory system, is particularly vulnerable at high altitudes. Although this opinion is generally held, the specific underlying mechanisms have not yet been elucidated. The question about brain edema formation at high altitudes is whether a deficiency in pulmonary oxygen uptake leads to cerebral hypoxia, or whether microcirculatory disturbances of the brain or other factors are instrumental. A clear answer to this problem does not seem available at the moment. However, since hypoxia and disorders of microcirculation occur at altitudes, the influence of both factors on brain edema formation is discussed in this paper.

Table 30-1 shows local and general conditions causing cerebral edema but without indicating what particular type of edema is induced. The table lists three conditions which have to do with hypoxia and microcirculatory disturbances: infarction, severe respiratory failure, and global cerebral ischemia.

The Influence of Hypoxia

Brain edema secondary to focal or global ischemia has become an extensively explored and rather well defined pathological entity (2,11), while it is uncertain whether anoxia without additional ischemia can produce brain edema. Bakay and Bendixen have maintained that hypoxia causes cerebral edema only when it occurs in conjunction with hypercapnia (3). However, more recently it was shown by Gilboe et al. and by Selzer that perfusion of isolated dog brain with anoxic blood, or of a single cerebral hemisphere of monkey in situ with venous blood (P_{O_2} 17–20 mm Hg) leads to development of marked brain edema (Table 30-2) (6, 16). The study of Gilboe and coworkers (6) led to the conclusion that brain perfusion with anoxic blood induces edema even more effectively than an equivalent period of cerebral circulatory arrest. Such an observation may be surprising because one assumes that perfusion of an organ, even with anoxic blood, brings it closer to physiological conditions than total

Table 30-1. Local and General Causes of Cerebral Edema. Infarction, Global Ischemia, and Severe Respiratory Failure Serve as Examples of Production of Brain Edema by Hypoxia or Circulatory Disorders.

Local	General
Trauma	Hypertensive encephalopathy
	Severe resp. failure
Tumor	Global ischemia
	Metabolic intoxication in
Infarction	1) uremia
	2) hepatic failure
Abscess	(e. g. Reye syndrome)
	3) diabetic coma
	Intoxications,
	e. g., hexachlorophene
	Triethyl-Tin
	Pseudotumor cerebri
	(endocrinologic)?
	Osmotic disequilibrium

ischemia, which not only deprives the organ of oxygen but also leads to an accumulation of metabolic waste. Nevertheless, tissue anoxia with flow may indeed be worse than ischemia. First, the presence of an ischemic component is difficult to rule out in "pure" anoxia, and second, maintenance of flow to tissue incapable of energy metabolism enhances the net influx of Na^+ and consequently water into cellular structures by providing extra fluid which is not available in complete ischemia.

This point is perfectly demonstrated by the data of Gilboe on anoxic brain perfusion (6). Observations of Norris and Pappius (13) on ventilation of experimental animals with different O_2-CO_2 mixtures may also relate to the problems encountered at high altitudes. Different types of respiratory hypoxia, or hypercapnia, were studied:

1) Asphyxiation for 5–7 min by turning off the respirator until the mean arterial pressure fell to 40 mmHg.
2) Hypoxic ventilation with 4–8% O_2 for $2^1/_2$–3 h.
3) Normoxic hypercapnia (10–20% CO_2) for 3 h.
4) Hypoxia plus hypercapnia by ventilation with 5–8% O_2 together with 10–20% CO_2 for $2^1/_2$ h.

Table 30-2. The Effect of Different Modes of Cerebral Hypoxia, or Anoxia Without Ischemia, on Formation of Brain Edema.

	Brain edema
1. Perfusion of CNS with anoxic blood (Gilboe, et al., 1976)	Yes
2. Perfusion of CNS with hypoxic blood (Selzer, et al., 1973)	Yes
3. Hypoxia + hypercapnia	
a) Respiratory arrest 5-7 min ($\rightarrow$ MAP 40 mmHg)	Transiently (yes)
b) Respiration with 5-8% O_2 2-3 h	No
c) Respiration with 10-20% CO_2 + 20% O_2	No
d) Respiration with 5-8% O_2 + 10-20% CO_2	No

Source: Norris, Pappius, 1970
In studies of Selzer et al. and Giboe et al., the presence of ischemia was not ruled out, however (6,13,16).

It is seen in Table 30-2 that except for a transient increase in brain water after asphyxiation, brain edema did not develop after the various procedures employed. It may be concluded from these studies that the hypoxic insult to the brain was not sufficiently large enough.

Kogure and coworkers attempted to define the hypoxic threshold of arterial oxygen tension, when brain tissue becomes edematous (12). The data are given in Fig. 30-1, where brain water, cyclic AMP, and CSF K^+ and Na^+ of rats were plotted against the arterial P_{O_2}. In these studies, the arterial P_{O_2} was lowered by ventilation with hypoxic gas mixtures for different intervals. Brain water started to rise if arterial P_{O_2} levels of less than 30 mmHg were reached. This level appeared to be critical, because at the same point the brain tissue concentrations of cyclic AMP started to rise

Cerebral Edema: The Influence of Hypoxia and Impaired Microcirculation

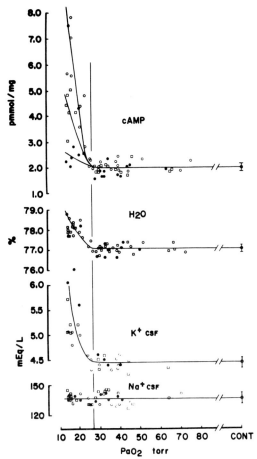

Fig. 30-1. Formation of hypoxic cerebral edema in rats in relation to the arterial P_{O_2} studied by Kogure et al. (12). The arterial P_{O_2} was lowered by ventilation with hypoxic gas mixtures. The authors feel that the rise in cAMP occurring in parallel with the increase in cerebral water has a role in hypoxic edema formation.

together with the CSF K^+ concentration, indicating that at arterial P_{O_2} levels lower than 30 mmHg energy metabolism of brain tissue is in jeopardy (12).

A threshold value of 25–30 mmHg Pa_{O_2} is in perfect agreement with corresponding studies of Siesjö et al. (18) who showed energy-rich phosphates, such as phosphocreatine, begin to decline at virtually the same arterial P_{O_2}, while glycolysis was turned on at somewhat higher P_{O_2} levels (Fig. 30-2). It is obvious from these data that a determinant of tissue swelling in hypoxic conditions is the failure of energy metabolism, rendering the tissue deficient in fuel for the active transport pumps.

The Concept of the Brain's Selective Vulnerability

Figure 30-3 provides evidence for the concept of selective vulnerability of the hypoxic brain. It shows the results of decompression experiments with monkeys conducted by Brierly. He exposed animals to a steady ambient pressure of 160 mmHg, corresponding to an altitude of 37,500 ft, or 11,430 m. This is somewhat above the altitudes accomplished by Messner and Habeler at Mount Everest. The animals were maintained at this pressure for 10–16 min until respiratory arrest became imminent. Then recompression ensued, but no special resuscitative measures were taken.

Figure 30-3 shows the distribution and severity of ischemic necrosis found in 5 of 18 animals so exposed. The neocortical areas involved were the boundaries of the territories of the anterior and middle or middle and posterior cerebral arteries. The animals in this subgroup survived the insult 3–20 weeks. In two other animals there were ischemic necrosis and demyelinization of the globus pallidus, putamen, and caudate nucleus (Fig. 30-4), while cortical tissue seemed to be less affected. A remaining group of 11 animals sustained no neuropathological damage. The pattern of selective vulnerability evolving in these animals is similar to that seen after a global ischemic insult, for example, after cardiac arrest. In this case too the boundaries of two adjacent vascular territories become selectively damaged. The necrotic foci which develop from such insults serve as port of entry for vasogenic brain edema on account of the breakdown of the blood–brain barrier.

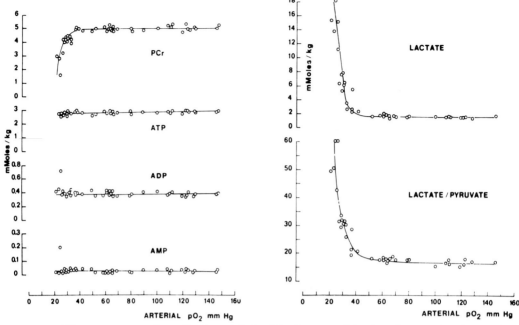

Fig. 30-2. Tissue concentrations (fresh weight basis) of energy-rich phosphates (phosphocreatine, ATP, ADP, and AMP) and glycolysis metabolites of supratentorial rat brains at various arterial P_{O_2} levels. The arterial P_{O_2} was lowered by respective changes in FI_{O_2} levels. Cerebral lactate, and lac/pyr ratios started to rise at a higher Pa_{O_2} than phosphocreatine started to fall (from B.K. Siesjö et al., 1971).

The Influence of Disturbed Microcirculation

The relationship of ischemia and edema development has been demonstrated in recent years in experimental models, such as limited global ischemia of the brain, explored by Hossmann, Betz and others, or focal ischemia, studied for example by Shibata, O'Brien, and Symon et al. (4,7,9,14,17, 19). Figure 30-5 demonstrates the changes in extracellular K^+ and Na^+ concentrations during and after 60 min of global cerebral ischemia. Extracellular K^+ concentrations increased immediately after the start of ischemia and leveled off at 60 mM while ischemia was maintained. Such an extracellular concentration is probably close to the passive concentration equilibrium between the intra- and extracellular space. Sodium ions were lost from the extracellular space. Note that the decrease in Na^+ from the extracellular fluid space exceeds the increase in K^+ resulting in a net loss of electrolytes from the extracellular compartment and leading to its shrinking.

Shrinking of the cerebral extracellular space following 8 min of ischemia is also demonstrated in Fig. 30-6 (6), which is an electron micrograph of mouse cerebellum obtained by freeze-substitution fixation (1). The material was obtained in studies on cytotoxic brain edema induced by a metabolic inhibitor. The preparation of the ischemic brain is shown at the right-hand side of Fig. 30-6. It shows swollen nerve fibers and glial processes with virtually no extracellular space between them, when compared to the preparation of the left-hand side (Fig. 30-5) (6) where swelling of only glial elements is seen, with nerve fibers and synaptic endings appearing normal. Here the extracellular space is well preserved and recognizable.

Employing the model of 1-h global cerebral ischemia by clamping the arterial cerebral blood supply, Hossmann studied brain water, intracranial pressure, tissue os-

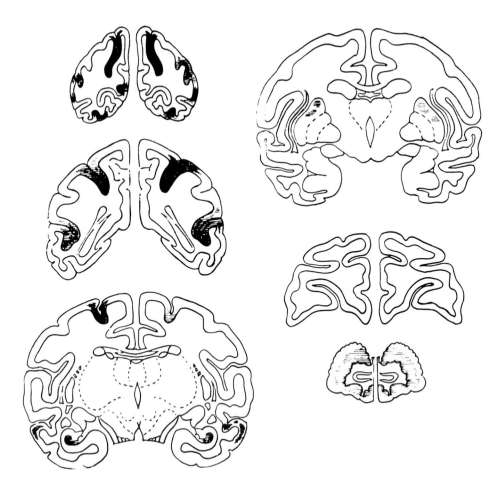

Fig. 30-3. Coronal section of a monkey brain showing schematically a pattern of selective tissue necrosis preferentially involving cortical areas in boundary zones between arterial supply territories. Hypoxia was induced by exposure of experimental animals to decompression down to 160 mmHg ambient pressure for 10–16 min. Out of 18 animals, 5 developed neuropathological damage as shown in the figure (from J.B. Brierly, 1971).

molality, and also the extracellular Na^+ concentrations of the brain (9). The data shown in Fig. 30-7 make it obvious that brain water and intracranial pressure remain unchanged during the ischemic phase, but start to rise with resumption of tissue perfusion. The observations demonstrate that brain edema cannot develop during total ischemia because no extra fluid is available. During ischemia, the available extracellular fluid is shifted into the intracellular compartment (Fig. 30-6) (6), or by the loss of Na^+ ions from the interstitial space observed by Hossmann (7–9). In Fig. 30-7, one sees that edema formation together with the intracranial pressure rise secondary to ischemia is reversible provided cerebral blood flow and oxygen supply meet the drastically increased requirements in the postischemic phase. If, however, postischemic hyperemia fails and cerebral blood flow remains subnormal, edema formation progresses. Then, disturbances of blood-brain barrier function may lead after a certain delay, to secondary vasogenic edema. In such conditions, the consequences of massive brain edema, for instance, a rise in intracranial pressure

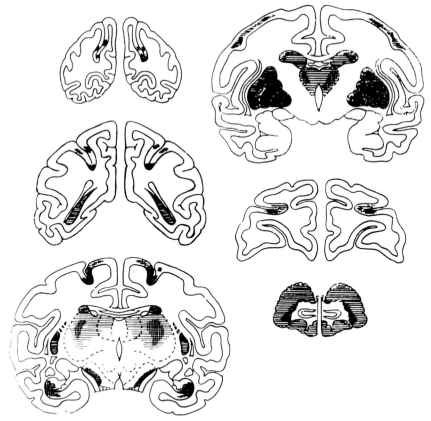

Fig. 30-4. Hypoxic damage of monkey brain secondary to decompression to 160 mmHg particularly affecting the basal ganglia. Such a pattern of cerebral tissue necrosis was found in 2 out of 18 animals studied (from J.B. Brierly, 1971).

causing brain tissue herniation, and finally arrest of the cerebral circulation, may ensue.

Disruption of the blood–brain barrier together with vasogenic edema influx into brain parenchyma is a prominent feature after focal ischemic insults. Ito et al. have studied the incidence of gross blood–brain barrier permeability changes (such as penetration of intravenously administered

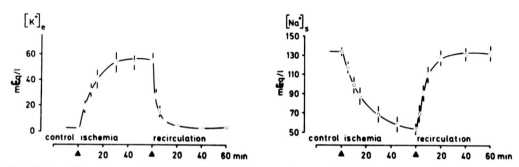

Fig. 30-5. Alterations of Na^+ and K^+ activity (caution sensitive electrodes) in extracellular or subarachnoid fluid of cerebral cortex of cats during and after 60 min of cerebral ischemia (from K.A. Hossmann et al., 1977).

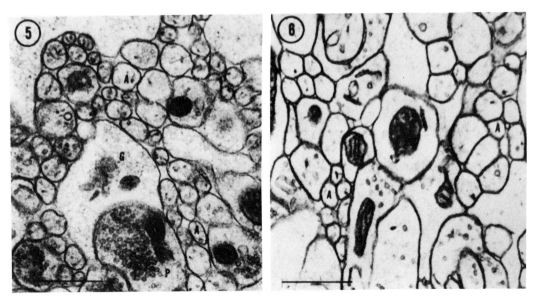

Fig. 30-6. Electron micrographs of the molecular layer of mouse cerebellum obtained by freeze-substitution fixation. On the left, a preparation of an experimental animal is shown which was administered for 2 days with a metabolic inhibitor, 6-aminonicotinamide, leading to swelling of glial elements (G), while the nerve fibers (A) or postsynaptic endings (P) appear normal. Extracellular space is clearly visible. On the right-hand side, the additional effect of 8 min of complete ischemia (decapitation) is shown leading to swelling of all the cellular structures with disappearance of the extracellular space. The calibration bars indicate 0, 5 μm (from A. Baethmann and A. Van Harreveld, 1973).

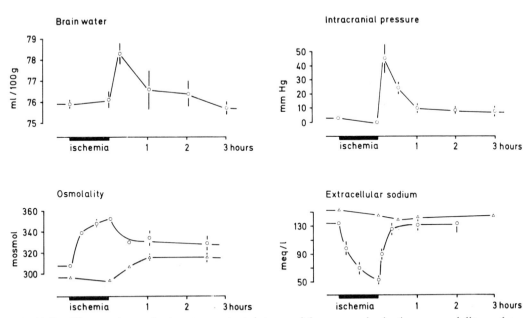

Fig. 30-7. Changes in cerebral water content, intracranial pressure, brain tissue osmolality and extracellular Na^+ in subarachnoid fluid of cats subjected to 60 min of cerebral circulatory arrest. Corresponding alterations in blood are shown as triangles. The increase in tissue osmolality during ischemia provides for the osmotic forces drawing fluid into the brain if cerebral circulation becomes established again. The increase in brain water, and consequently in intracranial pressure, were only transient since postischemic circulation allowed for recovery (from K.A. Hossmann, 1977).

Evans blue) in mongolian gerbils in relation to duration of flow obstruction and postischemic recovery. The experiments, which involved unilateral clamping of the common carotid artery, made evident that short periods of ischemia (15-30 min) did not cause any barrier leakage, or did so only in a fraction of experimental animals with neurological symptoms, while occlusion for more than 30 min always resulted in uptake of Evans blue. Moreover, the penetration of Evans blue into postischemic brain tissue occurred earlier the longer the period of vessel occlusion.

Figure 30-8 shows a coronal brain section of a gerbil obtained after 1 h of occlusion and 5 h of release, demonstrating the area of vasogenic edema by dark coloration. The area of dark coloration is representative of the ischemic infarct zone produced by clamping the carotid artery.

It should finally be mentioned that a particular form of experimental ischemia induced by cerebral microembolization which is used to study fat embolism is also a cause of vasogenic edema (15,20). However, microembolization initially does not produce ischemia or anoxia of the brain because the vessels not occluded by emboli dilatate compensatorily. Cerebral ischemia develops later as a secondary process after severe vasogenic edema has formed (15).

Summary and Conclusions

Cerebral anoxia and ischemia induce brain edema provided the insult is sufficiently large. Anoxic or ischemic brain edema develops in proportion to the preceding primary insult. The quality of cerebral blood flow during the postischemic period is an important determinant of secondary processes and the final outcome. Respiratory

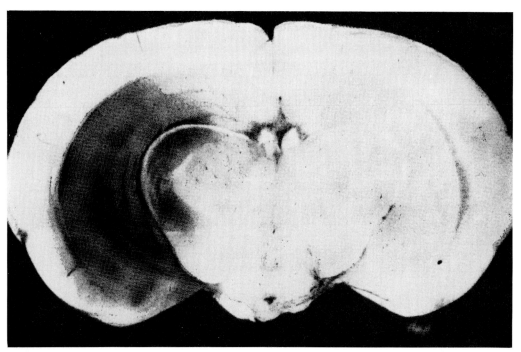

Fig. 30-8. Coronal section of mongolian gerbil brain 5 h after release of unilateral occlusion of a common carotid artery for 60 min. The dark area demonstrates the site of ischemic tissue injury causing blood-brain barrier dysfunction as shown by the uptake of intravenously administered Evans blue. The barrier marker was given prior to release of vessel occlusion (from Ito et al., 1976).

hypoxia alone may not always cause cerebral edema. Additional complications, as of the cardiovascular system, may be necessary to induce formation of edema.

Global anoxia or ischemia of the brain produces a pattern of selectively damaged brain tissue areas which are related to the vascular supply territories. From infarcted and necrotic tissue areas, the edema fluid enters the brain parenchyma through a malfunctioning blood–brain barrier. Amount and duration of a barrier dysfunction depend on the severity of the anoxic or ischemic insult.

Although the development of cerebral edema occurring at high altitudes is not yet fully understood, it is conceivable that components of disorders of cerebral microcirculation and oxygen supply play an important role.

Acknowledgments

The technical and secretarial cooperation of Hedwig Kuschke, Maxi Stempfle, and Mechthild Stein is gratefully acknowledged.

References

1. Baethmann, A. and Van Harreveld, A.: Water and electrolyte distribution in grey matter rendered edematous with a metabolic inhibitor. J. Neuropath. Exp. Neurol. 32:408, 1973.
2. Baethmann, A.: Pathophysiological and pathochemical aspects of cerebral edema. Neurosurg. Review 1:85, 1978.
3. Bakay, L. and Bendixen, H.H.: Central nervous system vulnerability in hypoxic states: Isotope uptake studies. In: McMenemey, W.H. and Schadé, J.F. (eds.): Selective Vulnerability of the Central Nervous System in Hypoxaemia. Oxford, Blackwell, 1963, pp. 63–78.
4. Betz, E.: Vascular reactivity and ion homeostasis in heart and brain. In: Zülch, K.J., Kaufmann, W., Hossmann, K.A., and Hossmann, V. (eds.): Brain and Heart Infarct. Berlin-Heidelberg-New York, Springer, 1977, pp. 10–16.
5. Brierley, J.B.: The neuropathological sequelae of profound hypoxia. In: Brierley, J.B. and Meldrum, B.S. (eds.): Brain Hypoxia. Clinics in Developmental Medicine, No. 39/40. London, Heinemann, 1971, pp. 147–151.
6. Gilboe. P.D., L.R. Drewes, and D. Kintner: Edema formation in the isolated canine brain: anoxia vs. ischemia In: Pappius, H.M. and W. Feindel (eds.): Dynamics of Brain Edema. Berlin-Heidelberg-New York, Springer, 1976. pp. 228–235.
7. Hossmann, K.A.: Development and resolution of ischemic brain swelling. In: Pappius, H.M. and W. Feindel (eds.): Dynamics of Brain Edema. Berlin-Heidelberg-New York, Springer, 1976, pp. 219–227.
8. Hossmann, K.A., Sakaki, S., and V. Zimmermann: Cation activities in reversible ischemia of the cat brain. Stroke 8:77, 1977.
9. Hossman, K.A.: In: Zülch, K.J., Kaufmann, W., Hossmann, K.A., and Hossmann V. (eds.): Brain and Heart Infarct. Berlin-Heidelberg-New York, Springer, 1977, pp. 107–122.
10. Ito, U., Go, K.G., Walker, J.P., Spatz, M., and Klatzo J.: Experimental cerebral ischemia in mongolian gerbils III. Behaviour of the blood-brain-barrier. Acta Neuropathol. (Berl.) 34:1, 1976.
11. Katzman, R., Clasen, R., Klatzo, I., Meyer, J.S., Pappius, H.M., and Waltz, A.: Brain edema in stroke. XV. Rep. of Joint Comm. for Stroke Resources. Stroke 8:509, 1977.
12. Kogure, K., Scheinberg, P., Kishikawa, H., and Busto, R.: The role of monoamines and cyclic AMP in ischemic brain edema. In Pappius, H.M. and Feindel, W. (eds.): Dynamics of Brain Edema. Berlin-Heidelberg-New York, Springer, 1976, pp. 203–214.
13. Norris, J.W. and Pappius, A.M.: Cerebral water and electrolytes. Effects of asphyxia, hypoxia and hypercapnia. Arch. Neurol. 23:248, 1970.
14. O'Brien, M.O., Jordan, M.M., and Waltz, A.G.: Ischemic cerebral edema and the blood-brain barrier. Arch. Neurol. 30:461, 1974.
15. Schuir, F.J., Vise, W.M., and Hossmann, K.A.: Cerebral microembolization. II. Mor-

phological studies. Arch. Neurol. 35:264, 1978.
16. Selzer, M.E., Myers, R.E., and Holstein, S.B.: Unilateral asphyxial brain damage produced by venous perfusion of one carotid artery. Neurology 23:150, 1973.
17. Shibata, S., Hodge, C.P., and Pappius, H.M.: Effect of experimental ischemia on cerebral water and electrolytes. J. Neurosurg. 41:146, 1974.
18. Siesjö, B.K., Nilsson, L., Rokeach, M., and Zwetnow, N.: Energy metabolism of the brain at reduced cerebral perfusion pressures and in arterial hypoxemia. In Brierley, J.B. and Meldrum, B.S. (eds.): Brain Hypoxia, Clinics in Developmental Medicine, No. 39/40. London, Heinemann, 1971, pp. 79–93.
19. Symon, L., Branston, N.M., and Chikovani, O.: Ischemic brain edema following middle cerebral artery occlusion in baboons: relationship between regional cerebral water content and blood flow at 1 to 2 hours. Stoke 10:184, 1979.
20. Zülch, K.J. and Tzonos, I.: Transudation phenomena at the deep veins after blockage of arterioles and capillaries by microemboli. Bibl. Anat. 7:279, 1965.

31
Physiologic Adaptation to Altitude and Hyperexis

J. Durand

In 1953 and in 1960, D.W. Richards published two papers in which he developed the concept of "hyperexis" (1,6,7): a state wherein a homeostatic response that ordinarily protects the body begins to have damaging effects. Later, this concept found several applications in pathology (1,4), and functional and adaptive, physiological responses to altitude may give other examples of hyperexis.

Usually adaptation has a connotation of teleologically wise transformations of various functions in order to maintain homeostasis with a minimal energy expenditure; however, as far as adaptation of human beings to altitude is concerned, some of the responses can scarcely be interpreted in this way.

Hypoxic vasoconstriction of the pulmonary arterioles is beneficial in patients suffering from obstructive lung disease, as local pulmonary vasoconstriction diverts blood flow to better-aerated alveoli. However, it becomes useless or even detrimental when hypoxia is generalized to the whole lung, as in hypobaria.

Reduced cutaneous vascular compliance directs blood from the skin toward the central circulation. In some subjects, es-

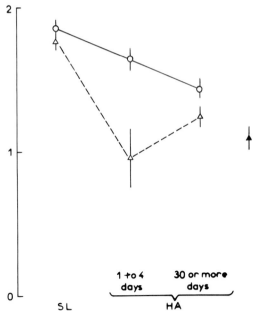

Fig. 31-1. Comparison of the reduction of cutaneous vascular distensibility during acclimatization at 3750 m in sea level residents (○) with altitude residents returning from sea level (△). Values obtained in resident highlanders (▲) are given for reference. Mean values and standard error. Measured are ΔV changes in volume of the right hand when transmural pressure is increased, ΔP, by 25 cm of water (2.45 kPa). Measurements were made at a local skin temperature (Θ) of 33°C.

pecially in altitude residents returning from a sojourn at sea level, the response to altitude exposure is greater than normal (Fig. 31-1) and may coincide with the development of high altitude pulmonary edema (3).

Increase in hemoglobin concentration is usually considered an appropriate response to low Pa_{O_2} as it maintains Ca_{O_2} at such a value that cardiac output may remain within the normal range. But this adaptive response defeats its purpose by increasing blood viscosity, which in turn enhances pulmonary hypertension, probably creates disturbances at the microcirculatory level, and may become a disease per se.

The increase in hematocrit and hemoglobin content which develops in some species (human, rat) exposed to altitude or chronic hypoxia ($Pa_{O_2} < 70$ torr or 93.3 kPa) does not exist in high altitude rodents (guinea pig, chinchilla, viscacha) or camelids (vicuña, llama, alpaca). In these animals nature chose another adaptive strategy, the shift to the right of the hemoglobin dissociation curve (8).

Furthermore, increase in hematocrit seems to grow with age (Fig. 31-2) (5,9), although this is not unanimously accepted (2), so that in human beings adaptation to altitude does not seem ever to reach a steady state.

Acknowledgments

This work was realized at the Instituto Boliviano de Biología de Altura, La Paz, Bolivia, and was supported in part by the Ministère des Affaires Etrangères (Coopération Technique), and by the C.N.R.S. (RCP 538).

References

1. Bradley, S.E. and Coelho, J.B.: Ellepsis and akairial disproportion in the hypothyroid rat. Bull. Physiopath. Resp., 15(5):707, 1979.
2. Chiodi, H.: Aging and high-altitude polycythemia. J. Appl. Physiol. (Resp. Environ. Exerc. Physiol.) 45:1019, 1978.
3. Durand, J. and Martineaud, J.P.: Resistance and capacitance vessels of the skin in permanent and temporary residents at high altitude. In Porter, R. and Knight, J.: High Altitude

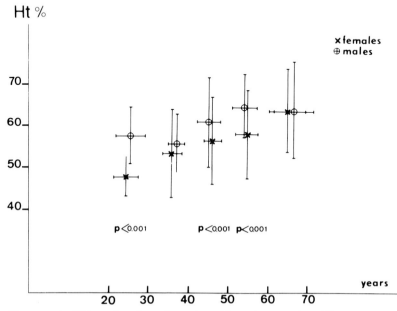

Fig. 31-2. Hematocrit (Ht) as function of age in subjects born and residing permanently between 3800 and 4200 m.

Physiology: Cardiac and Respiratory Aspects. Ciba Foundation Symposium. Edinburgh and London, Churchill Livingstone, 1971, pp. 159-170.
4. Fritts, H.W., Jr.: Hypoxic pulmonary vasoconstriction and the concept of hyperexis. Bull. Physiopath. Resp., 15(5):813, 1979.
5. Monge, C.C. and Whittembury, J.: Chronic mountain sickness. Johns Hopkins Med. J. 139:87, 1976.
6. Richards, D.W., Jr.: Homeostasis: its dislocations and perturbations. Perspect. Biol. Med. 3:238, 1960.
7. Richards, D.W. Jr.: Homeostasis and hyperexis. Scientific Monthly 77:289, 1953.
8. Turek, Z., Ringnalda, B.E.M., Moran, O., and Kreuzer, F.: Oxygen transport in guinea pigs native to high altitude. Pflüg. Arch. (in press).
9. Whittembury, J. and Monge, C.: High altitude hematocrit and age. Nature 238:278, 1972.

32
Eye Problems at High Altitudes

F. Brandt and O.K. Malla

One of the dangers to men at high altitudes was first described by Xenophon (15) in his *Anabasis* as "snowblindness." In 1802 Alexander von Humboldt reported eye changes which he noticed in himself and his companions during their attempt to climb the summit of the Chimborazo volcano in Ecuador (13). Now we know that they were symptoms of altitude sickness. Many reports have been published on this theme, especially in the last few years. In the following these eye changes will be discussed and our own observations added.

Extraocular Changes

A hyposphagma is a well-defined subconjunctival hemorrhage and may appear quite spontaneously, but it is also found in blood and blood vessel diseases and in altitude sickness. The hyposphagma itself is always harmless. When it appears at high altitudes, it may occur together with vascular changes and can be the first noticeable sign of altitude sickness. A hyposphagma requires no treatment as it is resorbed spontaneously within 10–14 days without leaving a trace.

The percentage of ultraviolet (UV) rays in sunlight increases from 1–2% at sea level to 5–6% at high altitude (3). When the cornea is exposed to UV light, usually the UV is absorbed by the cornea, sometimes injuring the epithelium. UV light reflected by snow, fog, and low-lying clouds at high altitudes is strong enough to cause damage to the epithelium.

Six to eight hours after exposure to UV light, blepharospasm, sharp pain, and epiphora occur. An examination of the eye shows strong conjunctival injection and fluorescein will stain the corneal epithelium (keratoconjunctivitis photoelectrica sive nivalis). The suggested therapy is a 24-h application of ointment and a patch, after which the cornea is generally healed without complication. Glasses with side protectors are recommended as prophylaxis because even light entering from the side can be strong enough to damage the corneal epithelium.

Intraocular Changes

Air pressure has no effect on intraocular pressure (6). It has been reported that intraocular pressure decreases with exercise (5) and it has been suggested that this phenomenon may explain the climber's propensity for intraocular bleeding (8).

Nausea, vomiting, and vision problems appear not only in altitude sickness, but they also are typical signs of an acute glaucomatous attack. The wide staring of the pupil, the cloudy cornea, the palpable high intraocular pressure, and the mostly unilateral appearance of these symptoms should be considered as reasons for a differential diagnosis of possible acute glaucoma. Therapy for the last is intravenous injection of Diamox (acetazolamide) and pilocarpine eye drops. As prophylaxis an ophthalmological examination before the expedition is recommended.

Sunlight, especially short exposure, has not been shown to influence the development of cataracts (2).

After long exposure to strong sunlight there may be retinal problems appearing to be adaptation problems as well as erythropsia (1). Generally snowblindness occurs at the same time. The problem disappears spontaneously and does not require therapy, but wearing protective glasses is suggested as prevention.

Changes in the Fundus

Disc Disturbances

Disc disturbances, in the sense of hyperemia or edema (8), have been described by many authors as symptoms of altitude sickness (9,10,14). Fluorescein angiographic investigations also have been done (12,14). A differential diagnosis must be obtained in the case of disc edema to determine if it is an expression of brain edema or if it appears as a direct result of tissue hypoxia with increased capillary permeability. On the other hand, an unsuspicious disc disturbance does not eliminate the possibility of brain edema with an increase of intracerebral pressure.

Because it is possible to confuse an anomaly of the disc with edema of the disc, we photographed the disc of expedition members before they set out, as a record. The pictures were added to the passport.

Vessels

Tortuositas vasorum, combined with a dilatation of the retinal vessels so often described in altitude sickness (7,9,14), is clinically similar to polycythemia, although vessel occlusion from altitude sickness has not so far been observed. Because blurred vision from central vein occlusion is usually not dramatic, and the branch vein occlusion is often not noticed by the patient, the occurrence may remain undetected until ophthalmologic examination. A central artery or artery branch occlusion is acute and results in immediate worsening or loss of vision. A patient with these symptoms will promptly seek medical attention.. In the past few years, no vascular occlusions in mountain climbers have been observed at the Nepal Eye Hospital or the Bir-Hospital Katmandu.

Retinal Hemorrhages

Retinal hemorrhage may occur noticed or unnoticed, may be either central in the macular area or peripheral. A central hemorrhage in the macular area results in a decrease of monocular vision and in distorted binocular vision, that is, loss of stereopsis. This condition can have fatal results for a mountain climber who loses the ability to judge distances and depths, for example how far he must jump to the next ledge. A climber with sudden loss of stereopsis should be handled as one with a serious injury until he is on level ground where a misjudged step will not result in a deadly fall.

Superficial retinal hemorrhages will disappear without leaving a retinal scar, and may not produce a scotoma (12). Should the hemorrhage occur in the nuclear layer itself, there is danger that the synapse of the ganglion cells may be destroyed, resulting in irreversible damage. An absolute scotoma must be expected as a result. Absolute scotomas have been observed after retinal hemorrhages (8).

A hemorrhage outside the macula is rarely noticed by the patient and is usually discovered only by ophthalmological examination. This bleeding usually is found in the middle periphery but may also occur in the outside periphery. Out of five European mountain climbers which we saw after returning from Everest base camp, two showed retinal hemorrhages throughout the periphery of both eyes; the other three had hemorrhages only in the middle periphery. No specific therapy is necessary since the bleeding is spontaneously resorbed in 4–12 weeks. Prophylaxis with Diamox has been discussed (4,11) as well as hemodilution (12).

Vitreous Humor

Very little has been reported on vitreous hemorrhages in altitude sickness (10).

Most fluorescein angiographic investigations to date (12,14) have been performed in order to discover a leakage. But in all these cases in which retinal hemorrhage had already occurred no leakage was found. The hemorrhage had covered or closed the leakage. Fluorescein angiographic investigations have not yet been done in patients who show only tortuositas vasorum and enlargement of the vessels but not retinal hemorrhage. This investigation should be made because a leakage from the retinal vessels could possibly be found before a hemorrhage appears.

Histological examination of eyes after altitude sickness has not been performed. We should like to suggest that all autopsies following death resulting from altitude sickness include histological investigation of the eyes, especially to obtain samples of reversible and irreversible damage to vision.

The Eye Clinic at the University of Munich stands ready to test fixed pathological material submitted by investigators lacking suitable facilities.

References

1. Axenfeld, P.M.: Lehrbuch und Atlas der Augenheilkunde. 11. Ed. Gustav-Fischer-Verlag, Stuttgart, 1973.
2. Chatterjec, A.: Cataracts in Punjab. In: The Human Lens in Relation to Cataract. Ciba Foundation Symp. 19, Excerpta Med. 265, 1973.
3. Duke-Elder, S.: System of Ophthalmology, Bd. XIV, Henry Kimpton, London, 1972, p. 918.
4. Frayser, R., Ch., Huston, A.Ch., Bryan, J.D., Rennie, D., and Gray, G.: Retinal hemorrhage at high altitude. New Engl. J. Med. 282:1183, 1970.
5. Lempert, P., Cooper, K.H., Culver, J.F., et al.: The effect of exercise on intraocular pressure. Am. J. Ophthal. 75:1673, 1967.
6. Leydhecker, W.: Glaukom. Ein Handbuch. II. Auflage Heidelberg, Springer-Verlag, S.170, 1973.
7. Rennie, D. and Morrisey, J.: Retinal changes in Himalayan climbers. Arch. Ophthal. 93:395, 1975.
8. Schumacher, G. and Petayan, J.: High altitude stress and retinal hemorrhage. Relation to vascular headache mechanisms. Arch. Environ. Health 30:217, 1975.
9. Shults, W. and Swan, K.: High altitude retinopathy in mountain climbers. Arch. Ophthal. 93:404, 1975.
10. Singh, I., et al: Acute mountain sickness. N. Engl. J. Med. 280:175, 1969.
11. Utz, G., Schlief, P., Barth, P., Linkert, P., Wollenweber, J.: MMW, 112:1122, 1970.
12. Vogt, C., Greite, H.-J., and Hess. J.: Netzhautblutungen bei Teilnehmern an Hochgebirgsexpeditionen. Klin. Mbl. Augenheilk. 172:770, 1978.
13. von Humboldt, A.: Südamerikanische Reise über den Versuch der Gipfel des Chimborazo zu ersteigen, 1802. Berlin, Safari Verlag, 1979.
14. Wiedmann, M.: High altitude retinal hemorrhage. Arch. Ophthal. 93:401, 1975.
15. Xenophen, Anabasis IV. 5. Zit. nach Duke-Elder.

33

Cotton-Wool Spots: A New Addition to High Altitude Retinopathy

PETER H. HACKETT AND DRUMMOND RENNIE

High altitude retinopathy, a term introduced by Dr. Charles Houston; presently includes retinal hemorrhages, retinal vascular engorgement, retinal edema, and perhaps papilledema (3–5). These conditions occur in lowlanders ascending to altitudes greater than 3000 m, and their pathophysiology is essentially unknown. The purpose of this report is to add to this category a new finding, that of cotton-wool spots.

Case Report

M.I., a 29-year-old Japanese male, had come to Nepal on a climbing expedition to the northwest ridge of Nuptse. He had been at altitudes between 5200 and 6600 m for 17 days when he noted over the period of a few hours changes in his vision, described as "blind spots" and "foggy vision," the latter affecting central vision. He descended with assistance to the base camp at 5200 m, and the next day reported to the Himalayan Rescue Association Clinic at 4243 m. He stated that his vision had improved while descending. He denied experiencing any symptoms of acute mountain sickness, and he had no history of visual problems. Examination at this time revealed normal visual acuity, and visual fields were grossly normal by the confrontation method. Funduscopy showed numerous typical retinal hemorrhages and lesions characteristic of cotton-wool spots (Fig. 33-1). We had examined his fundi 3 weeks earlier, as part of a study on retinal hemorrhages, and found them to be entirely normal. The rest of the ophthalmologic findings were normal, as was a complete physical examination. Specifically, he was normotensive, and urine dipstick was negative for protein and sugar. After a few days rest, he returned to the mountain and went to over 7000 m without further visual symptoms. On reexamination 20 days later, after successful completion of the expedition, the retinal hemorrhages were completely resolved, and the cotton-wool spots nearly resolved. Gross visual field testing and visual acuity were again normal, as was blood pressure. One year later, in a followup letter, M.I. was reported to be in good health.

Discussion

This is the first report of cotton-wool spots in a high altitude climber, although a short

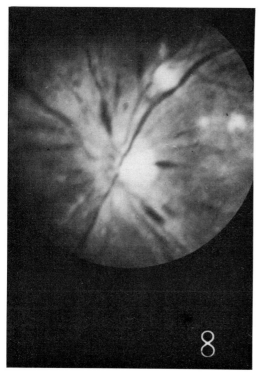

Fig. 33-1. Retinal photograph of subject M.I. Note the numerous hemorrhages and the cotton-wool spots above and to the right of the optic disc.

time later another case was observed by Dr. Murray McFadden and one of the authors in a woman at 5360 m on Mt. Logan. Cotton-wool spots are most commonly observed in hypertensive and diabetic retinopathies, although they have also been reported in collagen diseases, anemic conditions, toxemia of pregnancy, carbon monoxide poisoning, infection, and malignancy (2).

Pathologically, cotton-wool spots have been shown to be intracellular swelling of axons in the nerve fiber layer of the retina. It is generally agreed that they are caused by ischemia, that is, functional obliteration of the capillaries, and not solely from hypoxia (2). When induced experimentally in pigs by means of glass bead microemboli, leakage of fluorescein, which is characteristic of the lesions in humans, occurred only when the endothelium was damaged by the emboli (1). The period of ischemia necessary to produce the lesion was 24 h. In humans, they typically resolve in 4 to 12 weeks.

Although we could not detect visual field defects by the confrontation method, it is likely that careful testing, as with an Amsler grid, would have detected small defects, as it did in the second patient. In this case, we could not document that the subject's visual changes were actually caused by the cotton-wool spots, the retinal hemorrhages, or either. Neither do we know whether the hemorrhages and cotton-wool spots are related. To date, we have observed hemorrhages in more than 80 individuals, but cotton-wool spots in only two. We hope that this report will stimulate others to look for such lesions, so that their true incidence and significance can be determined.

References

1. Ashton, N., Dollery, C.T., Hending, P., Hill, D.W., Paterson, J.W., Romalho, P.S., and Shakid M.: Focal retinal ischaemia. Br. J. Ophthalmol. 50(6):2834, 1966.
2. Duke-Elder, S. and Dobree J.: Diseases of the Retina. In: Duke-Elder, S. (ed.): System of Ophthalmology, St. Louis, The CV Mosby Co., vol. 10, pp. 17–24.
3. Frayser, R., Houston, C.S., Bryan, A.C., et al.: Retinal hemorrhage at high altitude. N. Engl. J. Med. 282:1183, 1970.
4. Hackett, P.H. and Rennie D.: Rales, retinal hemorrhages, peripheral edema, and acute mountain sickness. Am. J. Med. 67:214, 1979.
5. Rennie, D. and Morrissey, J.: Retinal changes in Himalayan climbers. Arch. Ophthalmol. 93:395, 1975.

II. High Altitude Pulmonary Edema

34
High Altitude Pulmonary Edema: Analysis of 166 Cases

H.P. Lobenhoffer, R.A. Zink, and W. Brendel

Introduction

Ascent to altitude can cause a number of complaints: acute mountain sickness (AMS), high altitude retinal hemorrhages (HARH), high altitude pulmonary edema (HAPE), and high altitude cerebral edema (HACE). Whereas AMS is harmless and the importance of HARH is not completely understood, HAPE and HACE are life-endangering complications of altitude exposure. The interest in these complaints has increased since more and more people are reaching high altitudes in trekking and mountaineering tourism. Altitude complications occur and are nowadays recognized in mountain regions all over the world. HAPE seems to be the most frequent serious form and to date more than 900 cases have been reported in medical literature (1–43). Despite this great number of cases and intensive research work, there is disagreement about the environmental and individual factors causing this sickness as well as about its prevention and therapy.

This study has been made to obtain more data on patients and the course of their illnesses. Information from 166 detailed HAPE reports (142 from the literature, 24 from our investigations) were coded for computer processing and analyzed. A survey of characteristic findings follows.

Material

Our Investigations

We studied 24 German nationals who had experienced episodes of HAPE in the years 1967–1977. They all had had clear symptoms of HAPE (cough, rales, dyspnea), and most of them were treated in a hospital, where the diagnosis was confirmed. The patients answered a detailed questionnaire about their sickness and from some of them we got data from a medical checkup done after their recovery.

Literature

In the medical literature available to us, we found 142 case reports which were detailed enough to be included in this study (1–28). Cases were included if they were reported individually and with specifications on personal data, anamnesis, and clinical course. Global, summarizing case reports could therefore not be taken into account (29-32).

Methods

A standarized coding scheme for punch-card was used by which the information of the case history was divided into 57 variables. They were grouped into the following sections:
1) *personal data:* sex, age, nationality, height, weight
2) *anamnesis:* acclimatization, previous altitude complications, past and current diseases, ascent rate, altitude when taken ill
3) *sickness:* date, duration, course, symptoms, lab parameters, therapy
4) *comments*

Two arbitrary definitions were made to standardize the evaluations:
1) High altitude was regarded as above 2500 m
2) Acclimatization to high altitude was regarded as complete after more than 1 year of residence above 2500 m

Registration and analysis were done on the CDC Cyber 700 computer at the Leibniz-Rechenzentrum München by means of the SPSS[1] program package. Some of the reports were more detailed than others. Cases in which a specific variable was missing were excluded from statistics and diagrams for this variable.

Results

Geographical Distribution of Cases

Most of the episodes we studied occurred in the Andes, South America (52%), 29% were observed in the Himalayas, and 12%

[1]Statistical Package for the Social Sciences

in the United States and Canada; 6% took place in Europe and 1% in Africa.

Nationality

Natives of South America represent the largest proportion of population studied (31%), followed by North Americans (29%) and Europeans (28%). Other nationalities represented 12% of the total.

Sex

The majority of patients in the collected cases were men (84%). The proportion of women is 16% in the total, but is higher in several subgroups. It is 23% in all cases of natives from the Andes and 26% in the series of cases in high altitude residents. Women represent 20% of the patients from trekking tourism.

Age

HAPE was most common in persons from 20 to 29 years (26% of the total), but the incidence for the age range 0-9 and 10-19 years was also high (20, 12%). The mean

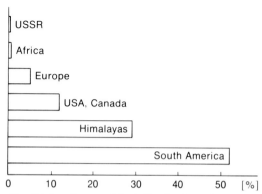

Fig. 34-1. Geographical distribution of the incidents studied. Percentage of cases by geographical regions. Eight cases with variable not given excluded.

age for all cases is 26 ± 14 years (Table 34-1).

Acclimatization

Sixty-seven percent of the analyzed cases were lowland residents visiting high altitudes ("entry" HAPE); 33% were persons living permanently at high altitude. Of the latter group, the majority (74%) got HAPE after reascent from a sojourn at sea level ("reentry" HAPE). The others moved to a place higher than where they lived and then fell ill.

Previous Altitude Complications

Thirty percent of the cases providing anamnesis had previously suffered serious altitude complications, most commonly HAPE. The frequency of repeated altitude complications is extremely high in children under 9 (48%) and decreases with age (40–49 years: 4%).

Reasons for Altitude Exposure

Of the patients who lived at low altitude and ascended, 49% of the total went for business or private reasons (such as work, army duty, visit to relatives), 22% for trekking, 22% for mountaineering and skiing, and 7% for expedition climbing.

There may be important differences between the cases among those expedition climbing, trekking, and mountaineering, and the others, made up mainly of natives of mountain regions. Genetic factors may influence frequency and prognosis of HAPE, as may different altitude experience, behavior at altitude, and therapeutic conventions in the two groups. This question has not been studied before, and therefore some data are reported separately for trekking tourists and mountaineers.

Altitude

The greatest frequency of HAPE is observed between 3000 and 4500 m (Fig. 34-2). Most of the cases reported from South America, Europe, and the United States occurred in this range. Only in the Himalayas did a marked number occur above 4500 m (Table 34-2). Below 3000 and above 5500 m, HAPE cases were rare (6; i.e., 7%).

Rate of Ascent

Sixty-five percent of all patients took less than 2 days to reach the altitude at which sickness set in. These patients used mechanical means of transport and ascended very little or no further on foot. Most of these cases occurred in the Andes, where roads and trains extend very high.

Table 34-1. Characteristics of HAPE Patients Studied.

	No. cases	%
Nationality		
South America	48	31
United States, Canada	46	30
Europe	43	28
Himalayas	14	9
Others	5	3
Sex		
Male	139	84
Female	27	16
Age, years[a]		
0–9	29	20
10–19	17	12
20–29	38	26
30–39	30	21
40–49	24	16
50–59	7	5
Acclimatization		
Lowland residents	104	67
Highland residents	51	33
Previous altitude complications		
No prev. HAPE or HACE	92	70
Previous HAPE or HACE	39	30

[a] Cases not reporting not tabulated

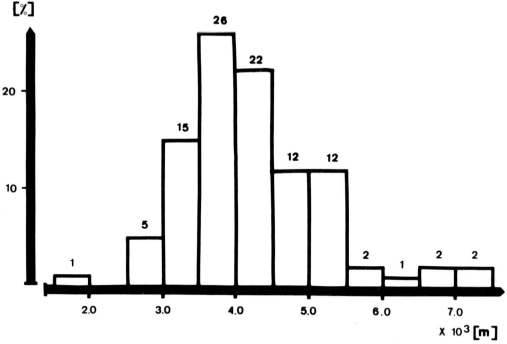

Fig. 34-2. Altitude at onset of sickness ($n = 166$). Frequency distribution of altitudes (classified per 500 m). Numbers in the figure are percentages.

Time Lag Between Ascent and Onset of Illness

Of the collected cases 149 occurred after ascent or reascent from low altitude. The majority of these people developed HAPE in the first week above 2500 m, and only 17% of all cases occurred after a longer period (Fig. 34-3).

Altitude Residents: Lowland Sojourn

Most cases of HAPE occurring in altitude residents came after a visit to sea level and reascent. Of these people, 25% spent less than one week at low altitude, 35% one to two weeks, 15% two to three weeks, and 25% more than three weeks.

Symptoms

Dyspnea, cough, and malaise were the most frequently mentioned symptoms. Cerebral complaints that took the form of ataxia, euphoria, apathy, confusion and unconsciousness, were also often reported, with a total frequency of 58%. All other

Table 34-2. Altitude at Which Symptoms of HAPE Were Manifested. Mean Altitude and Percentage of Cases Between 3000 and 4500 m, by Geographical Region.

Region	No. cases	Mean, m	Cases between 3000 and 4500 m, %
South America	83	4040	84
Himalayas	46	4950	47
United States and Canada	19	3500	89
Europe	9	3100	75
Others	7	—	—
Total	166	4216	69

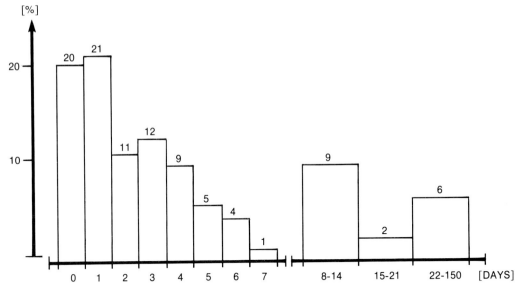

Fig. 34-3. Stay at altitude until onset of sickness. Days above 2500 m until HAPE-symptoms manifested ($n = 149$). 0 days means HAPE set in day of arrival above 2500 m. 149 cases with ascent from low altitude. Numbers in figure are percentages.

symptoms noted were much less frequent (Table 34-3).

The severe signs of cerebral edema (ataxia, confusion, coma) were observed mostly in trekkers and nountaineers (78%).

Vital signs and laboratory parameters are summarized in Table 34-4. Characteristic findings were very high respiratory rate, fast pulse, normal brachial artery blood pressure, and a slightly elevated body temperature.

Therapy

The most frequently used treatments were oxygen (61%) and descent (49%). Antibiotics were given to 25% of the patients, diuretics to 21%. Digitalis, steroids, and morphine were given to a smaller percentage (Table 34-5).

A cross-tabulation for treatment by oxygen and descent against deaths from HAPE

Table 34-3. Frequency of Symptoms in the Collected Cases of HAPE Patients.[a]

Symptoms	No. cases	%
Malaise	79	58
High altitude discomfort		
Headache	48	35
Nausea	31	23
Vomiting	23	17
Insomnia	16	12
Local edema	10	7
Cardiopulmonary symptoms		
Dyspnea	97	71
Cough	91	67
Cyanosis	53	39
Chest pain	29	21
Orthopnea	18	13
Heart complaints	14	10
Cerebral complaints		
Anxiety	6	4
Ataxia	8	6
Euphoria	3	2
Apathy	15	11
Confusion	13	10
Coma	33	24

[a] 30 cases not reporting excluded

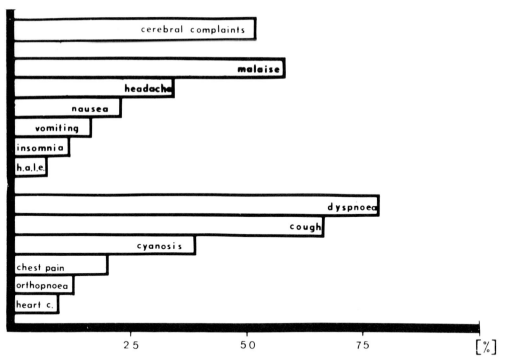

Fig. 34-4. Frequency of symptoms among incidents studied ($n = 136$). Cerebral complaints = ataxia, euphoria, apathy, confused and unconscious states; HALE = High Altitude Local Edema; heart complaints = palpitations, heart pain. Thirty cases with variable not given excluded.

shows that the mortality rate was very low among those treated by both measures and very high in the group which received neither (Table 34-6).

Table 34-4. Vital Signs and Laboratory Parameters in HAPE Patients.

	No. cases	Mean ± SD
Respiratory rate (min^{-1})	32	36 ± 13
Heart rate (min^{-1})	71	119 ± 24
RR systolic mmHg	44	119 ± 26
RR diastolic (mmHg)	45	73 ± 20
Temperature (°C)	49	37.7 ± 0.7
Hemoglobin (g/100 ml)	34	16.2 ± 2.4
Hematocrit (%)	38	48 ± 6
Erythrocytes (millions/mm^3)	10	5.7 ± 0.9
Leucocytes (thousands/mm^3)	50	13 ± 660

Mortality Rate

The mortality rate of HAPE, calculated for all the examined cases, is 11.4%. More males than females died of HAPE, more elderly people than younger ones, and more natives than tourists (Table 34-7).

Table 34-5. Therapeutic Measures and Drugs Used in the HAPE Patients Studied.

	No. cases	%
Oxygen	101	61
Descent	81	49
Medicines[a]		
Antibiotics	40	25
Diuretics	33	21
Digitalis	20	13
Steroids	18	11
Morphine	10	6

[a] 6 cases not reporting excluded.

Table 34-6. *Deaths from HAPE Among Patients Treated by Oxygen and Descent.*

Oxygen	Descent	Frequency		Mortality	
		No. cases	%	No. cases	%
Yes	Yes	39	24	1	3
Yes	No	62	37	3	5
No	Yes	42	25	5	12
No	No	23	14	10	44
Total		166	100	19	11

Discussion

Geographical Distribution

One reason for the high number of cases in the Andes is local geographical particularities. In this region there are permanent settlements at altitudes of more than 3500 m, which can be reached from the coast by car in only a few hours. There is much travel: highland residents come down and reascend, lowland residents ascend to these communities. These ascents are an important risk factor and most of the patients we studied fell sick after a rapid trip to high altitude.

As HAPE is relatively frequent among highland natives of South America, a high rate would also be expected among Sherpas, the Nepalese altitude population. But 70% of our cases from the Himalayas occurred in trekkers and mountaineers, the rest mainly in Indian mountain troops. We know of no reported case of HAPE in Sherpas and of only two instances of HACE (4). This may be caused partly by the lack of roads in Nepal, leading to a much lower ascent rate among Sherpas than among Andean natives. In addition, one assumes there are genetic differences between the two altitude populations, with a better altitude tolerance in Sherpas (45).

The different incidence for expedition climbers and trekking tourists, and for the Sherpas who accompany their party, may be explained in the same way: many tourists start their trip after a rapid ascent by plane, whereas their native porters and guides have to ascend by foot or are permanent residents of high altitude. Furthermore, as mentioned, Sherpas are supposed to have a better altitude tolerance and many of them have already experienced very high altitudes on several expeditions.

One may wonder why the number of cases in Europe is so small though there are mountains of considerable altitude in the Alps (Mont Blanc: 4807 m). One reason is that the existence of HAPE has been known in Europe for only a few years; before that episodes were mistaken for pneumonia. Also, fewer people ascend to altitude than, for example, in South America and, most important, they stay there a short time, often only a few daytime hours.

The typical onset of sickness is at night, when the patient is at rest, and many persons may escape an episode of HAPE by the short exposure time, descending before night. In addition, if hikers or mountaineers develop altitude problems, descent is relatively easy. In contrast to other mountain

Table 34-7. *Deaths from HAPE Among the Patients Studied by Sex, Age and Type of Cases.*

	No. cases	Mortality	
		No. deaths	%
Sex			
Male	139	17	12
Female	27	2	7
Age, years[a]			
0–19	46	3	7
20–39	68	8	12
40–59	31	5	16
Type			
Cases in mountaineers, trekkers, expedition climbers	77	6	8
Cases in natives and others	89	13	15

[a] 21 cases not reporting excluded.

regions, the horizontal distances in the Alps are relatively small compared to the vertical ones. Therefore descent below 2000 m seldom takes longer than a day, even under difficult conditions.

Acclimatization

The two main groups of HAPE are represented in our study, entry HAPE and reentry HAPE. The question has been raised whether highlanders who reascend are more likely to develop HAPE than are lowlanders who ascend (17,24,31,52). An epidemiologic study comparing the incidence of both types of cases has not been done before.

In the high altitude community of Leadville, Colorado, United States (3100 m, 8300 inhabitants), the number of HAPE episodes treated in the hospital were recorded over 7 years. Although there are about 100,000 tourists in the region every year, 30 of the 32 cases occurred among Leadville residents, all but one after a stay at low altitude and reascent. Only two patients were tourists from sea level (47). These findings and studies in India of larger numbers (24,29) suggest a higher risk for highlanders reascending after a stay at low altitude.

Sex

The small proportion of women in our cases is striking and a similar observation has been made by many authors (12,17,31,40,41,51). Men may be more susceptible, or perhaps many more men travel in the mountains. In order to interpret the data, the proportion of persons risking HAPE to persons developing HAPE has to be known for both sexes. Information on the risk group is available for two subgroups of our cases: the trekking tourists and the residents of high altitude communities.

Hackett and Rennie (9) tabulated all trekkers on the way to Everest Base Camp and found 29% females in 1975 and 30% in 1977. In the premonsoon season of 1978, 26% of all German trekkers in this region were women (our study). This was confirmed by German trekking organizations, which gave the proportion 25–30%. This proportion of female participants is roughly the proportion of female patients in the cases we collected from trekking, 20%. Hackett and Rennie also found both sexes at equal risk of HAPE. These results in trekkers there seem to demonstrate there is no difference in susceptibility between females and males.

High altitude communities where residents live with their families have also been studied in this respect. In a Peruvian study (La Oroya, 3750 m), the incidence of HAPE was equal for both sexes in a 2-year investigation (22).

Scoggin (47) in the cited Leadville study (3100 m, Colorado, United States) found among the HAPE patients 56% women and 44% men. In contrast to the studies of Hultgren and Scoggin, with about as many men as women, in the cases we collected from high altitude residents, the female proportion is only 26%, with no significant differences for young people or adults. We cannot explain this great discrepancy between the studies.

Age

Most of the patients younger than 20 years fell sick in the Andes. Young people often have to change altitude, for instance to visit school or relatives. The very high proportion of young people in the study from regions where they are exposed to altitude has raised the question of an age-dependent higher risk. This has been discussed in the literature, but a definitive answer can be obtained only by relating ascents to altitude and events of HAPE to age distribution in a given population. Hultgren studied a Peruvian high altitude community for 2 years (22). The incidence of severe HAPE was 8.1% for people 2-20 years and 0.6% for older people. This study gives data on the travel habits of the natives: 47% of the people who made descents and reascents were 2-20 years and 53% were more than 20 years old. The mean travel frequency was slightly higher for the younger group. We think these figures on travel habits can be applied to our cases in Andean altitude residents, as most of them were in the same region of Peru. Of these cases, 70% were younger than 20 years. It is obvious that this age group, which accounts for about one-half the ascents but about 70% of the HAPE episodes, is somehow at a greater risk. Other data in the literature support this opinion. In the 7-year Leadville study, 29 of 32 patients treated for HAPE in the local hospital were under 21 yr of age (47). Other large studies also suggest a high risk for young people (17,31,32,51). There are two possible explanations: the behavior of young persons coming to altitude or a specific biological susceptibility. Hultgren (17) has suggested that young people tend to run around and be active after ascent, whereas older ones may be more cautious. Mean pulmonary artery pressure (PAP) is raised by ascent to altitude and further increased by exercise (27,55). Excessive PAP has been observed in HAPE patients (7,21,25,43), so physical activity after ascent is thought to be a precipitating factor.

Altitude

Our study indicates HAPE is a typical complication of an altitude of from 3000 to 5500 m. The frequency of HAPE cases increases markedly above 2750 m. Other authors give similar figures: Hultgren 2750 m, Alzamora-Castro 2800 m (1,17). The value given by Indian investigators (49) is higher (3300 m), as cases generally occur at higher altitude in the Himalayas.

We conclude from our data that HAPE can be expected even at moderate altitude, one easily reached by a great number of people. Furthermore, since about 25 million people live permanently above 3000 m (41), the "risk group" for HAPE is quite large and widespread.

Several cases have been reported even below 2750 m. For some of these patients, organic problems disposing them to HAPE could be demonstrated: for example, absence of one pulmonary artery (16) and blunted respiratory response (23).

The reports of HAPE above 5500 m come from expedition climbers. They reached base camp and got sick later, while working on the mountain. It is interesting that these people ascended so high without problems and were only then struck by HAPE, as the majority of the cases occur far lower. The explanation could be the generally low ascent rate of expeditions on the way to base camp (transport problems), thus eliminating one precipitating factor, rapid ascent. Later, climbing at very high altitude and the difficulty of maintaining sufficient fluid intake often causes dehydration and HAPE may be produced by hemoconcentration and impaired microcirculation (45,46).

Time Lag Between Ascent and Onset of Illness

The risk of developing HAPE is greatest in the first days after reaching high altitude. When patients used transport facilities to 3500 m or more, onset of sickness within hours was very common. For example, 65% of the episodes reported from the Andes happened during the day of ascent and the following one. In contrast, only 17% of all HAPE in trekkers and mountaineers occurred in this period. These people ascended to above 2500-3000 m by foot, and the time spent walking or climbing at altitude before onset of sickness was longer than 3 days in most cases. Even in the cases with rapid onset of sickness there was a symptomless interval between exposure to altitude and onset. It was at least several hours in all cases.

Symptoms

Early HAPE symptoms are not very well defined: some degree of malaise, dry cough, dyspnea at rest, and more than normal elevation of respiratory rate and pulse are often also seen in newcomers to altitude. But these symptoms may progress within hours to massive malaise and dyspnea, orthopnea, cyanosis, wet cough with expectoration of pink frothy sputum, rales or bubbling in the chest, combined with brain symptoms such as ataxia, somnolence, or coma.

Table 34-3 shows that most of the patients we studied developed the severe symptoms and were not diagnosed in time. The high number of cases with cerebral complaints indicates how frequently brain edema complicates HAPE. Treatment of HACE is difficult and permanent brain damage has been observed (14). Patients with ataxia, somnolence, or coma had a far higher mortality rate than those without such symptoms (26% vs. 9%).

On the other hand, ataxia as an early sign of brain edema can be indentified before pulmonary symptoms are manifest (8). The Romberg or other coordination tests (walking on a line) can also be done even in the mountains and may enable the doctor to detect beginning HAPE/CE.

Therapy

Table 34-6 suggests a strong influence of oxygen supply and descent on the prognosis of HAPE. Both measures increase the blood and tissue oxygenation of the patient. For this table, we counted all descent, regardless of the stage of illness or amount of descent. This may explain the smaller effect of descent on mortality. It is well known and confirmed by our study that descent of 300–500 m can cure mild, beginning HAPE, but that the effect is less dramatic if the patient already has severe symptoms.

Almost all the cases in natives of South America and India were treated without descent only by oxygen and hospital bed rest. It has been shown in Indians that mild to moderate HAPE may even be treated by bed rest alone (32). However, it is doubtful that these experiences of natives may be applied to trekkers and climbers too. The safest therapy still seems to be rapid descent. If the symptoms are mild, this will be sufficient and recovery will be fast. In severe cases, the patient should receive oxygen (4–6 liters/min) on the way down. If descent is not possible, oxygen supply for at least 8 h may also cure the symptoms of HAPE, although the danger of HACE then will still be present if the patient stays at the same altitude (8). The role of drugs in the treatment of HAPE is obscure. There is growing evidence that diuretics, frequently used for treatment and even recommended for the prevention of HAPE, may be harmful. They cause a further reduction of the plasma volume, aggravating the shocklike hypotension (19) and impairing the microcirculation by increasing hemoconcentration (54,56). For morphine, another commonly used drug, there is also anecdotal information that it causes symptoms to worsen (3). There have been no controlled studies of drug therapy to prove or disprove its value.

Prevention

Some preventive measures arise from the results of our study:
1) *Slow ascent.* An acclimatization day should be planned for every 1000 m of ascent above 3000 m (8,57) and no ascent made by plane or car to above 3000 m, if avoidable. Now that tourists in the Mt. Everest region are

urged by the Himalayan Rescue Organization to ascend slowly, the number of deaths from altitude complications has gradually decreased from five in 1974 to none in 1977, this despite the increased number of trekkers (10).
2) *Rest after ascent.* Strenuous physical activity immediately after ascent should be avoided.
3) *Importance of early symptoms.* Cough, breathlessness, and malaise may be first symptoms of HAPE and signal the need for close observation of the person. If there is ataxia, or symptoms get worse, the patient should descend at once. Doctors on trekking tours often hurt the patient at this juncture (as our own experiences and those of Hackett and Rennie show) by trying drug therapy instead of descent. Important time is lost, the symptoms worsen, the patient has to be transported by litter, and his life is endangered.
4) *Screening.* People with a history of HAPE may be more susceptible to it. We do not know whether there is a transient or permanent tendency to altitude complications. Such people should ascend very carefully, in small steps. A simple physiological test has been published which identifies HAPE-susceptible individuals (53). We know of no study of this method.

Incidence

Data on the overall incidence of HAPE cannot be obtained from our approach. Figures given by other authors vary widely (Table 34-8). If one may generalize these data, the incidence for severe HAPE is at least 0.1–0.5% in lowlanders coming to high altitudes as well as in highlanders reascending. However, the personal and individual factors discussed above may increase dramatically the risk for a particular individual.

Prognosis

The prognosis is determined mainly by the available therapeutic facilities. Our cases

Table 34-8. Incidence of HAPE.

Risk group	Incidence, %	Authors (with literature references)
Climbers and trekkers		
7500 persons at Mt. Kenya	0.44	Houston 1976 (9)
648 persons at Mt. Rainier	0.5	Houston 1972 (9)
Approx. 55,000 persons at Mt. Kenya	0.1	Synder (41)
522 persons in the Mt. Everest region	2.3	Hackett and Rennie (42)
Indian soldiers		
Thousands of soldiers	2.3 to 15.5	Singh (30)
Thousands of soldiers	0.57	Menon (18)
Altitude residents		
97 inhabitants of La Oroya, Peru	0.6 (adults) 8.1 (youths)	Hultgren and Marticorena (43)

were collected from widely varied environmental conditions all over the world, and the mortality of 11% is only an average for severe HAPE. This figure shows that HAPE is a serious life-threatening complication of altitude exposure and deserves medical and scientific attention.

References

1. Alzamora-Castro, V., Garrido-Lecca, G., and Battilana, G.: Pulmonary edema of high altitude. Am. J. Cardiol. 7:769, 1961.
2. Arias-Stella, J. and Kruger, H.: Pathology of high altitude pulmonary edema. Arch. Pathol. 76:147, 1963.
3. Bates, T.: Pulmonary edema of mountains. Br. Med. J. 3(5830):829, 1972.
4. Clarke, C. and Duff, J.: Mountain sickness, retinal haemorrhages and acclimatisation on Mount Everest in 1975. Br. Med. J. 2(6034):495, 1976.
5. Dickinson, J.: Acute altitude sickness. J. Nep. Med. Assoc. 2(3):65, 1973.
6. Foster, J.C.: Letter to editor. Br. Med. J. 3(5820):232, 1972.
7. Fred, H., Schmidt, A., Bates, T., and Hecht, H.: Acute pulmonary edema of altitude. Clinical and physiological observations. Circulation 25:929, 1962.
8. Hackett, P.H.: Mountain sickness. Prevention, Recognition and Treatment. Albany, N.Y., Mountain Travel Inc., 1978.
9. Hackett, P.H. and Rennie, D.: The incidence, importance and prophylaxis of acute mountain sickness. Lancet 2(7996):1149, 1976.
10. Hackett, P.H. and Rennie, D.: Avoiding mountain sickness (letter). Lancet 2(8096):938, 1978.
11. Hasenhüttl, G. et al.: Die Höhenanpassung Ergebnisse der steirischen Karakorum-Expedition. Wiener Med. Wochenschr. 3:102, 1977.
12. Horrobin, D.F. and Cholmondeley, H.G.: High altitude pulmonary edema: pathophysiology and recommendations for prevention and treatment. East Afr. Med. J. 49:327, 1972.
13. Houston, C.S.: Acute pulmonary edema of high altitude. N. Engl. J. Med. 263:478, 1960.
14. Houston, C.S.: High altitude illness. Disease with protean manifestations. JAMA 236(19):2193, 1976.
15. Houston, C.S. and Dickinson, J.: Cerebral form of high altitude illness. Lancet 2(7938):758, 1975.
16. Houston, C.S.: Lessons from high altitude pulmonary edema. Chest 74(4):359, 1978.
17. Hultgren, H.N.: High altitude pulmonary edema. Adv. Cardiol., Basel/München/Paris/New York, Karger, 1970, Vol. 5, pp. 24–31.
18. Hultgren, H.N.: Furosemide for high altitude pulmonary edema. JAMA 234(6):590, 1975.
19. Hultgren, H.N., Spickard, W.B., Hellriegel, K., and Houston, C.S.: High altitude pulmonary edema. Medicine. 40:289, 1961.
20. Hultgren, H.N., Spickard, W., and Lopez, C.: Further studies of high altitude pulmonary edema. Br. Heart J. 24:95, 1962.
21. Hultgren, H.N., Lopez, C.E., Lundberg, E., and Miller, H.: Physiologic studies of pulmonary edema at high altitude. Circulation 29:393, 1964.
22. Hultgren, H.N. and Marticorena, E.A.: High altitude pulmonary edema. Epidemiological observations in Peru. Chest 74(4):372, 1978.
23. Kafer, E.R. and Leigh, J.: Recurrent respiratory failure associated with the absence of ventilatory response to hypercapnia and hypoxemia. Am. Rev. Resp. Dis. 106:100, 1972.
24. Khanna, P.K.: Pulmonary edema of high altitude. General aspects. Indian J. Chest Dis. 9:65, 1967.
25. Kleiner, J.P. and Nelson, W.P.: High altitude pulmonary edema. A rare disease? JAMA 234(5):491, 1975.
26. Kohli, P. and Stucki, P.: Das Hoehenlungenödem. Schweiz. Med. Wochenschr. 98:845, 1975.
27. Kronenberg, R. et al.: Pulmonary artery pressure and alveolar gas exchange in man during acclimatisation to 12470 feet. J. Clin. Invest. 50:827, 1971.
28. Lakshminarayan, S. and Pierson, D.J.: Recurrent high altitude pulmonary edema with blunted chemosensitivity. Am. Rev. Resp. Dis. 111(6):869, 1975.
29. Madan, L.: Clinical aspects of high altitude pulmonary oedema. Indian J. Chest Dis. 9:82, 1967.

30. Marticorena, E. et al.: Pulmonary edema by ascending to high altitude. Dis. Chest 45(3):273, 1964.
31. Marticorena, E. and Hultgren, H.N.: Evaluation of therapeutic methods in high altitude pulmonary edema. Am. J. Cardiol. 43(2):307, 1979.
32. Marpurgo, G., Arese, P., Bosia, A., et al.: Sherpas living permanently at high altitude: a new pattern of adaptation. Proc. Natl. Acad. Sci. USA 73:745, 1976.
33. Menon, N.D.: High altitude pulmonary edema. A clinical study. N. Engl. J. Med. 273:66, 1965.
34. Nanzer, A.: Koma und Höhenlungenödem. Aerztl. Praxis 27(99):3959, 1975.
35. Nayak, N.D., Roy, S., and Narayanan, D.C.P.: Pathological features of altitude sickness. Am. J. Pathol. 45(3):381, 1964.
36. Peñaloza, D. and Sime, F.: Circulatory dynamics during high altitude pulmonary edema. Am J. Cardiol. 23:369, 1969.
37. Pines, A.: Oedema of mountains (letter). Br. Med. J. 4(5938):233, 1974.
38. Radford, P.: High altitude oedema presenting as coma. Br. Med. J. 3(5874):294, 1973.
39. Ravenhill, T.: Some experiences of mountain sickness in the Andes. J. Trop. Med. Hyg. 20:313, 1913.
40. Remirez, G.F., Reinaldo, C.P., and Saenz, G.S.: Edema pulmonar de las alturas en la infancia. Bol. Med. Hosp. Infant 34(3):543, 1977.
41. Rennie, D.: Give me air-but not much. N. Engl. J. Med. 297(23):1285, 1977.
42. Rios-Dalenz, A.: Acute mountain sickness with cerebral edema (not published).
43. Roy, S.B. et al.: Haemodynamic studies in high altitude pulmonary edema. Br. Heart J. 31:52, 1969.
44. Roy, S.B. et al.: Transthoracical electrical impedance in cases of high altitude pulmonary hypoxia. Br. Med. J. 3(5934):771, 1974.
45. Samaya, M. et al.: Blood P_{50}2,3-DPG and Bohr effect in Sherpas and acclimated Europeans. Proc. Internatl. Union Physio. Sci. Paris, Comite National de Sciences Physiologiques, Vol. 13, p. 653, 1977. Cited from ref. 46.
46. Schaffert, W. et al.: Haemodilution bei hoehenbedingten Erkrankungen. Personal communication.
47. Shults, W.T. and Swan, K.C.: High altitude retinopaty in mountain climbers. Arch. Ophthalmol. 93(6):404, 1975.
48. Scoggin, Ch., Hyers, I.M., Reeves, J.T., and Grover, R.F.: High altitude pulmonary edema in the children and young adults of Leadville, Colorado. N. Engl. J. Med. 297(23):1269, 1977.
49. Singh, I. et al.: High altitude pulmonary edema. Lancet 1(7379):229, 1965.
50. Snyder, P.: Field aspects of experience, treatment and prevention of altitude sickness on Mt. Kenya. Proc. of the Hypoxia Symp., Banff, Canada, 1979.
51. Tejerina, Raygada M.: A propósito del edema agudo pulmonar de altura. Rev. Clin. Esp. 129:283, 1973.
52. Viswanathan, R., Jain, S.K., Subramanian, S., and Puri, B.K.: Pulmonary edema of high altitude. II. Clinical, aerohaemodynamic, and biochemical studies in a group with history of pulmonary edema of high altitude. Am. Rev. Resp. Dis. 100:334, 1969.
53. Viswanathan, R., Subramanian, S., Lodi, S.T., and Radha I.G.: Further studies of pulmonary edema of high altitude. Respiration 36:216, 1978.
54. Viswanathan, R.: Effect of furosemide on altitude tolerance. In: High Altitude Physiology and Medicine. New York-Heidelberg-Berlin, Springer Verlag, 1981.
55. Vogel, J., Goos, J., Mori, M., and Brammell, H.: PA pressure rise with high altitude and exercise. Clin. Res. 15:95, 1967. Abstract.
56. Zink, R.A.: Hämodilution bei Hochgebirgsexpeditionen als Erfrierungsprophylaxe. Aerztl Praxis 24(18):873, 1977.
57. Zink, R.A.: Aerztlicher Rat für Bergsteiger. Stuttgart, Thieme Verlag, 1978.

35
Hemodynamic Study of High Altitude Pulmonary Edema (12,200 ft)

G. Antezana, G. Leguia, A. Morales Guzman, J. Coudert, and H. Spielvogel

The high altitude pulmonary edema (HAPE) originally described by Hurtado (6) has been the object of numerous studies, some of them referring to clinical findings (1), others presenting physiological data and anatomical experiments which indicate that several aspects of this syndrome are not yet clarified.

HAPE is a vital problem for people living at high altitude. Its severity varies widely. HAPE is most common among people at high altitude for the first time, like tourists or mountaineers, or high altitude dwellers who return to high altitude after a stay at sea level or at the Bolivian "llano" (altitude 500 m). HAPE generally develops during the first 3 days at altitude, frequently in relation to unaccustomed exercise. Its onset reminds one of acute "soroche" (acute mountain sickness) (2,7), which is characterized by headache, shortness of breath, nausea, and vomiting. Then illness progresses to the respiratory syndrome signalled by progressive dyspnea and coughing which typically results in expectoration of blood and, in the most severe cases, frank hemoptysis. As Bolivia has a considerably expanded network of roads and improved communication, travel has increased, especially between eastern Bolivia and the cities of political and administrative importance situated in the Andes between 3600 and 4000 m. We believe that the HAPE developed by the high altitude native is rather severe, and that the risk increases with each additional change of altitude. These observations warrant a study of this subject. Fred (3) was the first one to study the dynamics of the syndrome; later Hultgren (5), Peñaloza (8), and Roy (9) contributed valuable data, and Hultgren and Grover (4) presented a very interesting study in this field.

Material and Methods

This study is divided into three parts: cardiocirculatory dynamics during acute HAPE, subsequent treatment and follow-up. General data of patients are shown in Table 35-1.

Five cases among high altitude natives were studied. The average age was 22.4 years. The patients had stayed in the Bolivian lowlands for longer than 4 weeks and returned by plane. HAPE developed during the second day after arrival, with the exception of case 3, who had returned by train. After clinical, radiologic, and elec-

Table 35-1. *General Data of Patients Studied During Acute High Altitude Pulmonary Edema.*

Case no.	Origin	Age (years)	Time at lowland (mo)	Onset of edema (h)	1st study (h)	2nd study (days)	3rd study (mo)
1	N	19	2	24	36		
2	N	35	48	42	48	7	
3	N	17	1	3	4		
4	N	22	2	25	35	7	3
5	N	19	2	24	35	6	
x̄		22.4		23.6	31.6	6.6	3
No.		5		5	5	3	1

N = High altitude native.

trocardiographic examination, right-heart catheterization was carried out in the routine manner. During the first several minutes pressures were registered and cardiac output was measured by the Fick method. In two cases alveolar oxygen tension was measured to determine the alveolar arterial oxygen gradient; this was done by the indirect method utilizing the formula of Rahn and Feen. Finally, the patients were studied during 20 min of oxygen breathing.

In three individuals a second heart catheterization was performed by the same team using the same method, 7 days after HAPE, as a control. The fourth case required a third right-heart catheterization 3 mo later.

A followup study was performed in four individuals who had HAPE repeatedly and severely; two were children with an average age of $9^1/_2$ years, and two were Bolivian soldiers.

It was not possible to measure the workload in the children. They exercised by leg raising. The adults, however, performed moderate exercise with a workload of 50 W.

Results

The data in Table 35-2, obtained during the acute stage of HAPE, show marked pulmonary hypoventilation, hypoxemia expressed by a drop in Pa_{O_2} and decreased arterial oxygen saturation, fall of alveolar oxygen tension, and increase in the alveolar-capillary oxygen gradient.

Table 35-3 shows a marked increase in heart rate together with a low cardiac output. An increase of right ventricular and pulmonary artery pressures with normal pulmonary wedge pressure was observed. We would like to point out the singular behavior of case 3, who was studied with few clinical symptoms except for a few rales in the left upper lung zone. This patient had already had numerous episodes of HAPE, one of which was described as severe. Hence the patient was transported directly from the railroad station to our laboratory. The blood gas findings were only slightly different, but the increase of pulmonary artery pressure was remarkably like that in the rest of our patients. Without these findings the chest x-ray could have been considered normal (Fig. 35-1) and would not have revealed HAPE.

Figure 35-2 shows the effects of breathing 100% oxygen, through a mask, which in two cases (1 and 2) caused a spectacular fall of mean pulmonary artery pressure to the normal mean range, whereas in the three other cases there was a drop of 30–50%. The rate of decrease was highest during the first 3 min. Case 4, clinically described as severe (Figs. 35-3 and 35-4), presents the smallest absolute decrease.

Table 35-4 shows the cardiocirculatory data registered during the control cathe-

Table 35-2. Respiratory Parameters of Four Patients with High Altitude Pulmonary Edema.

Case no.	VE (l/min) (BTPS)	VO_2 (ml/min) (STPD)	pH	Pa_{O_2} (mmHg)	Pa_{CO_2} (mmHg)	Ca_{O_2} (vol%)	Cv_{O_2} (vol%)	Capacity of O_2 (vol%)	Sa_{O_2} (%)	PA_{O_2} (mmHg)	$P(A-a)_{O_2}$ (mmHg)	Hct (%)	Hb (g%)
2	14.6	157	7.426	36	33.5	14.69	7.5	18.72	72			41	14.2
3	11.58	171	7.43	46	24	17.19	14.25	19.8	86			45	15
4	14.56	160	7.49	29.9	23.31	12.26	8.66	19.58	62	43.7	13.4	45	15.8
5	25	190	7.54	28.5	36.3	17.82	9.02	23.69	74	48.5	20	50	17
$\bar{x}$	16.43	169.5	7.471	35.1	29.27	15.49	9.85	20.32	73.5	46.1	16.7	45.25	15.5
No.	4	4	4	4	4	4	4	4	4	2	2	4	4

Table 35-3. *Circulatory Parameters in Acute High Altitude Pulmonary Edema.*

Case no.	Age (years)	BSA (m^2)	HR/min	SV (ml)	CI (l/min/m^2)	RA M	RV S	RV D_1	RV D_2	PA S	PA D	PA M	P_w M	SA S	SA D	SA M
1	19	1.70	120	—	—	2	82	0	1	82	53	62	2	110	65	80
2	35	1.80	110	29	1.21	3	72	0	3	72	43	60	7	110	67	80
3	17	1.74	130	86	3.34	4	80	0	4	80	50	64	9	138	84	100
4	22	1.55	100	68	2.86	6	87	1	5	87	56	70	2	98	80	87
5	19	1.62	115	30	1.33	10	83	0	7	83	45	54	—	113	70	86
$\bar{x}$	22.4	1.68	117	52	2.1	5	81	0.2	4	81	49	62	5	113	72	87
No.	5	5	5	4	4	5	5	5	5	5	5	5	4	5	5	5

BSA = body surface area; HR = heart rate; SV = stroke volume; CI = cardiac index; RA = right atrium; RV = right ventricle; PA = pulmonary artery; P_w = pulmonary wedge pressure; SA = systemic artery.

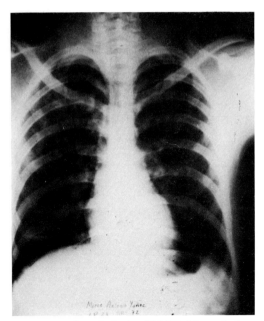

Fig. 35-1. Case 3 during HAPE. Normal chest x-ray. The patient had only a few rales in left upper lung zone and clinical symptoms. However, the hemodynamic study was typical for HAPE.

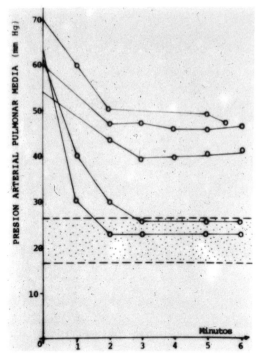

Fig. 35-2. Effects of breathing pure oxygen in five patients with HAPE. Cases 1 and 2 show a spectacular fall of mean PAP to the normal mean range. In cases 3, 4, and 5 the drop of mean PAP ranges from 30 to 50%. Case 4, clinically severe, presents the smallest absolute decrease in mean PAP. In all cases the greatest rate of decrease was observed during the first 3 min of oxygen breathing.

terization, performed only in three patients, only two of which showed return to normal values. Case 4, however, still had elevated pulmonary pressure and clinically recovery was slow. In this case and in view of the voluntary admission of the patient to our laboratory, a third diagnostic right-heart catheterization was performed after 3 mo to exclude cardiopathy. Right-heart and pulmonary vascular pressures were normal at rest. During exercise mean pulmonary artery pressure increased to 90 mmHg without decrease in arterial oxygen saturation. The workload imposed was moderate. Breathing pure oxygen decreased mean pulmonary artery pressure by 35%, and breathing an hypoxic mixture of 16% increased mean pulmonary artery pressure by only 6 mmHg.

These findings suggest that people with a history of HAPE behave differently, and that a followup study is required to confirm this assumption.

Table 35-5 shows resting (R) cardiocirculatory values considered normal for La Paz (3650 m) and for the age groups of the patients (two children and two adults). The exercise in the two groups is not comparable, because the children did much less exercise than the two young men. However, there was in all cases a moderate-to-significant increase in pulmonary artery pressure during exercise. Table 35-4 also shows that changes in oxygen saturation of the pulmonary artery are not significant. The findings of case 1 are difficult to interpret, whereas in the other three patients a discrete improvement of arterial oxygen saturation was found.

Please note case 2 (Fig. 35-5), whose

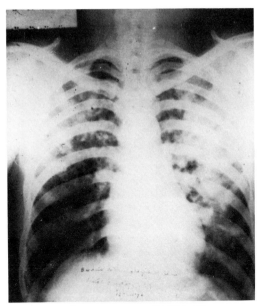

Fig. 35-3. Case 3, severe HAPE. Chest x-ray shows patchy lesions.

electrocardiogram shows severe right-heart overload and also a clear acute subendocardial lesion of the left ventricle, findings which have not been published before.

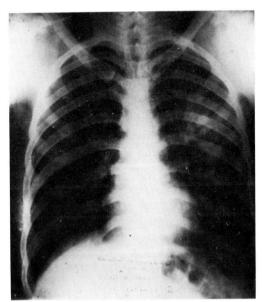

Fig. 35-4. Case 4 during recovery.

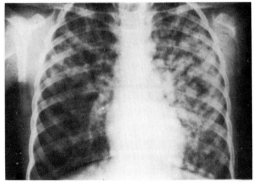

Fig. 35-5. Case 2 of group C (followup study) during HAPE. Chest x-ray shows patchy lesions typical of HAPE.

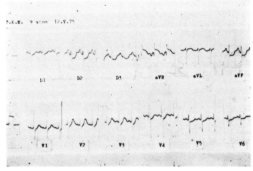

Fig. 35-6. Case 2 of group C (followup study) during HAPE. right-heart overload of extreme severity; acute subendocardiac lesion of the left ventricle.

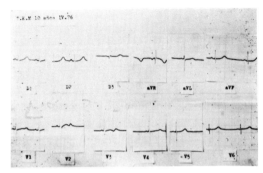

Fig. 35-7. Case 2 of group C (followup study) 11 months later. Normal ECG.

Table 35-4. Circulatory Parameters After Acute High Altitude Pulmonary Edema.

Case no.	Age (years)	BSA (m^2)	HR/min	SV (ml)	CI (l/min/m^2)	RA	RV			PA			P$_w$	SA		
						M	S	D$_1$	D$_2$	S	D	M	M	S	D	M
2	35	1.80	55	85	2.5	5	32	0	4	32	11	25	5	105	70	76
4	22	1.55	50	178	5.7	6	63	0	6	63	26	43	4	98	80	87
5	19	1.62	60	95	3.5	2	25	0	1	25	10	18	—	113	70	88

Pressure (mmHg)

Key: See Table 35-3.

Table 35-5. *Cardiorespiratory Parameters of Patients with Previous History of Acute High Altitude Pulmonary Edema.*

Case no.	Age (years)		BSA (m²)	HR/min	V_{O_2} (cc/min)	pH	Pa_{O_2} (mmHg)	Pa_{CO_2} (mmHg)	Sa_{O_2} (%)	CI (l/min/m²)	P_w (mmHg)	PAP (mmHg)	TPVR (dynes/sec/cm^{-5})
1	8	R	0.94	86	262	7.49	79	29.5	97	3.7	9	19	562
		E		150	345	7.45	58	27	92	4.7	10	40	705
2	11	R	0.81	90	160	7.43	41	36	78	7.4	7	30	372
		E		125	—	7.41	49	32	85	9.8	14	52	—
3	22	R	1.55	50	252	7.41	31	34	61	3.8	—	23	306
		E		120	884	7.39	46	35	81	5.7	—	90	758
4	19	R	1.71	59	176	7.45	57	31	92	2.4	7	19	361
		E		150	1221	7.30	70	36	93	9.3	12	75	354

Key: BSA = body surface area; HR = rate rate; CI = cardiac index; P_w = pulmonary wedge pressures; PAP = pulmonary arterial pressure; TPVR = total pulmonary vascular resistance; R = rest; E = exercise.

Discussion

Our findings are essentially similar to those in the literature, and closest to the findings of Peñaloza (8). Our figures were higher than those of Roy (9) and Hultgren (5), excepting his first case, an 8-year-old boy who probably had high altitude pulmonary hypertension. We have to point out that there exists no relation between the clinical picture and hemodynamic findings. Our case 3 is an example. He had few symptoms, but his hemodynamic values were typical for HAPE. When he arrived at La Paz by train, he had performed only minimal exercise before he was transferred to our laboratory (3650 m, PB-495 mmHg) by car. Study shortly after recovery in case 4 revealed abnormal behavior which suggested high altitude pulmonary hypertension.

The followup study demonstrated abnormal elevation of pulmonary artery pressure; however we did not find the significant changes in blood gases found by other authors (4). This fact suggested to us that perhaps the lungs of people with a history of HAPE overreact during exercise, even though they may not develop HAPE.

It seems that the cause of HAPE is an inability to acclimatize to high altitude hypoxia. It has been suggested that the thick pulmonary artery musculature does not involute uniformly at sea level (11), and that on return to the hypoxic environment, vasoconstriction causes pulmonary hypertension with uneven regional perfusion (10). Pulmonary hypertension can cause severe contraction of the muscular arterioles, distal occlusions within contracted arterioles, and increase of capillary permeability. Apparently the perivascular disorder represents an early stage of edema (12). The edema may start in the alveolus as the result of a transarterial leak (10) into the perivascular space, where the pressure is lowest. The excess pressure that cannot be relieved by elimination through the pulmonary lymphatic vessels is thought to cause progressive edema.

It is important for people who live at high altitude to be aware of the percentage of individuals who develop HAPE and are left with permanent pulmonary hypertension, as well as how many are taking a risk by continuing to live at high altitude. To underline this issue we offer the history of an 11-year-old boy in whose case high altitude pulmonary hypertension had been diagnosed and who was sent to sea level. He tired easily, there was significant accentuation of the second pulmonary heart sound, and a normal chest x-ray and electrocardiogram. After 6 mo of residence at sea level he consulted a physician who found no abnormal data, and the child was told he could return to high altitude. On arriving at the altitude of 3000 m the child died of severe irreversible HAPE.

References

1. Alzamora-Castro, V., Garrido-Leca, G., and Battiliana, G.: Pulmonary edema of high altitude. Am. J. Cardiol. 7:769, 1961.
2. Barcroft, J: The respiration of the blood—Lesson from high altitude. Cambridge, The University Press, 1925.
3. Fred, H., Schmidt, A., Bates, T., and Heecht, H.: Acute pulmonary edema of high altitude. Clinical and physiologic observations. Circulation 25:929, 1962.
4. Hultgren, H.N., Grover, R., and Howard, H.: Abnormal circulatory responses to high altitude in subjects with a previous history of high altitude pulmonary edema. Circulation vol. XLIV, 1971.
5. Hultgren, H.N., Lopez, C.E., and Lundberg, E: Physiologic studies of pulmonary edema at high altitude. Circulation 29:393, 1964.
6. Hurtado, A.: Aspectos fisiológicos y patológicos de la vida en la altura. Rimac Lima, 1937.
7. Monge, M.C.: High altitude disease. Arch. Int. Med. 59:32, 1937.
8. Peñaloza, D. and Sime, F.: Circulatory dynamics during high altitude pulmonary edema. Amer. J. Cardiol. 23:369, 1969.
9. Roy, S.B., Guleria, J.S., and Khanna, P.K.:

Hemodynamic studies in high altitude pulmonary edema. Brit. Heart J. 31:52, 1969.
10. Severinghaus, J.W.: London, Ciba Foundation, 1970.
11. Viswanathan, R., Jain, S.K., and Subramanian, S.: Pulmonary edema high altitude pathogenesis. Am. Rev. Rep. Dis., 1969.
12. West, J.B.: Cited by Severinghaus In Ciba Foundation, London, 1970.

36

Pathogenesis of High Altitude Pulmonary Edema (HAPE)

R. VISWANATHAN

Changes in physiologic functioning in high altitude environment are primarily due to the effect of low oxygen pressure and cold. Of these two hypoxic environment is the most important. Some of the acute effects of high altitude exposure which have been observed are pulmonary hypertension, coronary dilatation with normal blood flow, hyperventilation and consequent alkalosis, and polycythemia. Pulmonary hypertension is the most relevant to the pathogenesis of HAPE.

According to Green (1), the mechanisms which may underlie the development of HAPE are increased capillary hydrostatic pressure, increased capillary surface area, decreased plasma osmotic pressure, and increased alveolar surface tension. The question is, how are these conditions brought about?

It is recognized that only some of the people exposed to acute high altitude stress develop HAPE. During the Indo-China War of 1962, when thousands of Indian soldiers had to be flown to heights ranging from 10,000 to 14,000 ft and above, only 0.6% developed HAPE (6). Therefore, our first study was directed toward determining any constitutional susceptibility to HAPE in different species of experimental animals (8) (Table 36-1).

Our study shows that there is species difference in the development of HAPE, and within one species risk varies. These results have subsequently been confirmed by further studies.

During our experimental studies we saw two very significant changes in lung histology, one pertaining to the type of edema and the other to the appearance of constricted muscular arterioles, strikingly like a string of sausages.

That we saw contracted muscular arterioles only confirmed what had been described earlier by Weibel (9) and others. The above two observations contributed to my hypothesis regarding the mechanism of HAPE development.

Table 36-1. HAPE in Experimental Animals.

Species	No. exposed	Cases of HAPE No.	%
Dogs	12	2	16.7
Monkeys	28	7	25.0
Guinea pigs	66	20	30.3
Rats	168	40	23.8
Mice	258	171	66.2

Table 36-2. Location in Chest of Edema in HAPE.

	Upper	Middle	Lower	Total
Right	2	4	2	8 (3%)
Left	1	4	2	7 (28%)
Both sides	3	6	1	10 (40%)
Total	6 (24%)	14 (56%)	5 (20%)	25 (100%)

The appearance in man of a patchy edema has been observed by others, and we too observed it in the x-rays of a group of Indian soldiers with pulmonary edema.

The location of edema observed in the soldiers studied is summarized in Table 36-2.

Hemodynamic studies were done in 196 soldiers, of whom 101 had developed pulmonary edema 1 or 2 years before our study; the remaining 95 had not developed HAPE even though they had been at altitude for 2 years or more. We wanted to see whether the two groups differed significantly in any parameters. Both HAPE subjects and controls breathed a 10% hypoxic mixture (representing altitude of 6000 m) for 5 min. Hemodynamic parameters were recorded before and after hypoxic breathing (Table 36-3).

Table 36-3. Effect of Hypoxia on PAP in 101 HAPE Subjects and 95 Controls.

PAP (mmHg)	Subjects	Controls	Significance
Before hypoxic breathing	$\bar{x} \pm SD$	$\bar{x} \pm SD$	
Systolic	24.5 ± 6.3	21.3 ± 5.0	NS
Diastolic	9.5 ± 4.8	8.2 ± 2.4	NS
Mean	16.1 ± 6.7	13.9 ± 3.4	NS
After hypoxic breathing			
Systolic	39.5 ± 10.0	26.1 ± 5.8	HS
Diastolic	17.8 ± 5.2	11.4 ± 0.7	S
Mean	27.8 ± 8.4	17.5 ± 4.3	S

Table 36-4. Effect of Hypoxic Breathing on HAPE Subjects and Controls.

	HAPE (101)	Control (95)	Significance
Cardiac output			
Before	6.5 ± 2.1	6.2 ± 2.3	NS
After	6.6 ± 3.2	6.1 ± 2.2	NS
Wedge pressure			
Before	8.7 ± 2.8	7.6 ± 3.5	NS
After	8.8 ± 2.6	7.4 ± 3.4	NS

The results demonstrate no significant difference between subjects and controls in systolic, diastolic, and mean pressures, while breathing room air. After 5 min of 10% hypoxic breathing, however, there was significant a difference in the pressures of the two groups (Table 36-3).

There was no significant change in cardiac output and wedge pressure after hypoxic breathing. Oxygen concentration in arterial blood was lowered considerably in HAPE subjects as compared to control (Table 36-4).

While arterial blood serotonin levels did not rise (3), histamine levels rose in HAPE subjects and in controls after 10 min of hypoxic stress (7).

Another surprising observation concerned the cold pressor test. People who had had HAPE were hyporeactors to cold pressor tests while controls were normoreactors.

Hemodynamic studies have clearly shown that there is a significant difference in reactivity to hypoxic stress between HAPE subjects and controls.

Table 36-5. O_2 Saturation Hypoxia.

	Basal $\bar{x} \pm SD$	Hypoxia $\bar{x} \pm SD$	Significance
HAPE (57)	98.8 ± 20.0	70.1 ± 19.5	HS
Controls (47)	96.3 ± 8.4	84.6 ± 7.0	NS

Sig.—HAPE vs. controls—S

Table 36-6. *Effect of Hypoxia on Plasma Histamine Levels.*

	No.	Basal level $\bar{x} \pm SD$	No.	After hypoxia $\bar{x} \pm SD$	Significance
Normals	12	4.18 ± 0.85	12	13.4 ± 3.62	S
HAPE	12	4.32 ± 1.14	12	12.37 ± 3.54	S
Control	12	3.90 ± 0.99	12	11.57 ± 4.54	S

S, significant, $p < 0.05$

I believe that susceptibility to HAPE lies in the architecture of muscular arterioles, as well as in their reactivity to hypoxic stress. The patchy edema which occurs in different parts of the lung is obviously due to the variation in regional blood distribution caused by torrential flow through nonmuscular perpendicular arterioles branching from muscular ones, as represented in Fig. 36-2.

The top arteriole is muscular with a nonmuscular perpendicular arteriole branching from it. The bottom diagram represents the contracted arteriole with wider opening of the perpendicular arteriole. The character-

Table 36-7. *Cold Pressor Test.*[a]

	HAPE (101) $\bar{x} \pm SD$	Control (95) $\bar{x} \pm SD$	
Systolic	1.60 ± 3.45	19.85 ± 5.03	HS
Diastolic	1.51 ± 4.32	17.31 ± 7.15	HS

[a] BAP rise after 2 min immersion at 5°C.
HS, highly significant ($p < 0.01$).

istic appearance of the muscular arteriole (like a string of sausages) under hypoxic stress supports this hypothesis. It is likely that the torrential flow through the perpendicular arteriole is accentuated by sudden

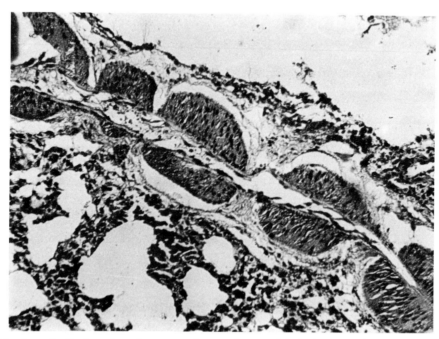

Fig. 36-1. Longitudinal section of contracted muscular arteriole in the lung of rats exposed to acute hypoxic stress.

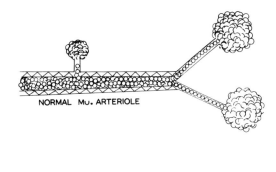

Fig. 36-2 Diagrammatic representation of muscular pulmonary arteriole before (above) and after (below) acute hypoxia. Note the wider opening of the perpendicular nonmuscular arteriole (below).

increase in pulmonary blood flow after acute hypoxic stress.

Figure 36-3 shows a laboratory model which I used for simulating a normal arteriole and a contracted arteriole. I have been able to show that with the same flow rate and pressure at the proximal end of the two glass tubes, the flow is remarkably higher through the perpendicular tube in the constricted glass tube. In fact, when the pressure is lowered from 20 to 15 cm of water, there is no flow through the perpendicular glass tube attached to the nonconstricted glass tube. On the other hand, bubbles of air were sucked into the flowing water. When the pressure head was raised from 20 to 25 cm, as under hypoxic stress, the flow of fluid through the perpendicular glass tube attached to the constricted one becomes more torrential. On the other hand, the flow decreases through the branches at the distal end of the glass tube

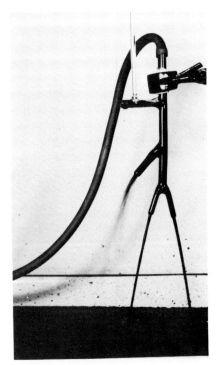
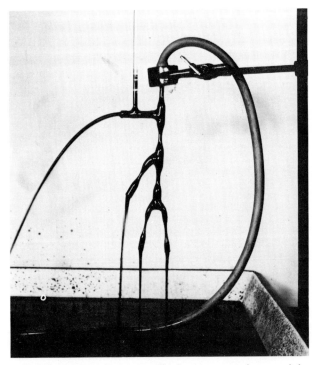

Fig. 36-3. Laboratory model to show torrential flow from the perpendicular nonmuscular arteriole (horizontal limb of glass tube). **Left** Normal. Note absence of flow from horizontal limb. **Right** On constriction of tube, flow from horizontal limb becomes copious.

representing the constricted muscular arteriole.

We have shown that the histamine levels in arterial plasma after hypoxic stress rose equally in HAPE patients and control subjects under hypoxic stress. It is postulated that higher reactivity of the pulmonary arterioles of HAPE subjects to mediators like histamine, rather than the actual rise in histamine level, is the cause of severe pulmonary vasoconstriction, resulting in more marked regional increase in perfusion of the arterioles. The muscular arterioles of HAPE subjects are hyperreactive to hypoxic stress.

The torrential flow of blood through perpendicular arterioles raises the pulmonary capillary hydrostatic pressure. There is also the increased permeability of the capillaries under hypoxic stress.

My hypothesis is essentially mechanistic and explains the development of HAPE in what is perhaps too simple a way.

I do not believe that the rise in pulmonary artery pressure is due to blocking of capillaries, either by edema vesicles in the capillaries [postulated by Heath (2)], or the intracapillary clotting or sludging of red blood cells, suggested by Singh and his colleagues (5), for the reason that such changes must be extensive to produce a significant rise in PAP.

Perivascular leakage due to microrupture of the arterial wall, postulated by Severinghaus, may not be the main factor in the development of alveolar edema (4).

Increased permeability of alveolar capillary membrane is an important factor in the development of pulmonary edema. The question is, what constitutes the permeability which results in blood constituents passing into the alveoli? I believe that it is a widened opening of intercellular junctions, in both the capillary and alveolar membranes, resulting perhaps from retraction of protoplasmic extensions, that allows seepage of blood constituent into the alveoli as is seen in the electronmicrophotograph (Fig. 36-4).

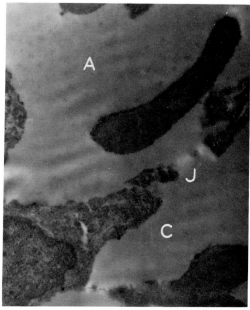

Fig. 36-4. Electronmicrophotograph of alveolar capillary membrane (*A-C*) in rat after acute hypoxic stress. Note the gap in the membrane (*J*) and the RBC in the alveolus.

References

1. Greene, D.C.: Pulmonary Edema, Handbook of Physiology. Sec. 2. Circulation. Vol. 2. Baltimore, Williams and Wilkens, p. 1985, 1964.
2. Health, D. and Williams, R.D.: High altitude pulmonary edema. In: Man at high altitude. 128. Churchill Livingston.
3. Radha, T.G., Venkatasubramanian, T.A., and Viswanathan, R.: Effect of acute hypoxia on blood serotonin in human beings and rats. Respiration 33:64, 1966.
4. Severinghaus, J.W.: Transarterial leakage: A possible mechanism of high altitude pulmonary oedema. High Altitude Physiology —Cardiac and Respiratory Aspects. p. 61. A Ciba Foundation Symposium. Churchill Livingstone, 1971.
5. Singh, I.H. and Chohan, I.S.: Reversal of abnormal fibrinolytic activity, blood coagulation factors and platelet function in high altitude pulmonary oedema with Furosemide. Int. J. Biometeorol. 17:73, 1972.
6. Sudhakaran, K. and Khanna P.K.: Pulmonary

oedema of high altitude. J. Chest Dis. 9:65, 1967.
7. Sudhakaran, K., Viswanathan, R., and Subramaniam, T.A.V.: Plasma histamine levels under hypoxic stress. Respiration 37:91, 1979.
8. Viswanathan, R., Jain, S.K., Subramanian, S., and Puri, B.K.: Pulmonary oedema of high altitude. 110:328, 1969.
9. Weibel, E.R., and Knight, B.W.: A morphometric study on the thickness of the pulmonary air-blood barrier. J. Cell. Biol. 21:367, 1964.

37
Subclinical Pulmonary Edema with Hypobaric Hypoxia

G. Coates, G. Gray, A. Mansell, C. Nahmias, A. Powles, J. Sutton, and C. Webber

Clinical pulmonary edema is a known but uncommon complication of exposure to high altitude (8,9). In a recent report by Hackett et al. (8) only 7 of 278 subjects exposed to an altitude of 5400 m developed clinical pulmonary edema. However, since clinical signs and symptoms are a late manifestation of an increase in lung water, it is possible that there is in fact a greater incidence of high altitude pulmonary edema (HAPE) than is suggested by the literature, but of a severity insufficient to cause clinical signs and symptoms. There is evidence to suggest that this is so. Gray (7) measured the washout of a single breath of nitrogen in 17 subjects during one week of exposure to 17,600 ft on Mount Logan, Yukon, Canada. There was a significant increase in the slope of the alveolar plateau in these subjects. This indicates that exposure to altitude produced an intraregional maldistribution of ventilation (3) which could have been due to an increase in interstitial lung water (13). The pupose of this study was to confirm the observed changes in the single-breath nitrogen washout at altitude and to measure changes in the volume of trapped gas. In addition, in order to obtain further evidence that these changes were caused by an increase in interstitial lung fluid, we measured changes in lung density.

Methods

The study was performed with 4 healthy male volunteers aged from 30 to 39 who were at a barometric pressure of 446 mmHg (equivalent altitude 4268 m) in a hypobaric chamber for 24 h. Decompression to this pressure took 28 min and the duration of exposure to altitude refers to time elapsed after this interval. During the 24 h before decompression and during the 24 h at altitude we monitored in each subject changes in lung volumes, single-breath nitrogen washout, lung density, and lung blood volume.

Lung Volumes and Single-Breath Nitrogen Washout

We made measurements at an altitude of 90 m approximately 20 h before decompression and at 4268 m 5 and 20 h after decompression. We used the single-breath nitrogen washout technique (1) to measure vital capacity (VC) and the alveolar plateau. A Med Science model 505 nitrogen analyzer and a Med Science model 270 Wedge spirometer were used and the expired nitrogen concentrations and volumes were displayed on an X-Y recorder. We performed at least three washouts on each subject on each occasion with 1 min between washouts. Although expiratory flow rates were not measured, they were kept below 0.5 liters/s, on the basis of total time taken for the expired VC maneuver. The best (largest VC and most linear alveolar plateau) washout was used for analysis.

We measured lung volumes by the closed-circuit helium dilution method. Helium concentrations were measured with a Warren Collins katharometer model 21211. A katharometer measures the thermal conductivity of a gas mixture which theoretically is independent of ambient pressure (11). We calibrated the katharometer in the hypobaric chamber at 90 m and at 4268 m. Two gravimetrically determined helium–nitrogen mixtures, one near full scale and the other near half scale on the apparatus, plus zero concentration produced linear readings that agreed within 0.5% helium at the two ambient pressures.

All volumes were corrected to BTPS. We analyzed the control values and the results after 5 and 20 h of decompression with Student's paired t-test.

Lung Density

We measured lung density with a Compton scatter technique previously described for use in lung (4) and bone (2,18). The technique was identical to that described by Garnett et al. (4). The procedure is based on the fact that when monoenergetic γ-rays impinge upon any tissue, the density of a defined volume of that tissue can be determined by measuring the scattered fraction of the incident γ-rays (Fig. 37-1).

From a 1.5 Ci source of radioactive samarium (^{153}Sm), 103 keV γ-rays are

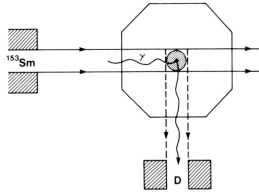

Fig. 37-1. Schematic drawing of Compton scatter device. Gamma rays from the samarium-153 source (^{153}Sm) are scattered by any object in their path. Our device detects the gamma rays scattered through 90° into detector D. The fraction of gamma rays scattered is directly proportional to the density of the object. The intersection of the beam of gamma rays with the sensitive volume of the detector defines a fixed volume in the object.

directed at the lower lobe of the right lung through a lead collimator. Compton scatter photons are detected with a collimated sodium iodide detector placed at right angles to the incident beam. In this way the detector measures ^{153}Sm γ-rays which have been scattered at 90° within a fixed volume or lung defined by the two collimators. We correct for attenuation of these scattered γ-rays (84 keV) with an 84 KeV source of radioactive thulium (^{170}Tm) placed opposite the detector of scattered photons. A second detector is placed in line with the ^{153}Sm source to measure transmitted photons. The sources and detectors are mounted on an adjustable horizontal C-frame. The details of the calculation of the results are similar to those already described for the measurement of bone density (12). Corrections for multiple scattering were made with the use of wooden phantoms of known density. The errors and reproducibility of the technique are described by Garnett et al. (4). Measurements were made with the subject sitting breathing quietly at 12 and 3 h before decompression, and after 2, 10, and 18 h at altitude.

The system was calibrated with a water phantom after each measurement.

Correction for Changes in Lung Blood Volume and FRC

Lung density is the mass of lung tissue, blood, and interstitial fluid per unit volume of lung. Clearly, an increase in the amount of air in the defined volume of lung would cause a decrease in density, while an increase in the amount of blood or interstitial fluid would increase the density. Thus in order to detect changes in density caused by changes in interstitial fluid, it was necessary to correct apparent changes in density for changes in lung blood volume and for changes in the volume of air within the lung.

Lung Blood Volume

We monitored changes in regional blood volume with a modification of the technique of Lindsey and Guyton (14). Ten days before decompression each subject received an intravenous injection of 10 μCi ^{59}Fe. By the time of decompression 90–95% of this radioactive iron would be incorporated into red cells, and thus any change in the amount of ^{59}Fe in an organ would reflect a change in the number of red cells within the organ. We monitored external counts from ^{59}Fe with a collimated sodium iodide detector over the upper lobe of the right lung, the liver, and the thigh. These measurements were made with the subjects supine, 26, 20, 12, and 3 h before decompression, and after 1, 9, and 19 h at altitude.

Corrections for Changes in Lung Volume

One week after decompression we measured lung density at sea level in two of the subjects (G.C., C.N.) while they breathed from a 9-l Collins spirometer. Lung density was measured twice while they breathed quietly at FRC and twice at 800 ml above FRC.

Results

Single-Breath Nitrogen Washout and Lung Volumes

Values for the measured lung volumes FRC, inspiratory capacity (IC), VC, and CV before and 5 h after decompression are presented in Table 37-1. Both FRC and IC suggested increases (19 and 24%, respectively), but each fell short of statistical significance. Between 5 and 20 h of decompression, VC fell significantly (10%, $p < 0.05$).

Mean values for the calculated lung volumes total lung capacity (TLC), residual volume (RV), and "closing capacity" (CC) are shown in Fig. 37-2. Each of these

Table 37-1. Lung Volumes Before and After Decompression.

Subject	Functional residual capacity (l, BTPS)		Inspiratory capacity (l, BTPS)		Vital capacity (l, BTPS)		"Closing volume" (l, BTPS)	
	Cont.	5 h	Cont.	5 h	Cont.	5 h	Cont.	5 h
A.M.	4.4	4.9	2.1	3.4	5.0	4.8	0.9	0.8
C.N.	4.1	4.6	2.9	4.0	5.5	5.9	0.7	0.9
G.C.	5.3	7.1	3.6	4.1	6.5	7.0	0.4	1.1
G.G.	3.5	4.0	4.1	4.2	5.7	5.7	0.7	0.5

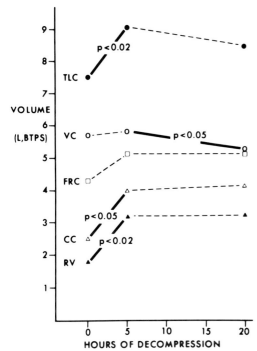

Fig. 37-2. Lung volumes before and during decompression. Points connected by lines represent mean values from the four subjects for total lung capacity (TLC), vital capacity (VC), functional residual capacity (FRC), "closing capacity" (CC), and residual volume (RV). Significant changes by the paired t-test are shown by solid lines with the associated p values.

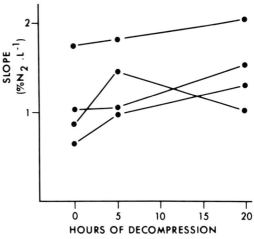

Fig. 37-3. Slope of alveolar plateau before and during decompression. Each set of points connected by lines represents individual values. Values at 20 h were significantly higher than control values by the paired t-test ($p < 0.05$).

increased significantly over control values after 5 hr of decompression. After 20 h, TLC was not significantly higher than the control value.

Slopes of alveolar plateaus appeared to have increased (23%, NS) at 5 h of decompression and the increase was statistically significant (36%, $p < 0.05$) after 20 h (Fig. 37-3). There was no suggestion of change in the amplitude of cardiac oscillations in the alveolar plateaus although these were not analyzed systematically.

Lung Blood Volume

There was no significant change in the external count rate from ^{59}Fe in the upper lobe of the right lung.

Lung Density

Values are presented in Table 37-2. The range determined in a separate group of normal subjects is 0.25–0.37 g · cm^{-3}. The apparent increase from a mean control of 0.30 g · cm^{-3} at 10 h of decompression fell short of statistical significance ($t = 2.55$, $0.05 < p < 0.10$).

Correction for Changes in Lung Volume (Apparent Lung Mass)

In both subjects density decreased by 15% as lung volume (FRC) increased by 800 ml.

Table 37-2. Lung Density Measured Before and After Decompression.

Time (h) before and after decompression	Lung density (gm · cm^{-3})				
	G.C.	G.G.	A.M.	C.N.	Mean
−12	0.27	0.26	0.32	0.34	0.30
−3	0.29	0.30	0.27	0.33	0.30
+2	0.27	0.34	0.46	0.36	0.36
+10	0.24	0.59	0.47	0.30	0.41
+18	0.36	0.37	0.34	0.36	0.36

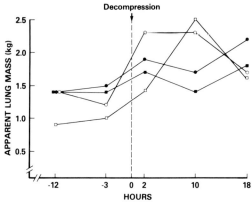

Fig. 37-4. "Apparent lung mass," the product of regional density and FRC, before and during decompression. Each set of points connected by lines represents individual values for "apparent lung mass." Values after 2 h of decompression were significantly higher than control values by the paired t-test ($p < 0.05$).

This represented a fractional increase in volume of 19% in one subject and 15% in the other. The product of lung volume and density (apparent lung mass) remained constant at sea level. However, at altitude apparent lung mass increased from a mean of 1.3 kg to 1.8 at 2 h and 2.0 after 10 h at altitude (Fig. 37-4).

Discussion

Changes in Lung Volume

The large (mean 21%) increases in TLC after 5 h of decompression were surprising. However, increases in both FRC and IC came close to statistical significance alone (Table 37-1). Previous studies have resulted in similar changes during exposure to altitude and hypoxia. Both Crusz-Jibaya (2) and Tenney et al. (16) found increases in FRC at altitude using nitrogen washout techniques. The latter study also suggested increases in TLC. More recently, Saunders et al. (15) used plethysmography to detect increases in TLC in normal subjects during acute isocapneic hypoxia. However, Goldstein et al. (5) found no change in TLC with hypoxia.

The increases in TLC in our study must reflect increases in distensibility of the lungs, inasmuch as lung compliance is very low near TLC in young adult subjects (17). This increase in distensibility cannot be explained by a reduction in total pulmonary blood volume, as our results indicate no change in regional pulmonary blood volume. A reduction in surface forces or a reduction in the tension of alveolar duct musculature may explain the increase in lung distensibility, although there is no evidence to support the existence of either of these mechanisms.

Gas trapping occurred at much higher lung volumes during decompression than at sea level. Figure 37-2 shows large increases in RV and CC but no suggestion of change in CV. Gray et al. (6) originally measured CV on Mt. Logan to detect interstitial pulmonary edema, which is associated with gas trapping in dependent lung zones (10), and which has been associated with measurable increases in CV (13). We also failed to find a change in CV at a comparable altitude in a hypobaric chamber. However, CV represents only that part of the VC where closure of dependent airways occurs during expiration. To be able to compare CV under different experimental conditions, one must assume that reserve volume remains constant; in fact we found major increases in reserve volume under hypobaric conditions.

Between 5 and 20 h of decompression there was a significant fall in VC associated with a slight decrease in TLC. Since RV and CC remained constant, the decrease in VC appears to represent a return of lung distensibility to normal unassociated with a decrease in gas trapping. This constitutes some evidence that the two changes were caused by separate mechanisms rather than by a simple loss of lung recoil and its subsequent return toward normal.

Distribution of Ventilation

The slope of the alveolar plateau in the single-breath nitrogen washout test increased significantly after 20 h of decompression and suggested an increase after only 5 h of decompression. However, all individual values for slope found using this technique were within normal limits throughout the study. The change in slope during decompression confirms the findings of Gray et al. (7) who used the same technique during an ascent of Mt. Logan. Their rate of ascent was similar to ours and they found a slightly smaller increase in the slope, after 24 h of exposure to 2988 m, than we found after 20 h exposure to 4268 m. The slope of the alveolar plateau reflects asynchronous emptying of lung units with uneven nitrogen concentrations and this can be detected within small lung zones (3). The slope is therefore thought to be a measure of intraregional maldistribution of ventilation and it is not clear why this worsens at high altitude. An increase in maldistribution of ventilation may be associated with an inhomogenous change in the resistance of small airway and/or inhomogenous change in the compliance of lung units.

Apparent Lung Mass

Apparent lung mass is the product of lung density and the lung volume at which the density is measured. At sea level lung density decreased as lung volume increased and thus apparent lung mass remained constant. At altitude lung volume increased but there was no corresponding decrease in density; in fact, density also increased slightly, consequently apparent lung mass increased. The increase in apparent lung mass at altitude was caused by either an increase in lung blood volume or an increase in extravascular fluid. Since we did not detect an increase in lung blood volume, we must conclude that the increase in apparent lung mass was caused by an increase in extravascular lung fluid. The estimates of lung mass at sea level (Fig. 37-4) are greater than the expected weight of normal lungs. This is probably because we have assumed uniform density throughout the lung when in fact our measurements reflect the greater density of the lower lobes.

Summary

None of the subjects developed symptoms of clinical pulmonary edema and yet we demonstrated a large increase in the volume of trapped gas, an increase in maldistribution of ventilation, and an increase in apparent lung mass. These findings are best explained by an increase in pulmonary interstitial fluid insufficient to cause clinical symptoms, that is, all the subjects had subclinical pulmonary edema. Although this study reports on only four subjects, it does suggest that interstitial pulmonary edema at altitude is a more common event than has previously been realized.

References

1. Anthonisen, N.R., Danson, J., Robertson, P., and Ross, W.: Airway closure as a function of age. Resp. Physiol. 8:58, 1964.
2. Cruz-Jibaya, J.C.: Respiration physiology on ascent to high altitudes. US Army Research Department (DAFA 19-68-6-0031), 1969.
3. Engel, L.A., Utz, G., Wood, L.D.H., and Macklem, P.T.: Ventilation distribution in anatomical lung units. J. Appl. Physiol. 37:194, 1974.
4. Garnett, E.S., Webber, C.E., Coates, G., Cockshott, W.P., Nahmias, C., and Lassen, N.: Lung density: clinical method for quantitation of pulmonary congestion and edema. Can. Med. Assoc. J. 116:153, 1977.
5. Goldstein, R.S., Zamel, N., and Rebuck, A.S.: Absence of effects of hypoxia on small

airway function in humans. J. Appl. Physiol. 47:251, 1979.
6. Gray, G.W., Rennie, I.D.B., Houston, C.S., and Bryan, A.C.: Phase IV volume of the single-breath nitrogen washout curve on exposure to altitude. J. Appl. Physiol. 35:227, 1973.
7. Gray, G.W., McFadden, D.M., Houston, C.S., and Bryan, A.C.: Changes in the single-breath nitrogen washout curve on exposure to 17,600 ft. J. Appl. Physiol. 39:652, 1975.
8. Hackett, P.H., Rennie, D., and Levine, H.C.: The incidence, importance and prophylaxis of acute mountain sickness. Lancet 2:1149, 1976.
9. Houston, C.S.: Acute pulmonary edema of high altitude. N. Eng. J. Med. 263:478, 1960.
10. Hughes, J.M.B., and Rosenzweig, D.: Factors affecting trapped gas volume in perfused dog lungs. J. Appl. Physiol. 29:332, 1970.
11. Jessop, G.: Katharometers. J. Sci. Instrum. 43:777, 1966.
12. Kennett, T.J. and Webber, C.E.: Bone density measured by photon scattering—inherent sources of error. Phys. Med. Biol. 21:770, 1976.
13. Lemen, R., Jones, J.G., Graf, P.D., and Cowan, G.: "Closing volume" changes in alloxan-induced pulmonary edema in anesthetized dogs. J. Appl. Physiol. 39:235, 1975.
14. Lindsey, A.W. and Guyton, A.C.: Continuous recording of pulmonary blood volume: pulmonary pressure and volume changes. Am. J. Physiol. 197:959, 1959.
15. Saunders, N.A., Betts, M.F., Pengelly, L.D., and Rebuck, A.S.: Changes in lung mechanics induced by acute isocapneic hypoxia. J. Appl. Physiol.: Resp. Environ. Exercise Physiol. 42:413, 1977.
16. Tenney, S.M., Rahn, H., Stroud, R.C., and Mithoefer, J.C.: Adaptation to high altitude: change in lung volumes during the first seven days at Mt. Evans, Colorado. J. Appl. Physiol. 5:607, 1953.
17. Turner, J., Mead, J., and Wohl, M.: Elasticity of human lungs in relation to age. J. Appl. Physiol. 25:664, 1968.
18. Webber, C.E. and Kennett, T.J.: Bone density measured by photon scattering—a system for clinical use. Phys. Med. Biol. 21:770, 1976.

38

Mechanism of Pulmonary Edema Following Uneven Pulmonary Artery Obstruction and Its Relationship to High Altitude Lung Injury

NORMAN C. STAUB

High altitude pulmonary edema (HAPE) is no longer considered a rare disease. Although it can be prevented by careful acclimatization and treated by descent to lower altitude or by breathing oxygen, death still occurs occasionally because of too rapid ascent of unacclimatized persons to high altitude, because the diagnosis is not made promptly, or because treatment cannot be given.

HAPE usually develops within 1 to 4 days of arrival at high altitude and is usually preceded by a period of physical exertion. The victims variously report dyspnea, headache, dry cough, fatigue, and other symptoms, and are often markedly apprehensive. Physical examination may reveal rapid breathing, crepitant inspiratory or bubbling coarse rales, cyanosis, increased heart rate, and low arterial blood pressure. The x-ray shows patches of exudate throughout the lung fields (Fig. 38-1). The x-ray pattern is quite different from that seen in the usual pulmonary edema due to congestive heart failure.

Cardiac catheterization studies *during* HAPE show extraordinarily high pulmonary artery pressure with normal pulmonary wedge pressure and normal or decreased cardiac output; that is, pulmonary vascular resistance is very high.

Postmortem examinations of numerous persons who have died of this disease have revealed severe confluent pulmonary edema with the alveoli flooded by what is described as protein-rich exudate. There may be evidence of obstruction of small pulmonary arteries and arterioles by clumps of white blood cells, platelets, and fibrinogen (13).

After decades of debate, there is now rather general acceptance of the facts that HAPE is not due either to direct hypoxic injury to the microvascular endothelium (4) or to left-heart failure (13). The initiating factor in HAPE is the active vasoconstriction of small muscular pulmonary arteries due to the direct effect of alveolar hypoxia (9, 10, 14). But how does vasoconstriction *proximal* to the fluid exchange vessels cause pulmonary edema; in other words, after pulmonary vascular resistance increases, what happens next? If a proposed mechanism is to be seriously accepted as the explanation of HAPE, it must be shown capable of producing pulmonary edema in its own right and must be present in every instance.

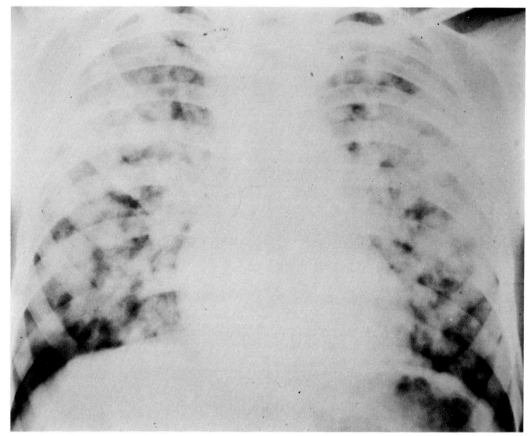

Fig. 38-1. Chest x-ray of HAPE in a 13-year-old Peruvian boy at 3200 m. It shows the characteristic pattern of focal exudates scattered throughout both lung fields with less edema at the bottom of the lung than at the top. (Courtesy of Dr. H. Hultgren, Palo Alto, Calif. See also Hultgren, 1978.)

Uneven Pulmonary Artery Obstruction

The Starling equation, in words, states that net fluid filtration is dependent upon the total transvascular driving pressure (both hydrostatic and protein osmotic) and the conductance (integrity) of the microvascular barrier. Indeed, this relationship defines two types of pulmonary edema: increased pressure edema and increased permeability edema (19).

Over the last several years, my associates and I have defined several experimental animal models of nearly pure increased permeability edema (*Pseudomonas* bacteremia, histamine, and multiple pulmonary emboli). Multiple pulmonary emboli belong to a group of acute lung injuries characterized by high pulmonary vascular resistance due to *uneven* pulmonary artery obstruction. I believe that HAPE also belongs to this group (21).

We have demonstrated that the uneven pulmonary artery obstruction phenomenon is easy to produce in sheep, which are our experimental animals of choice (16). We prepare sheep and catheterize the efferent lymph duct from the caudal mediastinal lymph node which drains nearly pure lung lymph (20). We do our experiments either in anesthetized sheep at the time of the lymphatic cannulation or in unanesthetized sheep that have recovered from surgery and have chronic lung lymph fistulas. The results in both groups are the same although the material presented here is based on work in the anesthetized, acutely operated sheep.

We have found that a variety of obstructing agents, including mineral oil, glass beads, fibrin, and continuous intravenous air infusions, will cause injury proportional to the degree of pulmonary vascular obstruction obtained. An example of this effect is shown in Fig. 38-2. We are also able to confirm the reports of Gibbon and Gibbon (8) and of Hultgren and associates (12) that surgical resection of 65% of lung mass leads to increased lung lymph flow and causes interstitial pulmonary edema in the remaining lung (Landolt, unpublished). This and other evidence indicates that the principal site of increased fluid filtration is in the open, perfused lung microvascular bed, not in the obstructed portion (16).

The exact mechanism of the injury is slowly being pieced together. At first glance, the logical explanation seems to be that the high pulmonary arterial pressure associated with the markedly decreased vascular cross-section would be transmitted to the microcirculation via the maximally dilated pulmonary arterioles, and that the increased distending pressure would make the microvascular endothelium more permeable to fluid and protein. This is the phenomenon of "pore stretching" that was proposed more than 20 years ago (18).

Although the increased pressure in the open, perfused vessels contributes to the rate of fluid leakage into the lung, there is no evidence that the increased pressure is the primary cause of the injury. Comparable or even larger increases in microvascular pressure produced by left atrial hypertension do not cause this type of increased permeability edema (6,16).

My own favorite theory for the injury to the endothelium is that it is associated with the high linear velocity of blood flow and increased wall shear stress that occurs in the restricted microvascular bed (21). Unfortunately, this is not an easy mechanism to demonstrate directly, although the lung resection experiments already mentioned support it.

Most investigators have directed their attention toward biochemical mediators.

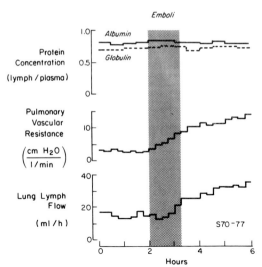

Fig. 38-2. Time course of the effect of uneven pulmonary artery obstruction in an anesthetized sheep. Pulmonary vascular resistance was increased by injecting siliconized glass microspheres over 90 min after a 2 h baseline observation period. Lung lymph flow increased in spite of a decrease in vascular surface area caused by the obstruction, and lymph protein concentration (expressed as ratio of lymph/plasma protein concentration, L/P, ratio) remained high, indicating an increase of protein flow proportional to the increase in lymph flow. (See also Ohkuda, 1978; Binder, 1979.) Reproduced by permission of J. Appl. Physiol.

There is strong evidence that, in the dog, the lung injury following microembolization is associated with increased fibrin deposit and subsequent fibrinolysis (17). Heparin appears to ameliorate the injury (15). However, Dr. Andrew Binder, working in our laboratory, found no evidence that either anticoagulation with heparin or virtually complete depletion of circulating fibrinogen affected the microembolism response in sheep. Figure 38-3 shows the time course of one experiment in a defibrinogenated sheep. The animal was depleted using viper

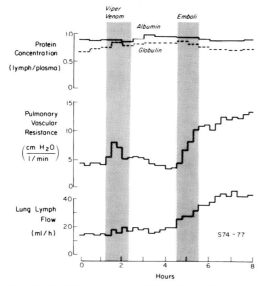

Fig. 38-3. Time course of the effect of uneven pulmonary artery obstruction in an anesthetized sheep in whom plasma and lung lymph had been depleted of fibrinogen in the second hour. There was no measurable fibrinogen in plasma or lung lymph. When pulmonary vascular resistance was increased by siliconized glass microspheres during the fifth hour, lymph flow increased and L/P remained high. (See also Binder, 1979.) Reproduced by permission of J. Appl. Physiol.

of nitrogen mustard. Figure 38-5 shows the time course of one of his experiments. Clearly, neutrophil depletion markedly attenuated the increased lymph flow associated with the microemboli (7).

In Fig. 38-6, I have compared all of the depletion regimes. The figure shows the effect of embolization on the highly sensitive lymph protein clearance (lymph flow × lymph/plasma protein concentration ratio). The animals depleted of fibrinogen and platelets show essentially the same response to embolization as do the control animals, whereas the sheep depleted of circulating neutrophils have a much reduced response.

Uneven Pulmonary Artery Obstruction, Neutrophils, and HAPE

Since the method of restricting the pulmonary microcirculation (various types of emboli or lung resection) does not appear to be critical to the acute lung injury that

venom and had no detectable fibrinogen in either plasma or lung lymph. The effect of embolization is no different than in the animal shown in Fig. 38-2 (3).

Platelet aggregation and release of vasoactive substances have contributed to increased lung microvascular permeability in isolated, perfused lungs (22). Dr. Binder depleted sheep of circulating platelets using specific antiplatelet serum. Figure 38-4 shows the time course of one experiment. The platelets were depleted by more than 95% but the embolism-induced lung microvascular permeability increase occurred as in the control sheep (2).

It has been shown recently that circulating polymorphonuclear leukocytes when activated by complement may be sequestered in the lung and lead to increased microvascular permeability (5). Dr. Michael Flick in our laboratory has successfully depleted sheep of more than 95% of circulating neutrophils using divided doses

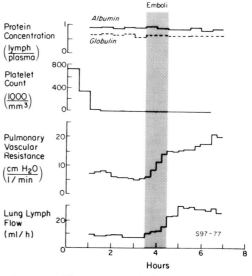

Fig. 38-4. Time course of effect of uneven pulmonary artery obstruction in an anesthetized sheep in whom circulating platelets had been depleted by more than 95% at t = 0. When pulmonary vascular resistance was increased in the fourth hour by siliconized glass microspheres, lung lymph flow increased and L/P remained high, as in the control sheep. (See also ref. 3.)

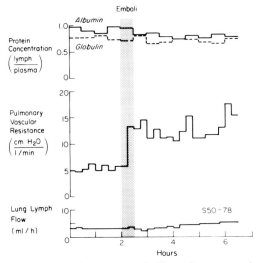

Fig. 38-5. Time course of effect of uneven pulmonary artery obstruction in an anesthetized sheep in whom circulating neutrophils had been depleted by more than 98% by nitrogen mustard. When pulmonary vascular resistance was increased by glass microspheres, lung lymph flow increased modestly and L/P decreased significantly. (See also Flick, 1979.)

follows, it may be that uneven pulmonary artery vasoconstriction that is reported to occur in high altitude hypoxia (13,23) belongs to the same group of acute lung injuries. Intense leukocyte infiltration is commonly reported in the lungs of patients dying of high altitude edema (1,11,24).

Summary

Because we lack an animal model of high altitude edema, my explanation of its pathophysiology is speculative. But my hypothesis has the advantage that it explains the clinical condition, including the characteristic chest x-ray, is clearly the cause of increased permeability in other types of uneven pulmonary artery obstruction, and that patients with HAPE show a marked increase in pulmonary vascular resistance.

Acknowledgment

This work supported in part by USPHS grants HL06285 (Program Project) and HL19155 (Pulmonary SCOR).

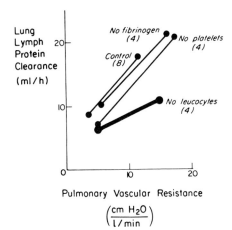

Fig. 38-6. Comparison of the effects of uneven pulmonary artery obstruction in sheep after depletion regimes. Pulmonary vascular resistance is used as the controlled variable to indicate the degree of pulmonary vascular obstruction. Protein clearance (lung lymph flow × L/P) is used as a sensitive index of vascular permeability. Embolization in sheep depleted of fibrinogen or platelets had no significant effect, whereas neutrophil depletion markedly attenuated the lung injury.

References

1. Arias-Stella, J. and Kruger, H.: Pathology of high altitude pulmonary edema. Arch. Path. 76:147, 1963.
2. Binder, A.S., Kageler, W., Perel, A., Flick, M.R., and Staub, N.C.: Effect of platelet depletion on lung vascular permeability after microemboli in sheep. J. Appl. Physiol. 48:414–420, 1980.
3. Binder, A.S., Nakahara, K., Ohkuda, K., Kageler, W., and Staub, N.C.: Effect of heparin on fibrinogen depletion on lung fluid balance in sheep after emboli. J. Appl. Physiol. 47:213–219, 1979.
4. Bland, R.D., Demling, R.H., Selinger, S.L., and Staub, N.C.: Effects of alveolar hypoxia on lung fluid and protein transport in unanesthetized sheep. Cir. Res. 40:269, 1976.
5. Craddock, P.R., Fehr, J., Brigham, K.L., Kronenberg, R.S., and Jacob, H.S.: Complement and leukocyte-mediated pulmonary

dysfunction in hemodialysis. New Eng. J. Med. 296:769, 1977.
6. Erdmann, A.J., III, Vaughan, T.R., Jr., Brigham, K.L., Woolverton, W.C., and Staub, N.C.: Effect of increased vascular pressure on lung fluid balance in unanesthetized sheep. Circulat. Res. 37:271, 1975.
7. Flick, M.R., Perel, A., Kageler, W., and Staub, N.C.: White blood cells contribute to increased lung vascular permeability after microemboli. Fed. Proc. 38:1327, 1979.
8. Gibbon, J.H., Jr. and Gibbon, M.H.: Experimental pulmonary edema following lobectomy and plasma infusion. Surgery 12:694, 1942.
9. Glazier, J.B. and Murray, J.F.: Sites of pulmonary vasomotor reactivity in the dog during alveolar hypoxia and serotonin and histamine infusion. J. Clin. Invest. 50:2550, 1971.
10. Hauge, A.: Hypoxia and pulmonary vascular resistance. The relative effects of pulmonary arterial and alveolar Po_2. Acta Physiol. Scand. 76:121, 1969.
11. Hultgren, H.N., Spickard, W., and Lopez, C.: Further studies on high altitude oedema. Brit. Heart J. 24:95, 1962.
12. Hultgren, H.N., Robinson, M.C., and Wuerflein, R.D.: Overperfusion pulmonary edema. Circulation. 34 (supl III):132, 1966.
13. Hultgren, H.N.: High altitude pulmonary edema, In Staub, N.C. (ed.): Lung Water and Solute Exchange. New York, Dekker, pp. 437–470.
14. Kato, M. and Staub, N.C.: Response of small pulmonary arteries to unilobar hypoxia and hypercapnia. Circ. Res. 21:426, 1966.
15. Malik, A.B. and van der Zee, H.: Mechanism of pulmonary edema induced by microembolization in dogs. Cir. Res. 42:72, 1978.
16. Ohkuda, K., Nakahara, K., Weidner, W.J., Binder, A., and Staub, N.C.: Lung fluid exchange after uneven pulmonary artery obstruction in sheep. Circ. Res. 43:152, 1978.
17. Saldeen, T.: The microembolism syndrome. Microvasc. Res. 11:227, 1976.
18. Shirley, H.H., Wolfram, C.G., Wasserman, K., and Mayerson, H.S: Capillary permeability to macromolecules: stretched pore phenomenon. Am. J. Physiol. 190:189, 1957.
19. Staub, N.C.: Pulmonary edema. Physiol. Rev. 54:678, 1974.
20. Staub, N.C., Bland, R.D., Brigham, K.L., Demling, R., Erdmann, A.J., III, and Woolverton, W.C.: Preparation of chronic lung lymph fistulas in sheep. J. Surg. Res. 19:315, 1975.
21. Staub, N.C.: Pulmonary edema due to increased microvascular permeability to fluid and protein. Circ. Res. 43:143, 1978.
22. Vaage, J., Nicolaysen, G., and Waaler, B.A.: Aggregation of blood platelets and increased hydraulic conductivity of pulmonary exchange vessels. Acta Physiol. Scand. 98:175, 1976.
23. Viswanathan, R., Jain, S.K., and Subramanian, S.: Pulmonary edema of high altitude III. Pathogenesis. Am. Rev. Resp. Dis. 100:342, 1969.
24. Wilson, R.: Acute high-altitude illness in mountaineers and problems of rescue. Ann. Intern. Med. 78:421, 1973.

39
Vasopressin in Acute Mountain Sickness and High Altitude Pulmonary Edema

P.H. Hackett, Mary L. Forsling, J. Milledge, and D. Rennie

While Forsling and Milledge (2) found that 4 h of exposure to 10.5% oxygen (equivalent to 5490 m altitude) had little effect on vasopressin (ADH) release in man, the result of prolonged exposure is not clear. Excessive fluid retention and hence possibly vasopressin could contribute to the various forms of edema seen at altitude. These include pulmonary, cerebral, and peripheral edema as defined previously (3). The purpose of the present study was to investigate the release of vasopressin in the human at altitude with a view to a greater understanding of the pathophysiology of acute mountain sickness (AMS) and altitude edema.

Methods

The observations were performed at Pheriche, Nepal (4243 m), in the Himalayas on 14 subjects. These included eight control subjects (five female) who showed no symptoms, four subjects with AMS (all female), and two subjects with high altitude pulmonary edema (HAPE), both male. On sampling, the blood was immediately centrifuged, and the plasma separated and maintained in liquid nitrogen until analyzed in London. Hormone determination was performed on extracted plasma, as described by Chard and Forsling (1).

Results and Discussion

The results are summarized in Fig. 39-1. The plasma vasopressin in the control group was 1.1 ± 0.15 μU/ml ($\pm$ SEM, $n = 8$) as compared with 0.7 ± 0.17 μU/ml ($n = 10$) in a previous study at sea level, and there was no difference in vasopressin concentration between the ascent and the descent from 5300 m. Thus altitude alone did not appear to affect the release of vasopressin. The results were similar to those seen with acute hypoxia, when a value of 1.1 μU/ml was recorded (2). In those subjects with AMS plasma vasopressin was significantly elevated to a mean concentration of 2.2 ± 0.5 μU/ml ($n = 4$, $p < 0.02$) and was 3.2 and 4.0 μU/ml in the two subjects with HAPE. The results of Singh et al. (4) suggest a similar trend. It was also noted that four of the climbers with AMS also suffered with peripheral edema (Fig. 39-1). However, mean ADH

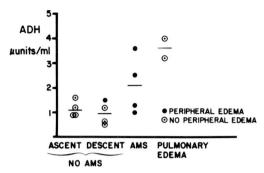

Fig. 39-1. Plasma ADH (vasopressin) concentration was elevated in those with AMS and those with HAPE. The mean concentrations in those with and without peripheral edema were similar (—, represents mean values).

concentrations in those with and without peripheral edema were similar. The present studies do not indicate whether the increased vasopressin secretion was a primary change and a major factor in the pathophysiology of AMS or secondary to a shift in fluid from the vascular bed or even to stress. The concentrations recorded would have been sufficient to cause a degree of water retention provided that the renal sensitivity to vasopressin was unchanged.

Acknowledgment

A grant from the Wellcome Trust is gratefully acknowledged as is Grant No. 47053 from the Rush Medical Center.

References

1. Chard, T. and Forsling, M.L.: In Antoniades, H.M. (ed.): Hormones in Blood. Cambridge, Mass., Harvard University Press, 1976, pp. 485–516.
2. Forsling, M.L., and Milledge J.S.: Effect of hypoxia on vasopressin release in man. J. Physiol. 267:22, 1977.
3. Hackett, P.H., Rennie, D., and Levine, H.D.: The incidence, importance, and prophylaxis of acute mountain sickness. Lancet II:1149, 1976.
4. Singh, I., Malhotra, M.S., Khanna, P.K., Nanda, R.B., Purshottam, T., Upadhyay, T.M., Radhakrishnan, U., and Brahachari, H.D.: J. Biometeorol. 18:211, 1974.

40

Hypoxic Pulmonary Vasoconstriction and Ambient Temperature

J. Durand, J. Coudert, J.D. Guieu, and J. Mensch-Dechene

It has long been known that hypoxia and arteriolar structural changes were the main factors in altitude pulmonary hypertension. However, other environmental factors such as cold may play a contributing role.

In order to test the effects of exposure to heat and cold on highlanders' pulmonary hemodynamics, 21 volunteers, born and residing between 3800 and 4200 m, were studied at 3750 m ($P_B = 493$ mmHg).

Cardiac output (measured by indocyanine green dilution), brachial and pulmonary (free and wedge) arterial pressures, heart rate (EKG), and blood gases were measured in different environments.

1) In 21 subjects at room temperature (Tgl 25°C, 40–45% humidity) breathing ambient air.
2) In 13 subjects after a 30-min exposure to radiant heat (5 kW).
3) In 9 of the proceding subjects for whom the heat stress was prolonged, and the hypocapnia induced by thermal hyperventilation was corrected by breathing hypercapnic mixture.
4) In 8 subjects after 30-min exposure to cold from ice bags on the limbs and the chest. In no case was shivering observed.

Heat exposure increases cardiac output with no significant change in pulmonary pressures. This situation results in a significant decrease in the so-called pulmonary arteriolar resistance; correction of the hypocapnia does not significantly affect the effect of heat.

Conversely, cold stress decreases cardiac output and induces an increase, although not significant, in pulmonary artery pressures. Pulmonary vascular resistance increases significantly (Fig. 40-1).

The effect of elevated ambient temperature on pulmonary hemodynamics has already been studied at sea level. As far as cold is concerned, very little has been published concerning humans; Viswanathan failed to find any change in pulmonary arterial pressure during cold pressure tests at sea level in subjects with a prior history of HAPE, whereas Atterhög et al. measured an increase central blood volume after exposure to a cold wind. However, there are several publications on the effects of both cold and altitude upon the

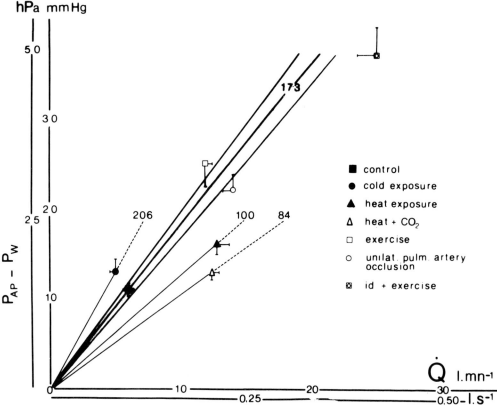

Fig. 40-1. Difference between mean pulmonary artery (P_{AP}) and pulmonary artery wedge pressure (P_w) as function of cardiac output ($\dot{Q}$). Data obtained during thermal stress are compared to those published by Lockhart et al. (2) during exercise and during unilateral pulmonary artery occlusion (rest and exercise). To make the comparison easier, values of cardiac output during pulmonary artery occlusion have been multiplied by 2 (mean and standard error). Isoresistance lines have been drawn with corresponding values expressed in $kPa \cdot ml^{-1} \cdot s$; shaded area corresponds to 95% of pulmonary arteriolar resistance measured in La Paz, Bolivia, in normal subjects at rest.

pulmonary circulation in animals (1,3) which show the potentiation of the vasomotor effect of hypoxia by cold.

Summary

Thermal environment has to be taken into account in the interpretation of changes in the pulmonary circulation induced by altitude hypoxia, even in resting conditions. For instance, at 3750 m, mean pulmonary artery pressure measured routinely in a catheterization room at 19°C was found to be 21 ± 4 mmHg ($n = 213$) whereas it is 19 ± 4 mmHg ($n = 118$) when measured at 25°C room temperature.

Acknowledgment

This work was realized at the Instituto Boliviano de Biología de Altura, La Paz (Bolivia), and was supported in part by the Ministère des Affaires Etrangères (Cooperation Technique) and by the C.N.R.S. (RCP 538).

References

1. Chauca, D. and Bligh, J.: An additive effect of cold exposure and hypoxia on pulmonary artery pressure in sheep. Res. Vet. Sci. 21:123, 1976.
2. Lockhart A., Zelter, M., Mensch-Dechene,

J., Antezana, G., Paz Zamora, M., Vargas, E., and Coudert, J.: Pressure flow-volume relationships in the pulmonary circulation of normal highlanders. J. Appl. Physiol., 41(4):449, 1976.

3. McMurtry, I.F., Reeves, J.T., Will, D.H., and Grover, R.F.: Hemodynamic and ventilatory effects of skin-cooling in cattle. Experientia 31:1303, 1975.

41

Pathophysiology of Acute Mountain Sickness and High Altitude Pulmonary Edema: An Hypothesis

J.R. Sutton and N. Lassen

Hypothesis

High altitude pulmonary edema and acute mountain sickness occur together more frequently than is generally recognized and have a common pathophysiological basis. Both are due to increased pressure and flow in the microcirculation, causing edema in the brain and edema in the lungs.

Introduction

Acute mountain sickness (AMS) and high altitude pulmonary edema (HAPE) are usually regarded as separate pathological entities. We believe that the two coexist and have a common pathophysiological mechanism. On exposure to altitude, there is a latent period between the time at which hypoxia begins and the onset of symptoms referrable to AMS or HAPE. This time interval varies, from several hours to days, as does the individual susceptibility and altitude of occurrence. Thus, hypoxia itself cannot directly produce the symptoms of AMS or HAPE and must initiate other pathophysiological events.

Brain Edema

Cerebral vasodilatation and an increased cerebral blood flow occur on exposure to high altitude and, when associated with high pressure in the cerebral microcirculation, will result in brain edema similar to that found in acute systemic arterial hypertension associated with hypertensive encephalopathy. In this situation, the normal autoregulatory vasoconstrictor response to increasing pressure breaks down and results in a forced vasodilatation of the arterioles and edema due to transcapillary and transarteriolar filtration. Exercise and cold, by producing transient elevations in systemic blood pressure, would contribute to the edema formation. Sleep may also aggravate cerebral edema by increasing the hypoxia. Thus, the factors identified as necessary to produce brain edema are hypoxia, followed by increased cerebral blood flow and acute transient systemic hypertension.

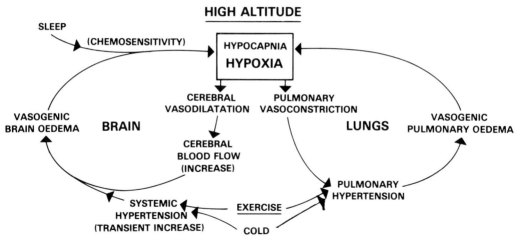

Fig. 41-1. Pathophysiology of high altitude lung and brain edema. Upon exposure to high altitude, all persons develop hypoxia and hypocapnia. In the lung, pulmonary vasoconstriction and pulmonary hypertension occur. Exercise and cold aggravate the pulmonary hypertension and areas of the lung which are overperfused develop pulmonary edema, thus worsening the hypoxemia and setting up a vicious cycle. In the brain, cerebral vasodilatation and an increase in cerebral blood flow occur. This, together with exercise and/or cold, which transiently elevate systemic blood pressure, results in brain edema. Sleep will decrease chemosensitivity, which will result in a decreased ventilation and worsening hypoxemia, which will aggravate the cycle. (Reproduced by kind permission of the editor of *Bulletin Européen de Physiopathologie Respiratoire*.)

High Altitude Pulmonary Edema (HAPE)

Pulmonary hypertension is invariable in HAPE. Pulmonary hyperperfusion will result from exercise and, with areas of hyperperfusion, pulmonary edema will result. In sleep, hypoxemia will contribute to the development of HAPE. Thus, the factors identified as necessary to produce HAPE are hypoxia, increased pulmonary blood flow (of a regional nature), and pulmonary hypertension.

Reference

1. Sutton, J.R. and Lassen, N.: Pathophysiology of acute mountain sickness and high altitude pulmonary edema: an hypothesis. Bull. Europ. Physiopath. Resp. 15:1045, 1979.

42

Use of Furosemide in Prevention of HAPE

S.K. KWATRA AND R. VISWANATHAN

Furosemide is a derivative of anthranilic acid. Its diuretic action is due to inhibition of salt and water resorption. Because it is fast acting and has minimal side effects, furosemide has been extensively used in a variety of edematous conditions, including pulmonary edema. Singh et al. found that soldiers who were prone to acute mountain sickness and pulmonary edema became oliguric on arrival at high altitude, whereas those immune to altitude effects passed urine freely. Working on this hypothesis, these workers gave dehydration therapy with furosemide prophylactically to soldiers before transporting them to high altitude. Singh has reported that HAPE was prevented completely and acute mountain sickness nearly so. Aoki and Robinson, however, failed to confirm the above results in a controlled study on 12 test subjects at a simulated altitude of 14,000 ft.

In view of the above conflicting reports, the present study was undertaken to evaluate the usefulness of prophylactic dehydration procedures.

Different batches of animals of different species (rats, mice, guinea pigs) were used for the purpose. Furosemide was given by injection 10 mg/kg body weight 2 h before exposure to simulated altitude of 10,000 m for 6 h. The number of survivors was noted. The results show that the animals' altitude tolerance was definitely reduced by giving the drug before exposure. Significantly more animals died in the group given furosemide than among the controls (Table 42-1).

Table 42-1. Death Rate of Furosemide-Treated Animals and Controls at High Altitude

	Control			Injected			
	Died	No.	%	Died	No.	%	Significance
Rats	9	32	28.1	18	32	56.2	HS
Mice	4	21	19.2	6	21	28	NS
Guinea pigs	3	11	27.3	8	11	72.7	HS

III. Chronic Mountain Sickness and Performance

43
Chronic Mountain Sickness: A Pulmonary Vascular Disease?

JULIO C. CRUZ AND SIXTO RECAVARREN

Introduction

Chronic mountain sickness (CMS), "soroche crónico," "eritremia de la altura," or "Monge's disease" was first reported in 1928 by Carlos Monge (25). Since then, concepts of and knowledge about this disease have evolved (3,17,19,26,27,29, 32,40,42) up to the point at which doubts have been raised as to whether this disease really exists, the doubt due to the lack of anatomical evidence (15). Most people with CMS are lifelong residents at high altitude (above 3000 m). They may be asymptomatic and live normally. Symptomatic patients may complain of blurred vision, headache, dyspnea, insomnia, and paresthesias. Physical examination may reveal clubbing of fingers, congestion of sclerae, and mucosa with cyanosis. In extreme cases, symptoms and signs of right-heart failure are seen (26, 29,32). Laboratory analyses show high hematocrit and hemoglobin above the range found in normal high altitude subjects. White cell counts and platelets are not different from normal subjects (21). Arterial blood shows low P_{O_2} and decreased arterial oxygen saturation, both abnormal for the altitude of residence, and some of them may have elevated Pa_{CO_2}. Many patients reveal no evidence of thoracic cage abnormalities or bronchopulmonary disorders. Symptoms and signs improve and may disappear with time when CMS patients move to lower elevations or to sea level (17,19,25–27,29,32). Differential diagnosis of CMS may become difficult when some known bronchopulmonary illness is also present because blood gas abnormalities may be due to the associated ailment rather than to CMS. A man with CMS should have excessive polycythemia, abnormal hypoxemia, and possibly hypercapnia without a recognized pulmonary disease. However, some published reports dealing with CMS patients have included patients with known pulmonary diseases or dysfunction (10,24,40), or pulmonary function testing was not performed at all (28,32). It is now recognized that 50% of people with excessive polycythemia living at high altitude have an associated pulmonary disease or dysfunction (23). What is the cause of excessive polycythemia and abnormal hypoxemia of the other 50%?

Hypothesis

We believe that high altitude dwellers with an excessive amount of smooth muscle in the lung arterioles (more than in normal subjects (1) at a given altitude) are candidates for CMS. The association of an increased lung arteriolar smooth muscle and the stimulus of chronic alveolar hypoxia leads to an increase in the pulmonary artery pressure (PAP), which is greater than that found in normal high altitude natives (1,4,16,33,41), but lower than that seen in primary pulmonary hypertension. This level of pulmonary hypertension opens up the preterminal arterioles (36) producing a venous-arterial shunt, which, in turn, decreases further the arterial oxygen tension and saturation. The accentuated hypoxemia will stimulate the production of more red blood cells leading to excessive polycythemia. With the passage of time, excessive polycythemia causes pulmonary dysfunction (8): airway obstruction and widening of alveolar arterial gas tension differences. Thus, according to our hypothesis, abnormal hypoxemia is due to a shunt increase, and hypercapnia, when present, is a consequence of ventilation-perfusion imbalance resulting from excessive polycythemia.

Supporting Evidence

Chronic mountain sickness patients have more severe pulmonary hypertension (10,32,38) than is seen in normal high altitude natives (4,16,33,41). Is this a consequence of alveolar hypoventilation? Our review of the literature (5,18,20,28,39,40,43) shows that mean $Pa_{CO_2} \pm 1$ SD in 22 cases of CMS patients was 38.1 ± 2.91 mmHg, while in 159 normal high altitude residents it was 32.5 ± 2.49 mmHg. In spite of this difference, 41% of CMS patients have Pa_{CO_2} within the normal range (Fig. 43-1). Moreover, when CMS patients descend to sea level, alveolar P_{CO_2} does not rise to abnormal limits (Table 43-1), as one would expect with a central or peripheral depression in chemoreceptor sensitivity. Furthermore, the calculated alveolar ventilation of CMS patients is not statistically different from that of high altitude residents (8,40) and their total ventilation may even be higher than normal (10,24). Thus, we may conclude that the high Pa_{CO_2} found in some CMS patients may be due either to an associated pulmonary illness or to uneven ventilation-perfusion caused by excessive polycythemia. Therefore, the hypoventilation theory originally proposed by Hurtado (19) and followed by others (32,40,42) does not appear to be the cause of the abnormal hypoxemia, excessive polycythemia, and pulmonary hypertension observed in CMS patients.

How, then, are we to explain the pulmonary hypertension in CMS patients? Pulmonary artery pressure and hypoxemia are correlated (7). However, pulmonary arterial hypoxemia alone, without airway hypoxia, does not produce an increase in PAP in mammals (9,14,37). Therefore, accepting that CMS patients do not hypoventilate, their elevated PAP, which is greater than that found in normal high altitude natives, must be due to other factor(s). We believe that the anatomical structure of the pulmonary vasculature is the main factor in determining the abnormally high pulmonary artery pressure. A linear relationship between the amount of smooth muscle in the pulmonary arterioles and the level of PAP and right ventricular hypertrophy when exposed to high altitude has been shown among several mammalian species, including man (44). Thus, it is expected that CMS patients with higher PAP than normal high altitude residents will have also more smooth muscle in their arterioles. Evidence for more vascular smooth muscle in CMS patients has been reported in some autopsy material (3,11,34). However, because of associated kyphoscoliosis and/or obesity found in those reports, excessive muscle in the pulmonary arterioles was attributed to those other ailments. We have also observed an increase in the amount of smooth muscle in the pulmonary arterioles (unpublished observations) in one patient with CMS. However, associated liver cirrhosis

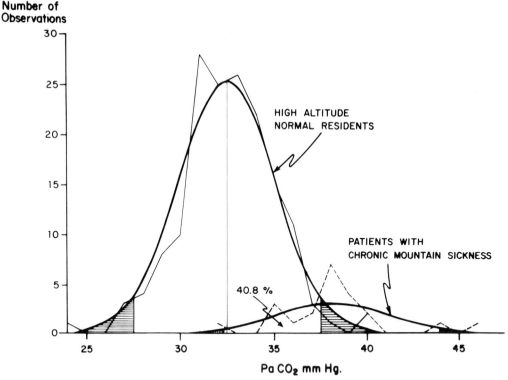

Fig. 43-1. Arterial P_{CO_2} distribution in high altitude normal residents (N) and patients with chronic mountain sickness (CMS). The best Gaussian curves have been fitted using the normal equation. Difference in curve height is due to the number of subjects studied. Mean and standard deviation of 159 N and 22 CMS patients were 32.5 ± 2.5 mmHg and 38.1 ± 2.9 mmHg, respectively. Data obtained from the literature.

prevented us from presenting him as a typical autopsy of CMS.

If we accept that CMS patients have increased lung vascular smooth muscle, the question then arises, how was this muscle developed? It is known that pulmonary and systemic arteries of the fetus at term have much the same smooth muscle mass (30). Furthermore, stillborn natives from high altitude and sea level show a similar structural pattern of the pulmonary arterial tree (2) and about the same right ventricular wall (35). These anatomical findings are in agreement with the lack of differences in electrocardiographic patterns (31) and PAP values (12) between newborns from high altitude and sea level. This anatomical cardiopulmonary feature regresses rapidly at sea level, acquiring an adult pattern at 3 to 6 mo of age (1,35). On the other hand there is a delay and incomplete involution of the fetal characteristics of pulmonary arterioles (1) as well as in the thickness of the ventricular wall (35) in high altitude native subjects. These anatomical differences are manifested by persistence of the rightward deviation of the mean AQRS axis in the electrocardiogram (31) and also in the high values of the PAP (41). The pulmonary arterioles resembling fetal structure observed in high altitude native adults are indicative of an increased amount of smooth muscle in the lung (1). That all the authors above reported normal findings suggests to us that, possibly, some people have more pulmonary vascular smooth muscle than is usually found in normals, that this is probably genetically determined, and that these people may then develop more severe pulmonary hypertension when living at high al-

Table 43-1. Arterial and/or Alveolar Gas Tensions (Mean ± SEM) Measured in Chronic Mountain Sickness Patients. Data Obtained From the Literature.

	Elevation (m)	No. subjects	PA_{O_2} (mmHg)	Pa_{O_2} (mmHg)	AaD_{O_2}	No. subjects	Pa_{CO_2} (mmHg)	PA_{CO_2} (mmHg)	aAD_{CO_2}
Monge et al. (28)	4,300					5	39.3 1.75		
Severinghaus et al. (40)	4,330	6	43.4 2.07	39.7 3.02	3.7 1.22	6	37.2 0.58	34.8[a] 0.91	2.4 0.37
Sørensen and Severinghaus (43)	4,360	3		41.1 1.06		3	37.0 1.27		
Lefrançois et al. (24)	3,660	9	54[b] 1			9		33 1	
Ergueta et al. (10)	3,650	20	56.5 0.77	47.9 0.86	8.6 0.54	20	43.5 0.59	35.4 0.61	8.1 0.3
Hurtado (unpublished observations)	4,540	10	46.1 1.06	37.1 1.03	9.0 0.81	7	37.4 1.03	33.5 0.77	3.9 0.93
1 day at	150	5	108.3 1.12			5		33.8 1.05	
10–16 days at	150	4	102.7 3.07			4		40.1 0.58	
20–36 days at	150	5	102.7 1.28			6		39.7 1.13	
Cruz et al. (8)	3,700	4	48.5 2.31	42.2 4.47	6.3 3.06	4	37.7 2.54	33.6 2.13	4.1 0.88

Key: PA_{O_2}, PA_{CO_2} = alveolar oxygen and carbon dioxide pressures. Pa_{O_2}, Pa_{CO_2} = arterial oxygen and carbon dioxide pressures. AaD_{O_2}, aAD_{CO_2} = alveolar-arterial oxygen and carbon dioxide tension differences.
[a] Measurements not simultaneously taken with Pa_{CO_2} in three cases.
[b] Results are from 44 values.

titude. Assuming that the amount of smooth muscle in the lung arterioles follows a Gaussian distribution among individuals, the upper extreme would predict (with an error of 5%) that 2.5% of the population will have a greater-than-normal amount of arteriolar smooth muscle in the lungs. We believe that this selected population when living at high altitude may develop CMS.

Is it possible that pulmonary hypertension may produce hypoxemia? One model of pulmonary hypertension without alveolar hypoxia is seen in primary pulmonary hypertension. A recent review on this subject shows that patients with primary pulmonary hypertension do have lower arterial oxygen tension than normals (45). A mechanism of intrapulmonary shunting through dilated capillary channels has been suggested (45). Our own observations in young calves show a good direct correlation between PAP and shunt (7) which has also been observed in dogs (22). Thus, it is quite possible that CMS patients with higher than normal PAP may have an increased shunt. It has been shown that CMS patients failed to reach arterial oxygen tensions as high as normals when they breathed oxygen (8). It should be emphasized that the moderate pulmonary hypertension observed in normal high altitude natives is not associated with abnormal hypoxemia. In fact, they have smaller alveolar arterial O_2 tension differences than sea level subjects (6,21). Furthermore, the shunt in high altitude natives is similar to that in sea level subjects dwelling at altitude (6). Thus, the hypoxemia of normal high altitude natives is the expected one for the altitude of residence.

Chronic Mountain Sickness: Animal Model

It is well known that some cattle develop pulmonary hypertension and right-heart failure when exposed to high altitude (13,47). Because of edema in the pectoral region, the ailment is called "brisket disease." However, "high mountain disease" has been suggested as being a more explicit term (47). Little effort has been made to link the high mountain disease in cattle with CMS in humans. Chronic high mountain disease in cattle is thought to be of vascular origin, while the one in humans has been thought to be due to a respiratory cause (32). Since our hypothesis of CMS proposes a vascular origin, the calf then seems to be a good animal model to study CMS. Furthermore, a breeding program established over the past 12 years has resulted in calves genetically susceptible or resistant to hypoxic pulmonary hypertension (48), and the distinguishing feature seems to be the amount of pulmonary arteriolar smooth muscle (46).

Summary

To explain the pathogenesis of Chronic mountain sickness, a new hypothesis is proposed. The basic anatomical feature for the development of this condition is the presence of an inordinate amount of vascular smooth muscle in the pulmonary arterioles. When an individual with excessive muscle in his pulmonary vasculature lives in a hypoxic environment at high altitude, pulmonary hypertension develops, and this hypertension is more severe than in normal high altitude natives. Pulmonary hypertension opens up the preterminal arterioles producing a venous–arterial shunt, decreasing further the arterial oxygen tension and saturation. The accentuated hypoxemia will stimulate the production of more red cells, causing excessive polycythemia. With the passage of time, excessive polycythemia produces pulmonary dysfunction: airway obstruction and increased alveolar arterial gas tension differences, leading to CO_2 retention and aggravated hypoxemia. Thus, according to our hypothesis, abnormal hypoxemia at high altitude results from a shunt increase, and hypercapnia, when present, is a consequence of ventilation–

perfusion imbalance due to excessive polycythemia.

Acknowledgments

We are indebted to Professor Hurtado for giving us his unpublished data obtained in normal high altitude residents and in patients with chronic mountain sickness. We wish to thank Dr. Robert F. Grover for kindly reviewing the manuscript.

References

1. Arias-Stella, J. and Saldaña, M.: The terminal portion of the pulmonary arterial tree in people native to high altitudes. Circulation 28:915, 1963.
2. Arias-Stella, J. and Castillo, Y.: The muscular pulmonary arterial branches in stillborn natives of high altitudes. Lab. Invest. 15:1951, 1966.
3. Arias-Stella, J.: Chronic mountain sickness: Pathology and definition. In Porter, R. and Knight, J. (eds.): High altitude physiology: Cardiac and respiratory aspects. London, Churchill Livingstone, 1971, pp. 31-40.
4. Banchero, N., Sime, F., Peñaloza, D., Cruz, J., Gamboa, R., and Marticorena, E.: Pulmonary pressure, cardiac output and arterial oxygen saturation during exercise at high altitude and at sea level. Circulation 33:249, 1966.
5. Cruz-Jibaja, J.C.: Alveolar gas exchange at altitude. DAMD-17-74-G-9383, US Army Med. Res. Rept 15, 1974.
6. Cruz, J.C., Hartley, L.H., and Vogel, J.A.: Effect of altitude relocations upon AaD_{O_2} at rest and during exercise. J. Appl. Physiol. 39:469, 1975.
7. Cruz, J.C., Russell, B.E., Reeves, J.T., and Alexander, A.F.: Relationship of venous admixture to pulmonary hypertension. Fed. Proc. 38:1379, 1979 (abstract).
8. Cruz, J.C., Diaz, C., Marticorena, E., and Hilario, V.: Polycythemia impairs pulmonary gas exchanges in chronic mountain sickness. Proc. Hypoxia Symp. Banff, Alberta, Canada, 1979.
9. Duke, H.N.: The site of action of anoxia on the pulmonary blood vessels of the cat. J. Physiol. (London) 125:373, 1954.
10. Ergueta, J., Spielvogel, H., and Cudkowicz, L.: Cardiorespiratory studies in chronic mountain sickness (Monge's disease). Respiration 28:284, 1971.
11. Fernán-Zegarra, L. and Lazo-Taboada, F.: Mal de montaña crónico. Consideraciones anátomo-patológicas y referencias clínicas de un caso. Revista Peru Patol. 6:49, 1961.
12. Gamboa, R. and Marticorena, E.: Circulación pulmonar en el niño recién nacido en las grandes alturas desde el nacimiento a las 72 horas de edad. Revista Peru Pediat. 29:5, 1971.
13. Glover, G.H. and Newsom, I.E.: Bulletin 204, Fort Collins, Colorado, Agricultural Experiment Station of the Colorado Agricultural Colleges, January 1915.
14. Hauge, A.: Hypoxia and pulmonary vascular resistance: the relative effects of pulmonary arterial and alveolar P_{O_2}. Acta Physiol. Scand. 76:121, 1969.
15. Heath, D.: Discussion. In Porter, R. and Knight, J. (eds.): High altitude physiology: Cardiac and respiratory aspects. London, Churchill Livingstone, 1971, p. 54.
16. Hultgren, H.N., Kelly, J., and Miller, H.: Pulmonary circulation in acclimatized men at high altitude. J. Appl. Physiol. 20:233, 1965.
17. Hurtado A: Sobre la patología de la altura. Revista Med. Peruana 2:335, 1930.
18. Hurtado, A. and Aste-Salazar, H.: Arterial blood gases and acid base balance at sea level and at high altitudes. J. Appl. Physiol. 1:304, 1948.
19. Hurtado, A.: Pathological aspects of life at high altitude. Milit. Med. 117:272, 1955.
20. Hurtado, A., Velásquez, T., Reynafarje, C., Lozano, R., Chávez, R., Aste-Salazar, H., Reynafarje, B., Sánchez, C., and Muñoz, J.: Blood gas transport and acid-base balance at sea level and at high altitudes. Randolph Field, Texas, USAF School Aviat. Med Rep, 56:104, 1956.
21. Hurtado, A.: Animals in high altitudes: resident man. In Handbook of Physiology. Adaptation to the environment. Washington, D.C., Am. Physiol. Soc., 1964, p. 843-860.
22. Kentera, D., Wallace, C., Hamilton, W., and Ellison, L.: Venous admixture in dogs with chronic pulmonary hypertension. J. Appl. Physiol. 20:919, 1965.
23. Kryger, M., McCullough, R., Doekel, R., Collins, D., Weil, J.V., and Grover, R.F.: Excessive polycythemia of high altitude:

23. role of ventilatory drive and lung disease. Am. Rev. Resp. Dis. 118:659, 1978.
24. Lefrancois, R., Gautier, H., and Pasquis, P.: Ventilatory oxygen drive in acute and chronic hypoxia. Resp. Physiol. 4:217, 1968.
25. Monge, C.: La enfermedad de los Andes. Sindromes eritrémicos. Anal. Fac. Med. Lima 11:1, 1928.
26. Monge, C.: Chronic mountain sickness. Physiol. Rev. 23:166, 1943.
27. Monge, C.: Syndromes biologiques et cliniques prodits par les changements d'altitude. Bull. Acad. Suisse. Sci. Med. 7:187, 1951.
28. Monge, C.C., Lozano, R., and Carcelen, A.: Renal excretion of bicarbonate in high altitude natives and in natives with chronic mountain sickness. J. Clin. Invest. 43:2303, 1964.
29. Monge, M.C, and Monge, C.C: High altitude diseases: Mechanism and management. Springfield, Ill., Charles C. Thomas Publishers, 1966, pp. 32–60.
30. Naeye, R.L. and Letts, H.W.: The effects of prolonged neonatal hypoxia on the pulmonary vascular bed and heart. Pediatrics 30:902, 1962.
31. Peñaloza, D., Dyer, J., Echevarria, M., and Marticorena, E.: The influence of high altitudes on the electrical activity of the heart. Electrocardiographic and vectorcardiographic observations in the newborn, infants and children. Am. Heart J. 59:111, 1960.
32. Peñaloza, D.: Corazón pulmonar crónico por desadaptación a la altura (mal de montaña crónico). Doctoral thesis. Lima, Univ. Peruana Cayetano Heredia, p. 1–114, 1969.
33. Peñaloza, D., Sime, F., Banchero, N., Ganboa, R., Cruz, J., and Marticorena, E.: Pulmonary hypertension in healthy men born and living at high altitudes. Am. J. Cardiol. 11:150, 1963.
34. Reategui-Lopez, L.: Soroche crónico. Observaciones realizadeas en el Cusco en 30 casos. Revista Peru Cardiol. 15:45, 1969.
35. Recavarren, S. and Arias-Stella, J.: Topography of right ventricular hypertrophy in children native to high altitude. Am. J. Pathol. 41:467, 1962.
36. Recavarren, S.: The preterminal arterioles in the pulmonary circulation of high altitude natives. Circulation 33:177, 1966.
37. Reeves, J.T., Leathers, J.E., Eiseman, B., and Spencer, F.C.: Alveolar hypoxia versus hypoxemia in the development of pulmonary hypertension. Med. Thorac., 19:561, 1962.
38. Rotta, A., Cánepa, A., Hurtado, A., Velásquez, T., and Chávez, R.: Pulmonary circulation at sea level and at high altitude. J. Appl. Physiol. 9:328, 1956.
39. Severinghaus, J.W. and Carcelén, A.: Cerebrospinal fluid in man native to high altitude. J. Appl. Physiol. 19:319, 1964.
40. Severinghaus, J.W., Baiton, C.R., and Carelén, A.: Respiratory insensitivity to hypoxia in chronically hypoxic man. Resp. Physiol. 1:308, 1966.
41. Sime, F., Banchero, N., Peñaloza, D., Gamboa, R., Cruz, J., and Marticorena, E.: Pulmonary hypertension in children born and living at high altitudes. Am. J. Cardiol. 11:143, 1963.
42. Sime, F., Monge, C., and Whittenbury, J.: Age as a cause of chronic mountain sickness (Monge's disease). J. Biometeorol. 19:93, 1975.
43. Sørensen, S.C. and Severinghaus, J.W.: Respiratory sensitivity to acute hypoxia in man born at sea level living at high altitude. J. Appl. Physiol. 25:211, 1968.
44. Tucker, A., McMurtry, I.F., Reeves, J.T., Alexander, A.F., Will, D.H., and Grover, R.F.: Lung vascular smooth muscle as a determinant of pulmonary hypertension at high altitude. Am. J. Physiol. 228:762, 1975.
45. Woelkel, N. and Reeves, J.T.: Primary pulmonary hypertension. In: Lung biology in health and disease. M. Dekker Edition (in press).
46. Weir, E.K., Will, D.H., Alexander, A.F., McMurty, I.F., Looga, R., Reeves, J.T., and Grover, R.F.: Vascular hypertrophy in cattle susceptible to hypoxia pulmonary hypertension. J. Appl. Physiol. 46:517, 1979.
47. Will, D.H., Alexander, A.F., Reeves, J.T., and Grover, R.F.: High altitude induced pulmonary hypertension in normal cattle. Circ. Res. 10:172, 1962.
48. Will, D.H., Hickes, J.L., Card, C.S., and Alexander, A.F.: Inherited susceptibility of cattle to high altitude pulmonary hypertension. J. Appl. Physiol. 38:491, 1975.

44
Predicting Mountaineering Performance at Great Altitudes

Hsueh-han Ning, Shao-yung Huang, Mei-chuen Gung, Chung-yuan Shi, and Shu-tsu Hu

With a view to developing some method of predicting sea level subjects' performance at great altitudes, groups of mountaineers and candidates were assessed chiefly for their cardiorespiratory functions in a hypobaric chamber and at the base camp (both at an intermediate altitude of 5000 m) during the 1975 Chinese Mountaineering and Scientific Expedition to Mt. Jolmo Lungma (Mt. Everest). The results of the studies were correlated with subsequent actual performance above the base camp up to the peak. Some test methods or indices showed better predictive value than others. This paper relates how the "cardiac pump function test," in conjunction with some other parameters, serves to predict the mountaineering performance at altitudes of 5000 m and higher.

Methods and Procedures

Cardiac Pump Function Test (CPF Test)

This composite test takes into consideration both inotropic and chronotropic properties of the heart. This noninvasive method involves measurements of systolic and diastolic brachial arterial pressure, ventricular systolic and diastolic time from cardiophonogram, and heart rate. From these, a *work index* (I_W) and a *driving force index* (I_{DF}) are computed. The results can be represented graphically and graded in order of degree of competence.

The Work Index

The work done by the left ventricle may be obtained by Eq. (1):

$$W = Q(\bar{P} - \bar{P}_a) \times 1332.8 \qquad (1)$$

where W = work done by the left ventricle (in dyne-cm/sec); Q = cardiac output (in ml/sec); $\bar{P}$ = mean arterial blood pressure (in mmHg); $\bar{P}_a$ = mean pressure in left atrium (in mmHg). The cardiac output Q may be computed:

$$Q = q \cdot R/60 \qquad (2)$$

where q = the stroke volume (in ml/beat); R = the heart rate (in beats/min). Since the stroke volume q bears a linear relationship, within a fairly wide range of physiological changes, (a) to the mean arterial blood pressure $\bar{P}$ (1-3); (b) to the pulse pressure or the difference of systolic and diastolic pressures $(P_s - P_d)$; (c) to the ventricular diastolic time t_d, (1-3). It is possible to write

$$Q = k \cdot \bar{P}(P_s - P_d)t_d \qquad (3)$$

where k is the proportional constant.

Substituting Eq. (3) for Q in Eq. (2), we have

$$Q = k \cdot \bar{P}(P_s - P_d)t_d \cdot R/60 \qquad (4)$$

Substituting Eq. (4) for Q in Eq. (1), we have

$$W = k \cdot \bar{P}(P_s - P_d)t_d \cdot R(\bar{P} - \bar{P}_d) \\ \times 133.28/68 \qquad (5)$$

Assuming the mean pressure in left atrium to be zero, then Eq. (5) may be written as

$$W = k \cdot \bar{P}^2(P_s - P_d)t_d R \times 1332.8/60 \qquad (6)$$

Let $\qquad K = k \cdot 1332.8/60$

Then $\qquad W = K \cdot \bar{P}^2(P_s - P_d)t_d R \qquad (7)$

The left ventricular work computed as such should not be considered identical with actual work done or directly determined, since some simplifying assumptions have been made in the computation. More accurately, it is a *work index* or *load index* (I_W). Thus,

$$I_W = K \cdot \bar{P}^2(P - P_d)t_d R \qquad (8)$$

In practice, the systolic and diastolic blood pressures (P_s, P_d) are taken from the brachial artery with sphygmomanometry. The mean arterial blood pressure is calculated as $\bar{P} = P_d + (P_s - P_d)/3$. The ventricular diastolic time t_d is taken from a phonocardiogram by measuring the time interval from the valvular component of the second heart sound (S_2) to that of the first heart sound (S_1). The heart rate is measured from ECG, averaged from 5-10 successive cardiac cycles. The work index I_W thus computed is expressed in K, or in K/m² body surface area to facilitate comparison between individuals. The body surface area (in m²) is calculated as

0.061 × height (cm) × 0.0128 × body weight (kg) − 0.1529

The Driving Force Index

The contractile property of the left ventricle is closely related to its systolic time, especially the ejection period. It takes a longer ejection period for a weaker myocardium to deliver a given stroke output, the peripheral resistance being constant. Thus the work index in conjunction with the systolic time could reflect a certain aspect of the ventricular contractility. To attain this, the I_W is first to be reduced to "stroke work index" by eliminating the factor of heart rate, and then dividing by the systolic time ($S_1 \sim S_2$ from PCG, in sec). The resulting index may be tentatively termed the *driving force index* (I_{DF}):

$$I_{DF} = \frac{I_W}{S_1 \sim S_2} \div \frac{R}{60}$$

Test Procedures

The CPF test has been done on 13 porter's team mountaineer candidates in a hypobaric chamber (simulated altitude 5000 m) as well as under sea level normoxia. In another group, 26 mountaineers were tested at the base camp (5000 m) during the 1975 Mt. Jolmo Lungma (Mt. Everest) Mountaineering and Scientific Expedition.

Each subject was trained to pedal a bicycle ergometer at a selected frequency controlled by a metronome. After the resting blood pressure, ECG, and PCG were taken, he performed the ergometer exercise for 4 min with a load of 450 kg-m/min (for male) or 300 kg-m/min (for female). In our experience these loads generally would bring about roughly equal cardiovascular response in the two sexes. Blood pressure, ECG, and PCG were taken at 150-180 sec from the end of exercise.

Values of P_s, P_d, $\bar{P}$, t_d, and R were taken or computed from the recordings for calculating I_W and I_{DF}. The I_W and I_{DF} obtained from recordings at 5000 m with exercise load may be related to those from recordings under normoxia at rest (I_{Wo} and I_{DFo}) by calculating the respective values of:

$$(I_W - I_{Wo}) \times 100/I_{Wo}$$
$$\text{and } (I_{DF} - I_{DFo}) \times 100/I_{DFo}$$

Representation Test Results

The I_W and I_{DF}, or $(I_W - I_{Wo}) \times 100/I_{Wo}$ and $(I_{DF} - I_{DFo}) \times 100/I_{DFo}$, as estimates of the ventricular inotropic property are in themselves a numerical scoring which may be used directly for evaluation; however, the chronotropic factor, the heart rate, may be incorporated in evaluating the results by graphing. A quadrant graph may be constructed by relating heart rate to any of the indices mentioned. Individual positions plotted in the graph are to be evaluated.

Study Design

The 13 porter's team mountaineer candidates did their sea level CPF tests both at rest and under exercise load before they were exposed to simulated altitudes of 5000 m, 6500 m, and 7500 m in a hypobaric chamber. At 5000 m each subject performed the CPF test under exercise load and had his resting Pa_{O_2}, PA_{O_2}, and PA_{CO_2} measured when he had been acclimated for at least 16 h at this altitude. If the subject tolerated 5000 m well for 24 h, he was elevated to 6500 m by slow ascent. He was further exposed to 7500 m if he tolerated 6500 m well for 12 h. All the subjects were under strict direct medical care.

The 26 mountaineers under investigation were given a CPF test at rest after they had acclimatized themselves for 2 mo at the base camp (5000 m). During the acclimatization period they went up and down between the base camp and intermediate altitudes which they could reach. The subjects were not told the results of the test, and had the same opportunities as any other members of the team.

Blood and Alveolar Gas Measurements

For Pa_{O_2} measurements, "arterialized" capillary blood samples were taken from the ear lobe. To collect alveolar air the subject blew through a classical Haldane tube, and the gas samples were withdrawn from the side tube near the mouth. All the gas analysis was done with Radiometer electrodes (Denmark).

The $PA_{O_2} - PA_{CO_2}$ were plotted in a Rahn-Otis diagram.

Results

Some combinations of the indices with the heart rate appear to be valuable in predicting hypoxic tolerance and mountaineering performance at altitudes above 5000 m. These results are presented.

CPF Tests at 5000 m Simulated Altitude and at Sea Level

1) The I_{DF} and heart rate of *resting* subjects under hypobaric hypoxia

Figure 44-1A is a quadrant graph of the I_{DF} and heart rate of resting subjects at 5000 m simulated altitude. Those who fell into the left upper quadrant had better tolerance at 6500 and 7500 m. Those in the right lower quadrant showed the contrary; they withdrew for reasons of intolerance from the hypobaric chamber kept at 5000 or 6500 m. In the central zone of the diagram, however, there is some mixing of the two cases.

2) The $(I_W - I_{Wo}) \times 100/I_{Wo}$ and heart rate of *loaded* subjects under hypobaric hypoxia

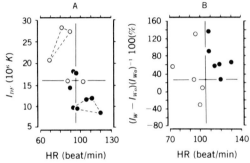

Fig. 44-1. A Quadrant graph of I_{DF} and heart rate (HR) of resting subjects at 5000 m simulated altitude. The horizontal and vertical lines crossing at the center represent the sample mean values of the I_{DF} and HR, respectively, and divide the area into four quadrants. Open circles (○), subjects who tolerated subsequent 7500 m exposure well in the same chamber. Filled circles (●), subjects who withdrew during exposure to 5000 or 6500 m for reasons of intolerance. Note the distribution of the two cases. **B** Quadrant graph of the index $(I_W - I_{Wo}) \times 100/I_{Wo}$ and heart rate. The meaning of the crossed lines and the circles is the same as in **A**. Note the distribution.

This $(I_W - I_{Wo}) \times 100/I_{Wo}$ signifies the degree of mobilization in response to hypoxia and exercise at 5000 m. The function, in conjunction with the heart rate, reveals that those who fell into the right upper quadrant (Fig. 44-1B) all failed to tolerate 5000 or 6500 m in the chamber.

3) The $(I_{DF} - I_{DFo}) \times 100/I_{DFo}$ and heart rate of loaded subjects *at sea level*

A CPF test was attempted with exercise load *at sea level*, in order to evaluate the mountaineering performance without resorting to a hypobaric chamber. It was found that this $(I_{DF} - I_{DFo}) \times 100/I_{DFo}$ together with the heart rate could, to a certain extent, distinguish candidates at risk from the others. Figure 44-2 shows that the five successful candidates who tolerated the 7500 m exposure gather around the left lower quadrant.

CPF Tests at the Base Camp (5000 m)

1) The I_{DF} and heart rate of *resting* subjects at base camp

Twenty-six mountaineers had their *resting* blood pressure, ECG, and PCG taken shortly before they left the base camp for an ascent of the peak. The I_{DF} and heart rate values derived from the recordings were plotted in a quadrant graph (Fig. 44-3). It is remarkable that the five mountaineers who reached the peak, together with four other companions who reached altitudes higher than 8200 m, all fell into the upper left quadrant. In contrast, those who could reach only 6500 m or lower were practically all confined to the right lower quadrant. It appears that the more competent climbers (Fig. 44-4) tend to occupy positions more toward the left.

In the upper left quadrant there is one (filled circle) who was withdrawn from the climbing party halfway to 6500 m on account of being ill.

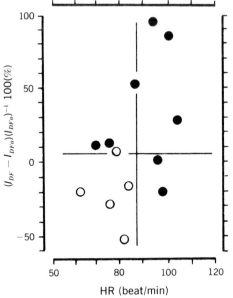

Fig. 44-2. Quadrant graph of the index $(I_{DF} - I_{DFo}) \times 100/I_{DFo}$ and heart rate of subjects with exercise load at sea level. The meaning of the crossed lines and the circles is the same as in Fig. 44-1. Note that the five competent subjects gather in the left lower quadrant, showing that better hypoxic tolerance requires both low index and low heart rate.

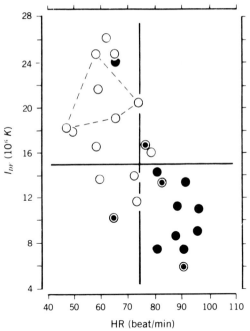

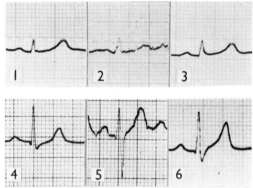

Fig. 44-3. Quadrant graph of the index I_{DF} and heart rate of 26 mountaineers at rest at 5000-m base camp. The meaning of the crossed lines is the same as in Fig. 44-1. The open circle (○), climbers who reached altitudes higher than 8200 m; filled circle (●), climbers who reached only 6500 m or less; doubled circle (◉), climbers who reached 7650 m. The four male climbers who made ascent to 8848-m peak are shown by joining them to form a quadrangle, with the only woman peak climber at its center. The crescent dot in the upper left quadrant, a subject who was withdrawn from the party halfway to 6500 m on account of being ill.

Fig. 44-4. ECG of two mountaineers who ascended to the 8848-m peak. 1,2,3: ECG (lead I) of Pan Do, leader of the women's climbing party; (1) recorded at 50 m before ascent; (2) telemetered at 8848 m and received in the base camp; (3) recorded at 50 m, 2 mo after returning. The ECG apparently showed no abnormality in spite of the fact that she worked energetically for 70 min without O_2 inhalation. HR = 99/min. 4,5,6: ECG (lead II) of a male climber Ping Tso; (4) recorded at 50 m before the ascent; (5) telemetered at 8680 m; (6) recorded 3 mo after returning to 50 m. ECG showed no evident abnormality other than right axis deviation and accelerated heart rate (101/min). No ECG telemetered at the peak.

Alveolar Gas Composition at 5000 m Simulated Altitude

Alveolar gas samples were taken from 10 subjects exposed to 5000 m for 16 h in the hypobaric chamber and analyzed for O_2 and CO_2 composition. The results were plotted in the Rahn-Otis $PA_{O_2} - PA_{CO_2}$ diagram (Fig. 44-5). It may be seen that those (open circles) who cross or approach the "acclimatized line" of the original Rahn-Otis diagram mostly tolerated still higher altitudes in the same chamber and later proved to be better climbers. However, there is one (filled circle) who crosses the "acclimatized line," yet apparently showed

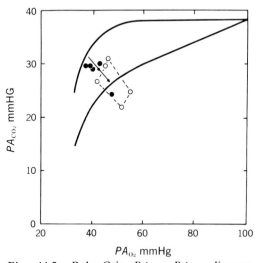

Fig. 44-5. Rahn-Otis $PA_{O_2} - PA_{CO_2}$ diagram showing the plotted position of alveolar gas composition of 10 mountaineer candidates exposed to 5000 m simulated altitude for 16 h. Open circle, subjects who showed good tolerance to subsequent exposures to 6500 and 7500 m; filled circle, those who failed to tolerate the subsequent 6500 and 7500 m exposures in the same chamber.

Discussion

This study shows that the resting or exercise loaded cardiac pump function test (done at sea level or intermediate height) does help predict potential hypoxic tolerance and/or mountaineering performance. We were limited in the number of subjects for investigation, yet the results of this preliminary study are encouraging as there is fairly close correspondence between prediction and actual performance.

The basis of the design of the CPF test is a consideration of both the inotropic and chronotropic factors of the heart. The choice of design is justified by the positive results of our study. The results also show that good functioning of the heart as a pump probably is a requirement for outstanding hypoxic tolerance as well as for climbing performance. We do not exclude the importance of other factors, for instance respiratory acclimatization as represented by the results of the Rahn-Otis $PA_{O2} - PA_{CO2}$ diagram, in which a certain degree of correlation between the plotted location and hypoxic tolerance and/or climbing performance is shown.

Most interesting is the predictive information from the I_{DF}-heart rate diagram of resting subjects at base camp (Fig. 44-3). Among the 26 test subjects there were 5 climbers who subsequently reached the peak, and all 5 had been indicated by the I_{DF}-heart rate diagram. They occupied a rather prominent position in the upper left quadrant, the four males joined to form a quadrangle with the only female "peak climber," Pan Do, at its center. Although she was known to us before the CPF test as three-time breaker of the women's world record, we did not know the prospects of the four males, all new recruits climbing for the first time. There was no motivational influence on the climbers, since none of the subjects were told their test results. Resting I_{DF} and heart rate at intermediate altitude may prove to be a useful objective noninvasive means for predicting climbing performance at great altitudes.

Summary

A noninvasive cardiac pump function test (CPF test), which involves measurements of systolic and diastolic arterial blood pressure, ventricular systolic and diastolic time from a cardiophonogram, and heart rate has been designed for and used as a method for predicting the hypoxic tolerance in a hypobaric chamber and of the climbing performance of mountaineers during the 1975 Chinese Mountaineering and Scientific Expedition to Mt. Jolmo Lungma (Mt. Everest). Several indices representing the inotropic properties of the cardiac pump were derived from the recording of the test done either at rest or during exercise. The indices were used with the chronotopic parameter heart rate and graphed. It was found that certain indices when used in conjunction with the heart rate were valuable in assessing hypoxic tolerance and climbing performance.

References

1. Ayres, S.M., Gregory, J.J., and Buehler, M.E.: Cardiology: A clinicophysiologic approach. London, Butterworths, Appleton-Century-Crofts, Educational Division/Meredith Corporation, 1971, pp. 107–127, 223–228, 243–257.
2. Berne, R.M. and Levy, M.N.: Cardiovascular Physiology, 2nd ed. Saint Louis, The C. V. Mosby Co., 1972, pp. 41–59, 84–99, 178–205.
3. Kelman, G.R.: Applied Cardiovascular Physiology, London, Butterworths, 1971, pp. 1–21, 42–62.

45

Effect of Ambient Temperature, Age, Sex, and Drugs on Survival Rate of Rats

R. VISWANATHAN

A number of studies have been made both in India and abroad of acute and chronic effects of cold, both in the plains and at high altitude (5,7,12). Altl and his colleagues (1) have observed that cold-acclimatized rats tolerate hypoxic environment less well than nonacclimatized rats. This is attributed to increased heat production in cold-acclimatized animals through "nonshivering thermogenesis."

The study presented here has a different but limited objective: to determine the optimum temperature for maximum survival rate of rats exposed to a simulated altitude of 10,000 m.

Effect of Ambient Temperature on Altitude Tolerance

Albino rats about 6 mo old were exposed to simulated altitude of 10,000 m for 6 h by keeping them in an altitude chamber with temperature control. Different batches of 24 animals each were kept at 5°, 10°, 15°, 20°, 30°, 35°, and 40°C in a temperature controlled exposure chamber. The number of survivors was noted, and survival rate calculated (Fig. 45-1).

The highest survival rate was obtained at 25°C. The survival rate decreased gradually with rise in temperature from 25°C, and also with lowered temperature.

Effect of Age and Sex on Altitude Tolerance

Batches of mice and rats, males and females of ages 15 days, 1, 3, 6, and 9 mo, as well as 1 year, were exposed. In all there were 393 rats and 251 mice. The animals

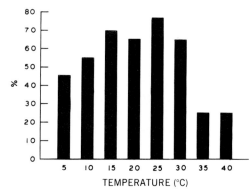

Fig. 45-1. Survival rates at different temperatures for rats exposed to 10,000 m simulated altitude for 6 h. Highest survival rate is at 25°C.

were kept in a desiccator in batches of four or six. The temperature of the air inside the desiccator was reduced by sending the air through a vacuum pump and maintaining a pressure simulating an altitude of 10,000 m. The inner tube for air inflow was connected with several coils of metal tube immersed in water at a constant temperature. The desiccator itself was in the same tank. This ensured a constant temperature of 25°C for the entering rarefied air used to keep the inside pressure constant. The results are represented in Fig. 45-2.

The survival rate is highest among male and female mice at 1 mo and at 9 mo of age. At 3 mo the survival rate falls and then increases up to 9 mo, falling again at 12 mo. The survival rate of female mice is uniformly higher than the males at all ages.

The survival rate of rats, male and female, was highest at 15 days and 1 year. The decline in survival rate at 3 mo was greater for male rats than for females. The survival rate of female rats was higher at all ages.

Statistical evidence shows that the probability of survival is age dependent among male and female mice as well as male rats. However, among female rats survival was independent of age. Sex is not correlated to the survival rate at different ages in mice. On the other hand, female rats had a significantly higher survival rate than the males at the ages studied.

Effect of Drugs on Altitude Tolerance

The use of drugs for preventing or reducing adverse effects of high altitude on men and animals has been studied (2–4,6,8,10,11). In studies the animals were exposed to simulated high altitude. To date there is conclusive evidence that any particular drug can increase altitude tolerance. It was decided to screen a number of drugs to determine their effect on altitude tolerance.

Batches of animals were given injection of the several drugs 2 h before they were exposed to simulated altitude of 10,000 m for 6 h. The quantity of the drug injected was ten times the amount per kg of body weight. An equal number of rats received distilled water injection and served as controls. The number of animals used varied from drug to drug. The survival rate at the end of the exposure period was noted. The results obtained are given in Table 45-1.

No drug improved altitude tolerance at all except the combination of niacin (Pelonin) and promethazine (Phenergan), which improved tolerance very significantly ($p < 0.001$) (Table 45-1).

We also studied the effect of glucose on altitude tolerance. Batches of animals were given 50 g of glucose in 100 cc water and no other food. Equal numbers of male and female animals were kept as controls. They were given only water, no food, for 24 h. There was no significant difference in survival rate between experimental female rats and controls. On the other hand, the difference in survival rate between the groups of male rats was highly significant ($p < 0.001$).

The action of the combination (niacin and promethazine) is not understood. Neither of these drugs given singly improved altitude tolerance. Niacin is a vasodilator and also functions in enzyme systems concerned with reversible oxidation and reduction by hydrogen transfer.

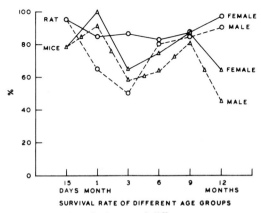

Fig. 45-2. Survival rate of different age groups in both sexes of rats and mice exposed to 10,000 m for 6 h at 25°C.

Table 45-1. Effect of Drugs on Altitude Tolerance

	Survival rate (%)	
Drugs	Treated	Controls
Acetylcholine	$\frac{2}{18}$ (11.1%)	$\frac{6}{18}$ (33.3%)
Aminophyline and Pelonin[a]	$\frac{8}{12}$ (66.5%)	$\frac{9}{12}$ (75%)
Antistine[a]	$\frac{58}{77}$ (75.3%)	$\frac{45}{77}$ (58.3%)
Ascorbic acid	$\frac{82}{110}$ (74.5%)	$\frac{74}{110}$ (67.4%)
Atropine	$\frac{3}{18}$ (16.6%)	$\frac{9}{18}$ (50%)
Caffeine and Phenergan[a]	$\frac{11}{14}$ (57.8%)	$\frac{11}{14}$ (57.8%)
Ciba 390 89 BA	$\frac{18}{21}$ (85.7%)	$\frac{20}{21}$ (95.2%)
Complamin[a]	$\frac{39}{39}$ (100%)	$\frac{38}{39}$ (97.5%)
Dexadrine[a]	$\frac{2}{16}$ (12.5%)	$\frac{8}{16}$ (50%)
Duvadilon	$\frac{28}{30}$ (93.3%)	$\frac{28}{30}$ (93.3%)
Dynesterol	$\frac{0}{18}$ (0%)	$\frac{13}{18}$ (16.6%)
Forhistal[a]	$\frac{114}{123}$ (92.6%)	$\frac{114}{123}$ (92.6%)
Malic acid	$\frac{39}{48}$ (81.2%)	$\frac{42}{48}$ (87%)
Nilamide	$\frac{44}{48}$ (91.6%)	$\frac{45}{48}$ (93.7%)
Phenergan[a]	$\frac{28}{30}$ (93.3%)	$\frac{28}{30}$ (93.3%)
Pelonin[a] + Phenergan[a]	$\frac{72}{90}$ (80%)	$\frac{21}{90}$ (28.3%)

Note: Only the combination Pelonin and Phenergan had highly significant effects ($p < 0.001$). Others not significant.

[a]Generic terms: Pelonin, niacin; Antistine, antazoline; Phenergan, promethazine; Complamin, xanthinol niacinate; Dexadrine, dextroamphetamine; Forhistal, dimethindene.

Promethazine is an antihistamine. While it does not prevent the release of histamine following antigen-antibody reaction, it may prevent histamine from acting on blood vessels. Hypoxic stress causes constriction of muscular arterioles in both lungs and brain. It has been postulated that hypoxia causes mast cells to release mediators like histamine, which produce vasoconstriction. Promethazine may prevent this action and thus may have a synergistic action on niacin, which is a vasodilator. It may be such counteraction of hypoxic effects that brings about increased altitude tolerance in rats.

Acknowledgment

The valuable assistance given by the technical assistants employed under the Research and Development Project of the

Ministry of Defense is gratefully acknowledged.

References

1. Altland, P.D., Highman, B., and Seller, R.G.: Tolerance of cold acclimatized and acclimatized rats to hypoxia at 1.7° C. Int. J. Biometeorol. 17:59, 1973.
2. Barach, A.L., Eckman, M., Ginsburg, E., Johnson, A.E., Brookes, R.D.: The effect of ammonium chloride on altitude tolerance. J. Aviat. Med. 17:123, 1946.
3. Brooks, M.M.L.: The effect of methylene blue on performance efficiency at high altitudes. J. Aviat. Med. 16:250, 1945.
4. Davis, B.D., and Jones, B.F.: Beneficial effect of estrogens on altitude tolerance of rats. Endocrinology 33:23, 1943.
5. Davis, T.R.A. Effect of cold on animals and man. Prog. in Biometeorol. I, part IA, 215, 1974.
6. Ilk, S.G., Segrin, J.J., Balraj, B., and Stevenson, J.A.F.: A criterion to assess the effects of various factors on altitude tolerance. Fed. Proc. 1961, vol. 20:211.
7. Locking, R.W., Zapata Ortiz, V., Hughee, F.D., and Tatum, A.L.: Effects of chronic ingestion of cocaine on tolerance to low oxygen. J. Pharmacol. Exp. Ther. 106:404, 1952.
8. Nair, C.S., Malhotra, M.A., and Tuoare, P.: Int. J. Biometeorol. 17:95, 1973.
9. Peterson, J.M.: Ascorbic acid and resistance to low oxygen tension. Letters to the Editor. Nature 148:84, 1941.
10. Pooger, S. and Dekareas, D.: Some effect of injected cytochrome C in myocardial and cerebral anoxia in man. J. Pediatr. 29:729, 1946.
11. Ward, M.: Mountain Medicine. A clinical study of cold and high altitude, London, Crossly Lockwood Staples, 1975.

IV. High Altitude Expeditions

46
Hemodilution: Practical Experiences in High Altitude Expeditions

R.A. Zink, W. Schaffert, K. Messmer, and W. Brendel

Increased hematocrit and evaluated hemoglobin concentrations are only two of many changes which occur regularly in high altitude climbers. Many authors have described these alterations, and enhanced erythropoesis due to chronic hypoxemia is the explanation usually given.

On high altitude expeditions it could be shown that this high altitude hemoconcentration is due to rapid and excessive water loss rather than to dysoxia. Most of this water is needed to humidify the cold, inspired air and is lost in expiration. When the hypoxia-induced hyperventilation is taken into consideration, the water lost through respiration can be approximated at 6 to 8 liters/day, according to the physical activity. Urinary output and perspiration play only a comparatively minor role in the total water turnover.

Figure 46-1 shows the course of temperature and the absolute water content of the air. The mean values at 7.00, 12.00, and 19.00 h, and the night's minimum are indicated. They were recorded in the base camp of the Kangchenjunga expedition in eastern Nepal during 50 days from April to May 1975, at 5500 m. These data are typical of the environmental situation of expeditionists in the Himalayas, since the ambient air is practically dry at temperatures below 0° C. (Oxygen from cylinders or cartridges is completely dry for technical reasons.)

The overall mean temperature was $-2.7°$ C and the average of the absolute humidity ($WC_{\bar{x}}$) was calculated as 2.6 g of water per m³ of air. Since almost 47 g of water are needed to saturate 1 m³ in the respiratory tract ($WC_{37°C}$), more than 40 g of water have to be withdrawn from the body resources to humidify this volume.

Fluid intake by mouth normally is limited in advanced camps due to the shortage of energy for melting sufficient amounts of snow. But also inexperience, ignorance of the dangers of dehydration, or exhaustion may let the wish to rest or to sleep dominate the feeling of thirst. Severe dehydration may result and hematocrit values up to 0.76 and over have been measured (1,11), whereas the hematocrit does not exceed 0.55 (maximum 0.60) even after several weeks at 5500 m when water intake is forced.

To substantiate these findings, total body water (TBW) was measured with tritiated water, the plasma volume with iodi-

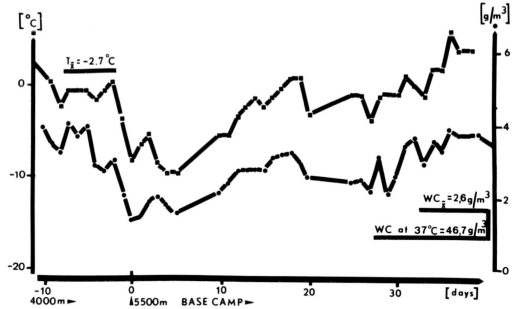

Fig. 46-1. Temperature (squares) and water content of the air (dots): Mean values at 7.00, 12.00, and 19.00 h, and night's minimum, measured in the Kangchenjunga base camp, eastern Nepal, 5500 m, from April to May 1975. Longitudinal averages of the temperature ($T_{\bar{x}} = -2.7°C$) and the water content of the air ($WC_{\bar{x}} = 2.6$ g/m³) are calculated; the amount of water needed for the saturation of 1 m³ of air at 37°C is indicated for comparison ($WC_{37°} = 46.7$ g/m³).

nated human albumin, and correlated with the hematocrit and body weight of 11 climbers during the 1975 Kangchenjunga expedition (Fig. 46-2). The hematocrit rose from an initial 100% (0.44) to 130% (0.57) during the first part of the expedition, when the participants climbed between 5500 m and 7200 m or rested in the base camp at 5500 m. The fluid intake during this period was never severely impaired. The hematocrit reached a maximum of 135% (0.59) after return from the summit (8438 m) 5–6 days later, though the climbers were hemodiluted prior to this.

Increasing hematocrit is associated with decreasing plasma volume, and the effect of hemodilution is seen in both parameters (indicated by arrow). TBW is reduced more than body weight, which indicates that the rapid weight loss is due to a loss of water and not of solid body mass.

Owing to the hemoconcentration an exponential increase in blood viscosity occurs, the coagulability rises and the afterload grows (3, 7). Hemoconcentration also causes a decrease in oxygen transport capacity (10,15) and tissue perfusion (2,8). This is equivalent to a reduced oxygen and heat supply in the peripheral circulation.

Figure 46-3 summarizes effects of hemoconcentration: blood with hematocrit double the normal (0.80; in untreated controls we found values up to 0.78) also showed a three-fold increase in apparent viscosity in high flow areas (such as arterioles) over low flow ones (such as venules), whereas the oxygen transport capacity dropped to only one-half of normal. It should be stressed that the same amount of oxygen can be transported after hemodilution to a hematocrit of 0.20, only one-fifth of the blood viscosity. Therefore in 1975 we started to hemodilute expedition climbers and to date have treated 23 alpinists with a stabilized human serum preparation[1] (Table 46-1). It

[1] Biseko. Biotest Institute, Frankfurt, West Germany.

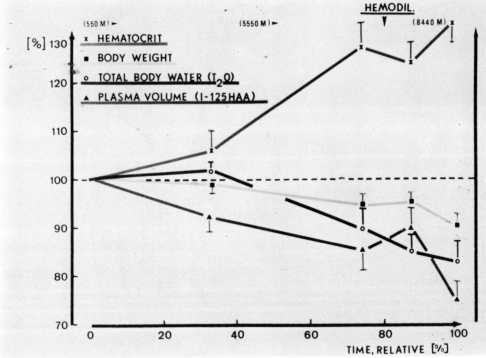

Fig. 46-2. Relative changes of hematocrit, body weight, total body water, and plasma volume, measured during the expedition to Kangchenjunga in 1975. The arrow indicates the time of hemodilution with stabilized human serum, prior to the summit attack, after adequate high altitude acclimatization.

combines high colloid osmotic activity with long intravascular half-life and good antisludge effect. The metabolic pathways of isogeneic proteins are also predictable under the conditions of chronic hypoxia, which is not possible with synthetic plasma expanders. A welcome side effect of the procedure is general antiinfectious prophylaxis due to the relatively high gammaglobulin content of the solution. After 3–4 weeks of acclimatization, just before the summit attack (8438 m and/or 8511 m), we infused 275 ml/m² (7.3ml/kg) in 14 cases (infusion group). In 9 more cases this treatment was accompanied by withdrawing an isovolemic amount of blood while hematocrit, blood pressure, and heart rate were monitored (Fig. 46-4).

Figure 46-5 shows hematocrit for climbers who were infused and those isovolemically hemodiluted (8438 m and/or 8511 m) 5–6 days later. There were no ill effects due to the treatment. One of the alpinists even returned from the top of Mount Lhotse (8511 m) with normal hematocrit (0.47) although he did not use additional oxygen. Infusion of human serum prepara-

Table 46-1. Contents of Stabilized Human Serum Preparation[a] Used for Hemodilution.

Protein	%	g/liter
Total protein	5.0	50.00
Albumin	64.0	32.00
Globulins	36.0	18.00
Electrolytes	mmol/	g/l
Na^+	157.0	3.60
K^+	3.9	0.15
Ca^{++}	2.1	0.08
Mg^{++}	1.0	0.02
Cl^-	100.0	3.50
Isoagglutinins	none	

[a]Biseko. Biotest Institute, Frankfurt, West Germany.

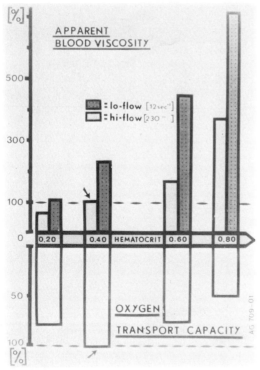

Fig. 46-3. Relative, apparent viscosity, and oxygen transport capacity of blood, depending on the hematocrit at different shear stresses (low flow = 12 sec^{-1}; high flow = 230 sec^{-1}). (After K. Messmer et al., Adv. Microcirc. 4:1, 1972).

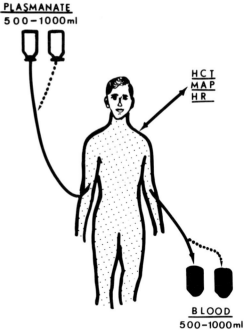

Fig. 46-4. Schematic drawing of the isovolemic hemodilution. Hematocrit (HCT), blood pressure (MAP) and heart rate (HR) are monitored during the withdrawing of 500–1000 ml of blood and its isovolemic replacement by stabilized human serum.

tion lowers the hematocrit sufficiently, but this effect is temporary. After return from the summit, hematocrit returned to earlier levels. However isovolemic hemodilution lowers the hematocrit significantly and even after ascent the readings are significantly below those prior to hemodilution.

Treatment gives high altitude climbers a longer period of relative safety; a mean hematocrit of 0.64 (0.58–0.78) was seen in untreated climbers ($n = 12$) after the same altitude exposure and a comparable schedule.

High altitude retinal hemorrhages (HARH) are frequently seen in high altitude climbers. We found HARH in 4 out of 13 funduscoped mountaineers;[2] on other expeditions the reported incidence was up

[2] We have to thank Dr. P. Hackett who performed these funduscopes during the 1977 Lhotse expedition.

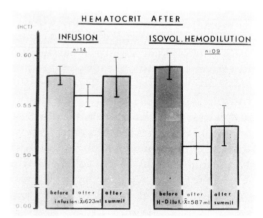

Fig. 46-5. Hematocrit in high altitude climbers before and after treatment and after the summit attack (8438 m and/or 8511 m).
Infusion group: intravenous infusion of 275 ml/m^2 of stabilized human serum.
Isovol. hemodilution: same procedure accompanied by withdrawing an isovolemic amount of blood.
Untreated controls ($n = 12$, not shown): 0.58 before and 0.64 (0.58 – 0.78) after summit attack.

to 60% (6). We think that local stases due to impaired microcirculation with (or because of) microembolisms, plus low flow hypoxia and hypoxia induced capillary membrane hyperpermeability causes local high blood pressure spurts which burst the capillaries.

In those members with retinal hemorrhages (Fig. 46-6, r), the hematocrit rises more steeply from the very beginning of altitude exposure (which is shown for comparison as the x line), as it does after the hemodilution. In those without bleeding the hematocrit rises relatively slowly and remains low after hemodilution in spite of ascent to the summit.

It is noteworthy that in this small group the "bleeders" were an average of 10 years younger than the "nonbleeders." Three weeks after hemodilution none of the "bleeders" demonstrated residual retinal changes.

We believe that climbers' high altitude cold injuries can be prevented much better by hemodilution than by any other medical procedure. The peripheral circulation, and therefore the heat supply to tissues, is necessarily restricted in chronically dysoxic high altitude climbers. Cold exposure jeopardizes the acral regions (fingers, toes, nose) even more, when hemoconcentration causes increased blood viscosity, increased

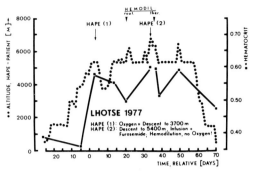

Fig. 46-7. Hematocrit (solid line) and mean altitude profile (dotted line) of one climber who suffered from recurrent HAPE (1,2). Isovolemic hemodilution (HEMODIL.) was routinely (rout.) performed on day 15 and therapeutically (ther.) on day 40. The additional HAPE therapy is indicated.

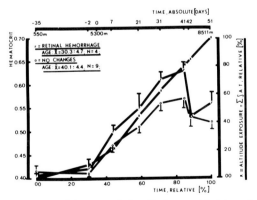

Fig. 46-6. Hematocrit in climbers with (r) and without (o) HARH in plotted against time (absolute and relative). The relative mean altitude exposure (x) is shown for comparison.

coagulability, and impaired oxygen transport capacity.

Hemodilution brings to tissues not only an improved oxygen supply, but also reduces washout of metabolites from them, and above all a much better heat flux through the tissues via sufficiently thin, warm blood.

None of our infused or diluted climbers suffered from frostbite even though the members of these two groups spent altogether 1.3 man years at altitudes between 5500 and 8500 m, in temperatures as low as $-36°$ C.

An increase in blood viscosity also seems to be of importance in the pathophysiology of high altitude pulmonary edema (HAPE). The case history of a climber who suffered from recurrent HAPE shows this relationship (Fig. 46-7). He was affected by HAPE at the beginning of altitude exposure, just after his arrival at the base camp at 5400 m (HAPE 1), when his hematocrit rose rapidly from 0.36 to 0.58. He was treated with oxygen and descended to 3700 m, where he recovered after a few days. His daily mean altitudes are indicated by the dotted line (Chapter 48, *this volume*). Twenty days after his return to base camp he again got pulmonary edema (HAPE 2) at 7200 m. This time he descended to 5400 m where 500 ml of stabilized human serum were infused; thereafter furosemide was given and later he was treated by isovolemic hemodilution. He did not need additional oxygen, recovered within 1 day, and subsequently reached altitudes above 7000 m without incident.

Note that his hematocrit rose very steeply not only at the beginning but also twice after hemodilutions on day 15 (routine) and day 40 (therapy), and that he also suffered from HARH.

To "autodilute" by increasing intake of fluids by mouth seems the easiest method of hemodilution, though the preparation of the necessary amount of water may be difficult during the climb. The ritual must be performed at least two times a day and it is doubtful that mere drinking of electrolyte solutions really covers the needs, though this should be done if no "real" hemodilution is possible or wanted.

During any long sojourn at high altitude a climber gets dysoxic in spite of his acclimatization. Various artificial processes cannot completely cope with the catabolic breakdown reactions, and lowered protein concentrations may result in the dysoxic, hardworking climber.

A decrease in total body waterbinding capacity may result from steadily lowered colloid osmotic activity. The water turnover is faster and even increased drinking may not prevent hemoconcentration.

Hemodilution with protein solutions does simultaneously add to protein stores and the water pool, however. We used human protein instead of other plasma expanders since its biological half-life covers about 1 week, generally enough for the most critical period of the summit attack.

In addition the breakdown of human proteins is certain to take place even in severe dysoxia, which is not certain in the case of synthetic plasma expanders and their metabolites. Several authors (8,12,13, cf. Chapter 15 and 16, *this volume*) assume that high altitude exposure necessarily causes more or less severe impairment of the microcirculation, which should be seen as a special form of hyperviscosity syndrome. We think impaired microcirculation is due to hemoconcentration, primarily caused by fluid loss. Hemodilution certainly decreases blood viscosity and counteracts microcirculatory disturbances safely (14). Furthermore it may prevent, treat, or lessen such high altitude complications as retinal hemorrhage, thromboembolism, cold injury; it is very likely to prevent chronic polyglobulinemia and probably pulmonary and cerebral edema as well.

According to many theoretical, experimental, and clinical results in our experiences hemodilution can be recommended as a feasible and safe method of reducing the risks of high altitude climbing, or to make them at least more calculable and predictable.

Summary

High altitude climbing is always associated with moderate-to-severe dehydration unless special precautions are taken. With much water lost by the humidification of inspired air (up to 7 liters/day), the resulting hemoconcentration may quickly become hazardous, since it causes increased blood viscosity, rise of the afterload, hypercoagulability of the blood, decrease in oxygen transport capacity, and disturbed microcirculation with an impaired heat supply, especially to the acral tissues.

We postulate that all these changes are of pathophysiologic importance in the etiology of high altitude complications such as HARH, cold injuries, chronic polyglobulinemia, thromboembolism, and pulmonary and cerebral edema.

Because of the basic importance of adequate hydration, we started in 1975 to hemodilute high altitude climbers, reporting on 23 cases. Hemodilution can safely lower hematocrit to 0.50–0.55 by the time acclimatization is accomplished. Isovolemic hemodilution, that is, simultaneously withdrawing blood and infusing the same amount of human stabilized serum, is more effective and its benefits longer lasting than those of mere infusion.

Hemodilution is a clinical routine procedure of proved safety, effective and feasible in high altitude climbers. We think that the method enables us to reduce considerably certain risks in high altitude climbing.

Acknowledgment

From the Documentation Center for High Altitude Medicine, Munich; Supported by the Deutsche Forschungsgemeinschaft (Zi 125/3).

References

1. Brendel, W: Höhenakklimatisation und Höhenkrankheit. Jahrbuch OAV 90:187, 1965.
2. Dormandy, J.A.: Clinical significance of blood viscosity. Ann. Roy. Con. Surg. Engl. 47:211, 1970.
3. Guyton, A.C. and Richardson, T.Q.: Effect of haematocrit on venous return. Circ. Res. 9:157, 1961.
4. Messmer, K., Sunder-Plasmann, L., Klövekorn, W.P., and Holper, K.: Circulatory significance of hemodilution: rheologic changes and limitations. Adv. Microcirc. 4:1, 1972.
5. Messmer, K. and Sunder-Plasmann, L.: Hemodilution. Prog. Surg. 13:208, 1974.
6. Rennie, D.R. and Morrissey, J.: Retinal changes in mountain climbers. Arch. Ophthalmol. 93:395, 1975.
7. Richardson, T.Q. and Guyton, A.C.: Effect of polychythemia and anemia on cardiac output and other circulatory factors. Am. J. Physiol. 197:1167, 1959.
8. Schmid-Schönbein, H.: Micro-rheology of erythrocytes, blood viscosity and the distribution of flow in the microcirculation. Int. Rev. Physiol. 9:1, 1976.
9. Schultz, W.T. and Swan, K.C.: High altitude retinopathy in mountain climbers. Arch. Ophthalmol. 93:404, 1975.
10. Sunder-Plasmann, L., Klövekron, W.P., Holper, K., Hase, U., and Messmer, K.: The physiological significance of acutely induced hemodilution. In Ditzel, J. and Lewis, D.H. (eds.): Proceedings of 6th European Conference on Microcirculation, Aalborg, 1970. Basel, Karger, 1971, pp. 23-28.
11. Ward, M.: Mountain medicine. A clinical study of cold and altitude. London, Crosby Staples Lockwood, 1975, p. 68.
12. Zink, R.A.: Ärztlicher Rat für Bergsteiger. Stuttgart, Thieme Verlag, 1978, p. 37.
13. Zink, R.A., Schaffert, W., Lutz, M., Messmer K., Bernett, P.: Hämodilution bei Höhenexposition. In Ehringer, H., Betz, E., and Bollinger, D. (eds.): Gefässwand—Rezidivprophylaxe—Raynaud—Syndrom. Cologne, G. Witzstrock Verlag, 1979, p. 126.
14. Zink, R.A., Brunner, W., Stuck, E., Holper, K.: Hypothermic and normothermic total body washout with cellfree aqueous medium in the pig. Europ. Surg. Res. 8, Suppl. 2:177, 1976.

47

How to Stay Healthy While Climbing Mount Everest

OSWALD OELZ

The Austrian Mount Everest expedition of 1978 was one of the smallest and most successful expeditions attempting to climb the highest peak in the world (8848 m; 29,028 ft). The summit was reached by 8 of the 12 Caucasian members and by 1 Sherpa. For the first time Everest was climbed by two men who did not use supplementary oxygen.

Two simple preventive measures were used to reduce high altitude deterioration, dehydration, and their possibly deleterious consequences: first, after staying on the mountain for 2–10 days, climbers regularly returned to base camp (BC) at 5300 m for full recovery. The final summit attempts started after a rest period in BC and proceeded within 3–4 days to the south col, from there in 1–2 days to the summit, and back to BC within 48 h. Second, meticulous care was taken to assure adequate fluid intake. Climbers were advised to estimate their daily urine output and to drink enough beverages and soups to achieve a daily urine volume of at least 1.5 liters. The high rate of fluid loss from the lungs associated with increased ventilation in the dry cold air above 6400 m made it necessary to consume 4–7 liters of fluid every 24 h. For instance, after an unbroken 10-h climb from the south col (7986 m) to the summit and back, I had to drink 5 liters of tea before adequate urine flow started again. In spite of the difficulties involved in melting that much snow and ice under often miserable conditions and temperatures below $-30°$ C, the climbers generally followed the "fluid prescription."

Polycythemia of high altitude and hydration status were monitored by regular hematocrit (Hct) and hemoglobin (Hb) determinations at BC using a minicentrifuge Compur M 1100 and a miniphotometer Compur M 1000 from Zeiss, Zürich. Control measurements in Europe demonstrated an excellent reproducibility and were identical to data obtained with a conventional microhematocrit centrifuge and a Coulter Counter respectively. The hematocrit levels are shown in Table 47-1. Before leaving Europe Hct was $44.3 \pm 1.0\%$, Hb, 15.4 ± 0.3 g% (mean $\pm$ SEM; $n = 12$). After a 10-day hike from Lukla (2850 m) to BC and a 6-day stay there, Hct rose to $51.2 \pm 1.9\%$ and Hb to 18.3 ± 0.7 g% ($n = 12$). Within the next 9 days, while climbing up to 7400 m, both Hct and Hb reached a maximum of $55.6 \pm 2.2\%$ and 19.2 ± 0.8 g% ($n = 10$), respectively. Owing to low temperatures, Hb measurement reagents decomposed and only Hct determinations could be continued. Hct remained on a plateau: after 2 more weeks of climbing up to the south col (7986 m), it was 55.2 ± 1.9% (n = 8) and after another 10 days 55.4 ± 1.5% (n = 10). Climbing to the summit did not change Hct levels; the Hct of all climbers who reached the summit were determined 48 h later upon arrival at BC, and were 55.7 ± 3.1%. The "oxygenless" climbers R.M. and P.H. had a Hct of 58 and 54%, respectively, the successful Sherpa 55%.

Thus, the maximal adaptive increase of red cell mass occurred after 16 days' exposure between 2850 and 5300 m and 9 additional days up to 7200 m. These findings of a maximal Hct increase of 25.7% are at variance with reports from other Himalayan expeditions (1,2). Cerretelli studying a group of Italian climbers found a rise in Hct from 44.7% to 63.8% after 7 weeks of high altitude exposure up to 6400 m, and a further increase to 66.3 ± 4.1% thereafter. The maximum Hct of climbers returning from above 8000 m were 72 to 74%. There was a concomitant impairment of maximal oxygen consumption, most likely a consequence of impaired O_2 diffusion in the hyperviscous state. This excessive rise in Hct was almost certainly caused in part by a reduction in plasma volume due to dehydration. This may account for a 20–30% increase in red blood cell counts and Hb concentration (2). We avoided dehydration and blood hyperviscosity by adequate fluid intake and regular rest periods at 5300 m. Whether the frequent previous high altitude exposure and the excellent physical fitness of our climbers limited the overproduction of red cells and Hb is subject to speculation.

Climbing Everest without supplementary oxygen but in a well-hydrated state did no harm to R.M. and P.H. On a physical examination at 7400 m, 20 h after their summit success, I found a mild conjunctivitis in R.M. and otherwise two perfectly healthy, somewhat tired individuals. The neurologic and mental status were also normal. No retinal hemorrhages could be detected.

Careful ophthalmoscopy in a dark tent was performed regularly in all members except myself. Before attempts to reach the summit, retinal hemorrhages were observed in two subjects, one of them climbing to the summit 2 weeks later. Eight of nine summit climbers were examined 20–48 h after climbing Everest. There were minor fresh bilateral retinal hemorrhages in two subjects who did not have any symptoms. This $^2/_8$ incidence of retinal hemorrhage in Everest climbers is comparable to the observations of Rennie and Morrissey who found a $^5/_{15}$ incidence in Americans on Dhaulagiri, Nepal (8167 m), after each climber had descended from his highest point to 5883 m (3).

No frostbite, pulmonary edema, cerebral edema, obvious neurologic abnormalities, or other serious diseases occurred in the Caucasian members. However a 42-year-old, heavily smoking Sherpa had a stroke with aphasia and complete left-side hemiplegia at 6400 m. His previous fluid intake

Table 47-1 Hematocrit of Everest Climbers.

	Total days above 2850 m	Mean ± SEM	Range
Sea level (n = 12)		44.3 ± 1.4	40–46
10 days from 2850 m to 5300 m and 6 days at 5300 m (n = 12)	16	51.2 ± 1.9	45–60
7–9 days climbing up to 7400 m (n = 10)	23–25	55.6 ± 2.2	49–63
11–14 days climbing up to 7986 m (n = 8)	36–39	55.2 ± 1.9	51–62
7–10 days climbing up to 7986 m (n = 8)	46–49	55.4 ± 1.5	49–60
48 h after Everest summit (8848 m) (n = 9)	51–61	55.7 ± 3.1	48–64

was apparently insufficient, resulting in clinical dehydration and a Hct of 72%. The patient was rapidly rehydrated with 4 liters of saline given intravenously and oxygen. Aphasia gradually disappeared; however, hemiplegia persisted even after an emergency evacuation to BC and later by helicopter to Katmandu. This is the sixth reported case of stroke occurring in healthy Himalayan climbers; two patients died and three recovered completely (4).

Our experience strongly indicates that dehydration, hyperviscosity of blood and their possible consequences (such as high altitude deterioration, high altitude mountain sickness, pulmonary and brain edema, thrombosis, frostbite) can obviously be avoided by adequate fluid intake and regular descents to 5300 m for full recovery. Even after climbing Everest, mean Hct did not rise above a mean level of 55%. Austrians like to drink a lot and this apparently keeps them healthy when climbing Mount Everest.

References

1. Cerretelli, P.: Limiting factors to oxygen transport on Mount Everest. J. Appl. Physiol. 40:658, 1976.
2. Pugh, L.G.Ċ.E.: Animals in high altitudes: Man above 5000 m—mountain exploration. In Dill, D.B. (ed.): Handbook of Physiology: Adaptation to the environment. Washington, D.C., American Physiological Society, 1964, pp. 861–864.
3. Rennie, D. and Morrissey, J.: Retinal changes in Himalayan climbers. Arch. Ophthalmol. 93:395, 1975.
4. Ward, M.: Mountain Medicine. London, Crosby Lockwood Staples, 1975, pp. 289–292.

48

Proposals for International Standardization in the Research and Documentation of High Altitude Medicine

R.A. ZINK, W. SCHAFFERT, AND H.P. LOBENHOFFER

High altitude research has always been hampered because only small populations could be investigated in the field, and most of the severe high altitude complications show up only in reports of a single case. When one attempts to group cases or reports in order to achieve statistics with more meaning, not much is gained. Various authors use identical terms to denote different high altitude reactions and their symptoms. The lack of common terms and of a commonly applicable index for grouping people according to altitude exposure could be remedied by systematic terminology and a high altitude exposure index. The Documentation Center for High Altitude Medicine has such a system, which in the past several years has proved to be practical.

The Course of Adaptation to High Altitude and its Disturbances

In Fig. 48-1 high altitude induced reactions of an organism are ranked according to severity of symptoms, urgency of therapy, and—as the superior pathophysiological hypothesis—the extent of microcirculatory disorder.

Grade I: Normal Adaptation to High Altitude (HANA)

Adaptation to high altitude (HANA) is normal if ascent is slow enough that adaptation processes have time to organize functioning at the given new level. Neither respiratory nor macro- or microcirculatory or other complaints occur in HANA, though there

I. High Altitude Normal Adaptation	HANA
No problems, no therapy	
II. High Altitude Discomfort	HADI
Harmless, but warning symptoms No therapy, but time for acclimatization e.g. "Acute Mountain Sickness" High Altitude Local Edema	AMS HALE
III. High Altitude Complications	HACO
a. Neither hazardous to health, nor to life High Altitude Retinal Hemorrages	HARH

b. Hazardous to health, potentially to life
     High Altitude Cold Injuries
     High Altitude Thromboembolisms
	HACI
	HATE

c. Hazardous to life
     High Altitude Pulmonary Edema
     High Altitude Cerebral Edema
     High Altitude Chronic Polyglobulinemia
	HAPE
	HACE
	HACP

Fig. 48-1. Systematic terminology of high altitude complaints.

are alterations in the involved systems (1,13).

Grade II: High Altitude Discomfort (HADI)

High altitude discomfort (HADI) includes all the well-known mild symptoms of classical acute mountain sickness (AMS). HADI is harmless per se, but should be understood as a warning of insufficient or incomplete adaptation to the new environment. It indicates that the various physiological systems which have to be tuned to altered surrounding are not yet readjusted.

No therapy is required except more time, maybe at lower altitude, where the acclimatization processes can be completed for the given altitude. That is, the same symptoms may occur after complete acclimatization to one level and subsequent ascent to a higher one. As a rule of the thumb, these ranges are in steps of 1500 m, starting from 3500 m above sea level.

Reactions of the "fast" systems, first of all, of respiratory regulation, seem to have priority over changes of the microcirculation.

High altitude local edema (HALE), manifested as peripheral, lid, or facial edema, is a form of HADI and usually does not require therapy. Its etiology remains unknown, or rather is the subject of speculation. Some hormonal (dis-) regulations may be involved since HALE seems to be more common in females than in males.

Grade III: High Altitude Complications (HACO)

Grade III includes all high altitude complications (HACO) which constitute more or most serious complaints. All these severe forms may be seen as the result of seriously disturbed microcirculation in dysoxic man.

Category A

Category A consists only of high altitude retinal hemorrhage (HARH). This is a harmless condition, if the fovea is not affected (2,3). HARH generally requires no therapy (3,9) but is the visible sign of a serious impairment of perfusion, not only at the retina, where it can be seen easily, but also in such other tissues as the lungs and the brain, to which only pathologists have access to detect such bleeding (2).

Category B

Category B is made up of ailments hazardous to health and potentially life-threatening. It consists of high altitude induced cold injuries (HACI) and high altitude thromboembolisms (HATE). Both are due to the hyperviscosity of hemoconcentrated blood. HACI occurs as a combination of cold exposure and hypoxemia-induced vasoconstriction plus impaired blood supply to tissues caused by the slow perfusion of hyperviscous blood (13,14). HATE, fairly frequent in expedition climbers (11), is due to increased aggregability of slowly floating, packed erythrocytes and cold-induced platelet aggregation, as well as the hypercoagulability of condensed blood in which the clotting factors are also more concentrated.

Category C

Category C illnesses are hazardous to life, as is well known about high altitude pulmonal edema (HAPE) and high altitude cerebral edema (HACE). Both often lead to rapid death if appropriate therapy is not launched immediately. Both coincide with, among other symptoms, pronounced impairment of microcirculatory processes (10). The high altitude induces chronic polyglobulinemia (HACP), which is a special form of the hyperviscosity syndrome. The HACP syndrome is seen only in Indios who live permanently in high areas such as the Altiplano or Chinese lowlanders living for years in Tibet. Hypoxia induced erythropoiesis increases the hematocrit and therewith the blood viscosity. This worsens the microperfusion (10) and the oxygen transport capacity of the blood (11-13). A further enhancement of hypoxemia results and closes the vicious cycle. HACP patients die of pulmonary complications if this fatal cycle cannot be stopped by descent, phlebotomy, or hemodilution (4,6,7).

Altitude Exposure Index (AI)

The duration of an altitude sojourn is commonly used as the independent variable in plotting the course of any dependent parameter such as respiratory or hematologic values. This plot is valid for one particular expedition or trekking tour, though greater differences between individuals may remain. It is of doubtful validity to compare such courses with the corresponding ones for another expedition or trekking tour. The effective altitude exposure may vary considerably, even if the summit height is identical: Use of the altitude exposure index (AI) will overcome some of these problems and in addition make it possible to compare differences such as differences of approach between various expeditions and trekking tours.

Evaluation of the Daily Mean Altitude ($\bar{A}$)

The $\bar{A}$ is based upon the individual sleeping altitude (A_s) of a given day n and requires only under special circumstances the minimal (A_{min}) or maximal (A_{max}) altitude reached during day n.

The index day n starts with the beginning of the "active period" at the sleeping altitude of the preceding day $n-1$ ($= A_{s(n-1)}$) in the morning and terminates with the end of the resting phase at the sleeping altitude of day n (A_{sn}) in the morning of the subsequent calendar day.

Starting point for the evaluation of mean altitude of the given index day n ($\bar{A}n$) is the sleeping altitude of day n (A_{sn}), since a person dwells under such circumstances approximately two-thirds of the time (about 16 h) in the area of the campsite where the night is to be spent. The remaining time is required for ascents and descents from or to the sleeping altitude of the previous day ($A_{s(n-1)}$). In mathematical terms this would be a differential equation ($\frac{d\ \text{Altitude}}{d\ \text{Time}}$). But as a rough approximation one-quarter of the difference between two adjacent sleeping altitudes [$0.25 \times (A_{sn} - A_{s(n-1)})$] may be used as an empiric weighting factor with satisfying accuracy (Fig. 48-2). In some special cases, as when the difference between the sleeping altitude (A_{sn}) and the minimal or maximal altitude gained during day n is greater than 300 m absolute (300),

DAILY MEAN ALTITUDE

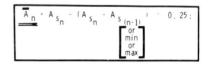

A_s = sleeping altitude
$\bar{A}_n$ = mean altitude of day n
A_{min_n} = minimal altitude of day n
A_{max_n} = maximal altitude of day n

EXAMPLES:

TIME e.g.:	7.00	12.00	19.00	–
MOVEMENT	$A_{s(n-1)}$ [m]	$A?_{min\ max}$ [m]	A_{s_n} [m]	$\bar{A}_n$
↗	5000	(5500)	6000	5750
↘	5000	(4500)	4000	4250
↗	5000	6000	5000	5250
↗	5000	6000	4000	4500
↘	5000	4000	5000	4750

Fig. 48-2. Formulas and examples for the evaluation of the mean altitude ($\bar{A}_n$) for a given day n.

A_{minn} or A_{maxn} is used in the equation instead of the preceding sleeping altitude ($A_{s(n-1)}$).

In formulas the whole procedure seems much more complicated than it really is. Some examples may make it clearer (Fig. 48-2):

1) Ascending from 5000 m ($=A_{s(a-1)}$) in the morning of the index day a, reaching 6000 m in the afternoon, and spendin the night there ($=A_{sa}$) makes a mean altitude for day a ($=\bar{A}_a$) of 5750 m.

2) Starting point of day b is again 5000 m ($=A_{s(b-1)}$), descent to 4000 m, and sleeping there ($=A_{sb}$) results in a mean altitude for day b ($=\bar{A}_b$) of 4250 m.

3) Ascending from 5000 m ($=A_{s(c-1)}$) in the morning, to 6000 m ($=A_{maxc}$) and returning to sleep at 5000 m ($=A_{sc}$) gives a mean altitude for day c ($=\bar{A}_c$) of 5250 m.

The remaining examples in Fig. 48-2 are analogous.

Evaluation of the Daily Altitude Exposure Unit (DU)

The surface below an altitude profile (Fig. 48-3) represents for each day n a given, individual altitude exposure. The surface of this area could be obtained precisely by an integration according to the following formula (Fig. 48-4):

$$DU = \int_{n-1/2}^{n+1/2} f(n)\, dn$$

But as in the calculation of the mean altitude ($\bar{A}$) a sufficient accuracy can be obtained by a geometric approximation, which simplifies the calculation considerably. The small sides of the "DU squares" (Fig. 48-3) are always "1" and their long sides are 1/2 of the preceding ($\bar{A}_{(n-1)}$) plus

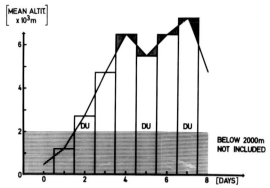

Fig. 48-3. Altitude profile, gained from the daily mean altitudes ($\bar{A}$), indicating the daily altitude exposure units (DU). Triangles below and above the curve indicate the error of the geometrical integration. Hatched area, below 2000 m, may be disregarded since this range is not relevant to altitude adaptation processes.

1/2 of the subsequent mean altitude ($\bar{A}_{(n+1)}$). Prior to the addition of $\bar{A}_{(n-1)}$ and $\bar{A}_{(n+1)}$), 2000 m is substracted from each value in order to cope here with the altitude range between 0 m and 2000 m above sea level,

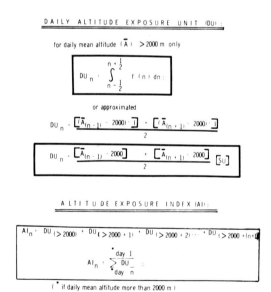

Fig. 48-4. **Upper** Exact and approximative formulas for determining the daily altitude exposure unit (DU). **Lower** Accumulation formulas for arriving at the altitude exposure index (AI).

which is not significant for adaptation processes (Fig. 48-4). This is a second weighting step which helps to avoid overrepresentation of lower altitude ranges.

If $\bar{A}_n$ is smaller than 2000 m, the corresponding DU_n becomes "0" by definition and may be disregarded. The values of DU_n are free of dimensions and given in square units (SU) (Fig. 48-4):

$$DU_n = \frac{(\bar{A}_{(n-1)} - 2000) + (\bar{A}_{(n+1)} - 2000)}{2}$$

The error of the above approximation is minimal, as indicated by the small shaded triangles, and is still largely compensated for by the accumulation process of the following step.

Evaluation of the Altitude Exposure Unit (AI)

The third and last step in calculating the individual altitude exposure for a given day n (AI_n) consists only of adding all consecutive altitude exposure units (DU) from day 1 to day n (Fig. 48-4).

$$AI_n = \sum_{day\ n}^{day\ 1} DU$$

For computational reasons it may be necessary that the DU of days with $\bar{A}_n$ less than 2000 m is 1, 2, or 3 instead of 0 SU. Otherwise there could be trouble in plotting programs where for software reasons division by DU_n might be inevitable. This step is impossible if $DU_n = 0$. That error, however, is negligible since ordinary trekking tours come to a total of 20,000 to 50,000 SU and some reach 200,000 SU.

Future studies will determine whether the kind and extent of the weighting factors are really representative of the physiological phenomena, or whether others could better facilitate our understanding of the adaptation processes. However, this seems to be of secondary importance and would be easier to work out if more material were available in comparable form.

Summary

As a means of avoiding idiomatic confusion, this study offers a system of terminology of high altitude complaints ranked according to severity of symptoms, urgency of therapy, and superior pathophysiologic consideration.

An altitude exposure index (AI) is introduced which is easy to calculate from the altitudes where an individual sleeps during the time of altitude exposure. AI is free of dimensions and given in square units (SU). The index enables the user to compare directly with each other data from different expeditions or trekking tours. This may help to avoid statistical problems in high altitude research, which are commonly due to small numbers in the investigated groups.

References

1. Hannon J.P. and Vogel, J.A.: Oxygen transport during early altitude exposure. Eur. J. Appl. Physiol. 36:385, 1977.
2. Houston, C.S.: High altitude retinopathy. Proc. Hypoxia Symp., Banff, 1979.
3. Houston, C.S.: Long term effects of altitude on the eye. Lancet 49, 1979.
4. Hurtado, A.: Some clinical aspects of life at high altitude. Ann. Intern. Med. 53:247, 1960.
5. Monge, M.C. and Monge, C.C.: High Altitude Diseases; Mechanism and Management. Springfield Illinois, Charles C. Thomas, Publisher.
6. Peñaloza, D. and Sime, F.: Chronic cor pulmonale due to stress of altitude acclimatization (chronic mountain sickness). Am. J. Med. 50:728, 1971.
7. Peñaloza D., Sime, F., and Ruiz, L.: Cor pulmonale in chronic mountain sickness: present concept of Monge's disease. In Porter, R. and Knight, J. (eds.): High Altitude Physiology: Cardiac and Respiratory Aspects. CIBA Foundation Symposium, Edinburgh/London, Churchhill Livingstone, 1971.
8. Rennie, D.R. and Morrisey, J.: Retinal

changes in mountain climbers. Arch. Ophthalmol. 93:395, 1975.
9. Shults, W.T. and Swan, K.C.: High altitude retinopathy in mountain climbers. Arch. Opthalmol. 93:404, 1975.
10. Schmidt-Schönbein, H.: Microrheology of erythrocytes, blood viscosity and the distribution of flow in the microcirculation. International Rev. Physiol. 9:1, 1976.
11. Weingart, H., Zink, R.A., and Brendel, W.: High Altitude Complaints, Diseases, and Accidents in Himalayan High Altitude Expeditions (1946-1978), Chap. 29, *this volume*.
12. Wilson, R.: Acute high altitude illness in mountaineers and problems of rescue. Ann. Intern. Med. 78:421, 1970.
13. Zink, R.A.: Ärztlicher Rat für Bergsteiger. Stuttgart, Thieme Verlag, 1978.
14. Zink, R.A.: Hemodilution in high altitude mountain climbing: A method to prevent or treat frostbite, high altitude pulmonary edema and retinal hemorrhage. In: Proc. of Am. Soc. Anesthesiol. Ann. Meeting (1978).

49
Equipment Requirements for High-Altitude Studies (Personal Experiences)

W. Schaffert and R. A. Zink

As the sport of mountain climbing becomes increasingly popular, more and more persons are spending prolonged periods of time at altitudes above 3000 m. Unfortunately, there has been a corresponding increase in the incidence of altitude-related illnesses. Severe forms with pulmonary and brain edema can become acutely life-threatening even for young, well-trained climbers. As a result, both the physiology and pathophysiology of adaptation to prolonged high-altitude exposure has become an important area of research in tourist medicine.

The preservation of working capacity and avoidance of injurious influences are the subject of scientific field studies.

An important prerequisite for such studies is optimum equipment for conducting performance tests and laboratory analyses in the field. However, such instruments tend to be quite delicate and suitable for use only in a stationary and climatically stable situation. The proper functioning of these devices depends in part on reliable and often costly maintenance by trained personnel.

Our experience, based on two large-scale expeditions (Kanchenjunga, 1975; Lhotse, 1977), has shown that it is possible to utilize even highly delicate instruments in large, carefully planned field studies, provided that certain conditions are met:

1) An accurate understanding of the design, operation and maintenance of the instruments.
2) Strong, shockproof packing units in a size convenient for carrying, weighing no more than 30 kg.
3) Allowance for low ambient pressure for items packed in airtight containers in order to avoid package rupture (vacuum-packing preferred!).
4) Allowance for subfreezing temperatures, even in the cargo area during intercontinental jet flights, in the case of liquid electrolyte-filled devices, thermolabile liquid quartzes, or other cold-sensitive chemicals or components.
5) A heated on-site working area.
6) Sufficient spare parts, tools, and instruments capable of operating independently of conventional supply facilities.
7) A reliable portable power supply in the form of a gasoline-driven generator. The generator must deliver an adequate output even at one-half barometric pressure, taking into account the starting-current peaks of the principal current-

consuming devices. The generator engine must be designed for extreme environmental conditions, i.e., for a reduced oxygen supply (special carburetor jet or compressor blower, spark plugs with reduced heat range). The generator must also be able to withstand extreme cold ($-30°C$) and varying atmospheric humidities, and have good cold-start properties and reliable long-term operation with minimum fuel consumption. We have had poor experience with the 2500-W Honda generator; during 20 days of operation at 5500 m, it had to be repaired 17 times, until finally it ceased to operate. However, we have had very good experience with the Kirsch (Kirsch, Trier, West Germany) 1.5-kW generator with a two-stroke Fichtel & Sachs engine with special carburetor jet attachment and modified spark plug heat range.

The following laboratory instruments, properly packed, arrived undamaged at our base camp at an altitude of 5500 m after having been transported for 28 days and were fully operational when set up in a special double-walled thermal tent:

1) Semiautomatic blood-gas analyzer: Corning, Model 165, U.S.A.
2) Portable ECG recorder: Cardiostat, Siemens, Munich, West Germany.
3) Mini tape storage ECG recorder: Ricard.
4) Various centrifuges (laboratory, hematocrit, manual): Hettich and Heraeus-Christ, Osterode, West Germany.
5) Pulmonary-function recorder: Spirotron, Draeger Werke, Luebeck, West Germany.
6) Gas-analyzer tublets for measurement of end-expiratory CO_2: Draeger Werke, Luebeck, West Germany.
7) Photometer: Eppendorf Model 6120, Hamburg, West Germany.
8) Glucose analyzer: Reflomat Laboratory, Boehringer Mannheim, West Germany.
9) Portable freeze-dry equipment: WKF-Forschungsgeräte, Model 2000, Brandau, West Germany.
10) Immersion freezer: Kryofix, WKF-Forschungsgeräte, Brandau, West Germany.
11) Ultravacuum pump: Model LD2, Brand, Wertheim, West Germany.
12) Various laboratory equipment: Brand, Wertheim, West Germany.
13) Oxygen analyzer: Lex-O_2-Con, Lexington, U.S.A. (This device worked, but too slowly—up to 20 min per sample—to be feasible for use at high altitude. The problem may have been low barometric pressure.)

The most important discovery from experiences during the well-organized Kangchenjunga expedition is that with proper planning and execution, it is possible even under extreme conditions to produce "ordinary" laboratory conditions (though at risk of a certain failure rate) at modest expense.

Nevertheless, the following problems remain:

1) Lack of instrument mobility, which makes it difficult to monitor subjects in the field.
2) Disruptions in lines of supply, which can jeopardize the entire program (causes may lie in carriers, climate, objective hazards such as avalanches, rockslides, or floods).
3) The possibility of instrument malfunction due to instrument-specific properties often cannot be eliminated despite the presence of well-trained personnel.

The following problems must be dealt with in every case:

1) Geographic and climatic problems: Remote sites with poor roads at high altitudes under extreme climatic conditions. Altitudes above 5000 m, temperatures between $+30$ and $-30°C$, low or rapidly fluctuating atmospheric humidity, intense solar radiation.

2) Transport-specific problems: Limited pack size and weight (max. 30 kg); instruments should be packed in strong, shock-resistant but nonbulky containers with good thermal insulation and, if necessary, negative interior pressure (to avoid pressure buildup when ambient pressure falls).
3) Instrument-specific problems: Low power consumption limit and low starting current due to dependence on generator power; fast operational readiness (no prolonged warm-up); reliable, problem-free recalibration; simple specimen preparation; high operating reliability with modest maintenance.
4) Methodological considerations: Separation between practicable on-site determinations and the safe, reliable perservation of specimens for accurate at-home analysis (deep freezing, freeze drying); determinations should be limited to a few representative quantities that can be measured without undue difficulty.

In accordance with the principles and experiences described above, the following characteristics describe ideal high-altitude field-research equipment:

1) Robust, portable construction: small, lightweight, resistant to shock, cold, moisture and corrosion.
2) Low-energy operation from portable power supply: alkaline-manganese cells or, for subfreezing temperatures, lithium batteries.
3) Ease of operation: rapid setup regardless of time or location, easy calibration, high measuring accuracy, low susceptibility to malfunction.
4) Representative selection of measured quantities: limited to those that can be simply measured on site, with means for specimen preservation for at-home analysis.

The instruments described below were tested during the Lhotse expedition of 1977 and have a great many of the characteristics listed above:

1) Miniphotometers, battery run, with single-use cuvette sets: Compur (M1000) for hemoglobin and RBC; Compur (M1001) for glucose (GOD-PAP method), total bilirubin (DPD method), total bilirubin (DPD method), and cholesterol (ChOD-PAP method).
2) Minicentrifuge, battery run, for hematocrit and plasma samplings: Compur (M1100), Compur Werke, Munich, West Germany.
3) ECG miniscope and minigraph: Schiller (MS2), West Germany.
4) Immunodiffusion plates for TRF, fibrinogen, and protein components.
Partigen plates: Behring Werke, Marburg, West Germany.
Quantiplates: Biotest Institute, Frankfurt, West Germany.

The following devices are still under development and will be of great interest for use in future investigations:

1) Erythrocyte aggregometer, by Schmid-Schönbein, Department of Physiology, RTW University, Aachen, West Germany. This equipment was successfully tested during the Kangchenjunga expedition in 1975, then in combination with a compensation recorder. An improved battery-run version with tape recording unit has been successfully used in field studies since 1978.
2) Epicutaneous oxygen-measuring electrodes devised by F. Luebbers (Max Planck Inst. f-Work Physiology, Dortmund, West Germany), have proved feasible in the laboratory over several years. Modified versions for use in the field are under development.

Index

A

Accidents, 193–197
  classification of, 194
Acetazolamide
  acute mountain sickness and, 7
  sleep hypoxemia and, 4–7
Acute mountain sickness, see Mountain sickness, acute
Adaptation
  course of, 301–303
  normal, 301–302
Adenosine triphosphate (ATP), maximal exercise and, 86
α-Adrenergic blockade, phentolamine, 158–161
β-Adrenergic blockade, propranolol, 159, 166
Aerobic power, maximal, see Oxygen consumption, maximal
Age, pulmonary edema and, 220–221, 226, 233
Air breathing, and shift to oxygen breathing, 9–10
Albumin, capillary leakage of, 176–177
Alkalosis, blood, 74
Alpine accidents, 193–197
  classification, 194
Altitude exposure index, 303–305
Altitude residents, see Highlanders
Alveolar-arterial oxygen gradient, and lowland acclimatized highlanders, 22–24
Alveolar-capillary membrane
  diffusion across, 16–18
  lung model and theory, 16–18
  oxygen uptake limit and, 16–20
  permeability of, 246
AMS, see Mountain sickness, acute
Anaerobic alactic power, maximal, 88–92
Anaerobic (lactacid) metabolism, 94–101
  buffer value in, 98–99
  exercise and, intermittent, 99–100
  hematologic characteristics of, 95
  hydrogen ion concentration and, 97–98
  and lactacid oxygen debt and exercise, 99–100

  lactate, blood, 95, 97–98
  lactate disappearance during recovery, 100–101
  metabolic characteristics of, 95
  pyruvate in blood and, 95
Andes, 81
Anoxia, cerebral edema and, 199–200, 206–207
Anoxic zone, oxygen supply and, 48–49
Arterial blood, ventilation and gas in, 22
Arterial crossover P_{O_2}, 56
Arterial oxygen content, 151
Arterial oxygen pressure, 70
Arterial oxygen tension, hypoxic threshold, 200–201
Arterio-venous oxygen difference, 70
Arterioles, pulmonary edema and architecture of, 244–246
Atmospheric pressure, 137

B

Barbiturates, and ischemic brain insults, 39
Barometric pressure
  low, 137
  rapid increase of, 12
Blood-brain barrier, ischemic insults and, 204–206
Blood electrolytes, 183–185
Blood fluidity
  cell aggregation and, 113
  factors influencing, 115
  plasma viscosity and, 113
  red cell fluidity, 110
  stasis phenomenon, 113
  vasomotor influence and, 114–115
Blood infusion, aerobic power and, 14
Blood oxygen dissociation curve, see Oxygen dissociation curve
Blood rheology, in hemoconcentration, 109–115
  blood and red cell fluidity, 110–112
  disadvantages of, 112–115
  in microcirculation, 113–114

Blood rheology, in hemoconcentration *(cont.)*
   risk factors, 112–115
   slowly flowing blood, 109–110
Blood viscosity, 109; *see also* Blood rheology
   high clinical importance, 118
   pulmonary edema and, 295–296
Blood volume, 82
Body water, total (TBW), 291–292
Bradycardia, and low altitude, 24
Brain edema, *see* Cerebral edema
Breathing
   air to oxygen, shift from, 9–10
   periodic, 1–2, 6–7
   pulmonary edema and pure oxygen, 233, 236
Buffer value, during acclimatization, 98–99
Buffering, lactate, 74

C

Calcium, plasma, 184
Capacitance coefficient, β_{O_2}, 20
Capillary density, heart, 153
Capillary leakage, albumin, 176–177
Capillary network, flow regulation in, 49–50
Carbon dioxide, and oxygen dissociation curve, 73–77
   blood investigations, 74–76
   exercise testing, 76–77
Carbon monoxide, pulmonary diffusing capacity for, 19
Cardiac output, 129–137
   distribution of, 129–135
   functional circulations and, 131–132
   heat and cold stress and, 263
   hemoconcentration and, 117–118
   in highlanders and lowlanders, 129–137
   of lowland acclimatized highlanders, 24
   nutrient circulations and, 132–135
   oxygen transport capacity and, 118–120
   pulmonary circulation and, 145
Cardiac pump function test (CPF test), 278–280, 283
Cardiac weight, 151–153
Cardiovascular circulation, 82
Cataracts, 213
Cell metabolism, oxygen depletion and, 37–39
Cellular oxygen utilization, components of, 34–36
Cerebral blood flow, 82, 102
Cerebral edema, 199–207
   anoxia and, 199–200
   causes of, 200
   hypoxia and, 199–201, 206–207
   ischemia and, 203–206
   microcirculation and, disturbed, 203–206
   pathophysiology of, hypothesis on, 266
   selective vulnerability of brain, 201
Chloride, plasma, 184
Chronic mountain sickness, *see* Mountain sickness, chronic
Climbing expeditions, *see* Expeditions

CMS, *see* Mountain sickness, chronic
Cold injuries, high altitude (HACI), 196, 197
   hemodilution and, 295
Cold pressor test, high altitude pulmonary edema and, 243
Cold stress, cardiac output and, 263; *see also* Thermal conditions
Complaints, high altitude, 195–196
Complication classification, high altitude, 194, 302–303
Corneal epithelium, 212
Coronary blood flow, 153, 163
Cortisol, 166, 169
Cotton-wool spots, 215–216
Critical O_2 extraction, definition of, 55–56
Critical peripheral (unloading) P_{O_2}, definition of, 56
Cutaneous circulation, 131–132
Cutaneous vascular compliance, reduced, 209–210
Cycloergometric exercise, measurement of maximal power, 92

D

Decompression, altitude, 179–180
Dehydration, 291–300
   hemoconcentration and, 121
   hemodilution and, 291–297; *see also* Hemodilution
   prevention of, 298–300
2,3-Diphosphoglycerate (2,3-DPT), 54, 74–76, 86
   maximal exercise and, 86
Discomfort, high altitude, 302
Diseases, high altitude, 195, 196
   definition of, 194
Driving force index, 279
Driving pressures, 115
Drugs, altitude tolerance and, 285–286
Dysoxia, 33–39
   causes of, 33–34
   definition of, 33
   hyperoxic, 34
   long sojourn at high altitude and, 296
   mechanisms for coping with, 37–39
   stages of, 36
   substrate (glucose) supply and, 36–37, 39
   types of, 33–34

E

Electrolyte changes, in climbers, 183–185
Endcapillary, definition, 56
Energy expenditure, whole body, 88
Energy requirements, oxygen depletion and, 38–39
Equipment, high altitude studies, 307–309
Erythrocyte function, respiratory and metabolic changes, 82
Exercise, high altitude, 81–86; *see also* Trekking
   carbon dioxide dissociation curve, 74
   cell metabolism and, 37–38

chronic, 37–38
hormonal responses, see Hormonal responses, and exercise
intermittent, 99–100
lactacid oxygen debt and, 99–100
maximal, 74, 82–86
oxygen debt and, 103–106
oxygen depletion and, 37–38
oxygen return and, 156
parameters of, biologic, 81–82
respiratory values in, 18–82
submaximal, 103–106, 129, 156
sustained rhythmic, 123–127
Exercise testing, postaltitude, 76–77
Exhaustion, 195
Expeditions, high altitude
alveolar gas measurements, 280, 282–283
cardiac pump function tests, 278–280, 283
definition of, 193
driving force index, 279
electrolyte changes during, 183–185
hemodilution and, see Hemodilution
performance prediction, 278–283
preventive measures for, 298–300
work index, 278–279
Exposure index, altitude, 303, 305
Exposure unit, daily altitude, 304–305
Eye problems, 212–216; see also Retinal hemorrhage

F

Fahraeus-Lindqvist effect, 111, 124
Fluid intake, 291, 296
Free fatty acids (FFA), 166, 169, 170, 173, 175
Furosemide, and high altitude pulmonary edema, 268

G

Gas
  in arterial blood, 22
  transport, 82
Glaucoma, acute, 213
Glucose, 166–167, 169
  blood, 173, 175
  supply, and oxygen depletion, 36–37, 39
  tolerance test, 166–167
Glycolytic capacity, and oxygen depletion, 38
Growth hormone
  plasma, during exercise, 172–175
  serum levels, in high altitude dwellers, 166–167, 169–170

H

Heart, see also cardiac entries; Heart rate
  capillary density, 153
  muscle fiber density, 153
  muscle fiber diameter, 153
  oxygen supply to, 150–154
Heart rate
  in lowland acclimatized highlanders, 24
  maximal, at high altitude, 82
Heat exposure, cardiac output and, 263
Hematocrit levels, high altitude, 109, 113–114
  blood flow changes and, 124
  chronically elevated, 127
  control of, 120–121
  hemodilution and, 292–293
  high, 117, 118
  increase in, 210, 292
  maintenance of, 115
  maximal exercise and, 84
  in microcirculation, 114
  optimal, 123–127
  oxygen delivery and, 124
  oxygen extraction and, 124–125
  pulmonary edema and, 295–296
  rhythmic exercise and, 123–127
Hemoconcentration, see also Hemodilution
  acute, 117–118
  blood rheology and, 109–115
  cardiac output and, 117–118
  dehydration and, 121, 296–297
  microcirculatory impairment and, 296
  oxygen transport capacity and, 117–118
Hemodilution
  cold injuries and, 295
  expedition experience with, 291–297
  fluids by mouth for, 296
  hematocrit and, 292–293
  intentional, 120, 121
  isovolemic, 293–294
  limited, 120
  maximal performance and, 14
  normovolemic, acute, 119, 120
  protein solutions in, 296
Hemoglobin
  concentration, increase in, 210
  maximal exercise and, 84
  maximal oxygen consumption and, 12, 14
  oxygen affinity of, 84
Hepatosplanchnic blood flow, 131
Highlanders (high altitude natives)
  cardiac output in, 129–137; see also Cardiac output
  lowland acclimatized, see Lowland acclimatized highlanders
  pulmonary circulation of, 124–149
  regional blood flow in, 129–137
  risk, 240
Hiking, see Trekking
Histamine levels, in high altitude pulmonary edema, 246
Hormonal responses, and exercise, 165–170, 172–175; see also specific hormones
  glucose tolerance test, 166–167
  growth hormone, see Growth hormone
  maximal exercise, 166

Hydrogen ion buffering, 98–99
Hydrogen ion concentration, blood, 97–98
17,21-Hydroxy-20-ketosteroids, urinary elimination of, 24
Hypercapnia, acute mountain sickness and, 30–31
Hyperexis, 209–210
Hyposphagma, 212
Hypoxemia
  acute mountain sickness and, 7
  sleep, see Sleep hypoxemia
Hypoxia, see also hypoxic entries
  acute mountain sickness and, 30–31
  hypobaric, 248–253
  cerebral edema and, 199–201, 206–207
  lactic acid and, 94
  oxygen consumption and sudden, 10, 12
  polycythemia and, 119
  pulmonary edema and, 243
  pulmonary hypertension and, 275
  ventilation and, 22
Hypoxic brain, selective vulnerability of, 201
Hypoxic dysoxia, 33, 38–39
Hypoxic pulmonary hypertension, see Pulmonary hypertension, hypoxic
Hypoxic threshold, 137
  and arterial oxygen tension, 200–201
Hypoxic vasoconstriction, pulmonary arterioles, 209
Hypoxic zone, oxygen supply and, 48–49

I

Infections, 195
Insulin, 166, 169
International standardization of altitude documentation, 301–305
Intestinal disease, 195
Ischemia
  barbiturates and, 39
  cerebral edema and, 203–206
  necrosis from 201

K

Krogh model of oxygen supply, 45–51, 153–154

L

Lactacid anaerobic metabolism, see Anaerobic metabolism
Lactacid oxygen debt, intermittent exercise and, 99–100
Lactate, blood, 94, 95, 97–98
Lactic acid, 173, 175
  blood concentration, 90–92
  disappearance rate, 105

maximal exercise and, 90–92
oxygen deficit and, 105
Lethal corner, definition of, 56
Lowland acclimatized highlanders, 21–26
  alveolar-arterial oxygen gradient, 22–24
  gas in arterial blood of, 22
  heart rate, 24
  lung volumes, 21–22
  maximal aerobic capacity, 24, 26
  ventilation, 22
Lowlanders
  cardiac output in, 129–137
  regional blood flow in, 129–137
Lung arterioles, pulmonary edema and architecture of, 244–246
Lungs, see also pulmonary entries
  arterioles, 244–246
  blood volume, 250, 251
  diffusion in, 16–20
  injury, 257–259
  mass, 251–253
  microvascular permeability, 258
  platelet survival, 179–180
  volumes, 248–252
Lymph protein clearance, pulmonary edema and, 258

M

Magnesium plasma, 185
Medroxyprogesterone, sleep hypoxemia and, 7
Metabolic characteristics, anaerobic, see Anaerobic metabolism
Microcirculation
  blood rheology in, 113–114
  cerebral edema and disturbed, 203–206
  hematocrit in, 114
  hemoconcentration and impairment, 296
Microembolization, pulmonary edema and, 257–259
Microvessels, blood rheology in, 115
Mixed-venous oxygen content, 69–71, 151
Monge's disease, 271
Mountain sickness, 195; see also Mountain sickness, acute; Mountain sickness, chronic
Mountain sickness, acute (AMS)
  acetazolamide and, 7
  hypoxemia and, 7
  hypoxia and, 30–31
  pathophysiology of, 266–267
  susceptibility to, 28–32
  vasopressin and, 261–262
Mountain sickness, chronic (CMS), 271–276
  animal model, 274
  differential diagnosis, 271
  hypothesis on, 271–272
    supporting evidence for, 272–275
  pulmonary artery pressure in, 272–275
  symptomatic patients, 271
  as vascular disease, 271–276

Mountaineering performance, see Expeditions
Muscular exercise, see Exercise
Muscular work, energetics of, 88–89
Myocardium, vascularity of and oxygen transport, 119

N

Neutrophils, circulating, pulmonary edema and, 258–259
Niacin, 285, 286
Nitrogen washout, single breath, 248–251
Non-rapid eye movement sleep, 4
Normovolemic hemodilution, acute, 119, 120
Normoxic dysoxia, 33–34
Nutrient circulations, 132–135

O

Ophthalmological problems, 212–216; see also Retinal hemorrhage
Oxidative phosphorylative capacity, oxygen depletion and, 37
Oxygen affinity, hemoglobin, 84
Oxygen breathing, air breathing shift to, 9–10
Oxygen consumption, maximal (V_{omax}), 9–14, 82
  barometric pressure, rapid increase of, 12
  blood infusion and, 14
  factors influencing, 9
  hemodilution and, 14
  hemoglobin concentration and, 12, 14
  hyperoxia and, 10, 12, 14
  in hypoxia, 9–14
  limits, 13–14
  in lowland acclimatized highlanders, 24, 26
Oxygen content of blood, oxygen transport capacity and, 118–120
Oxygen debt
  asymptotic value of, 104
  calculation of, 105
  intermittent exercise and, 99–100
  lactacid and, 99–101
  submaximal exercise and, 103–106
  temperature and, 105–106
Oxygen deficit, 103–106
  lactic acid disappearance rate and, 105
  slow component of, 105
  steady state data on, 103
  submaximal exercise and, 103–106
  transient phase data on, 104
Oxygen delivery, hematocrit and, 124
Oxygen depletion, see Dysoxia
Oxygen dissociation curve, 54–62
  alterations to, 62
  blood investigations for, 74–76
  crossover phenomena, 60–62
  discrepancies, 66
  exercise testing and, 76–77
  implications of, 58–69, 62
  mixed-venous oxygen pressure and, 69–71
  optimal, 62
  original position, influence of, 67
  oxygen extraction, effect on, 56–59, 62
  oxygen supply to tissues, influence on, 65–71
  parameters of, 76
  peripheral P_{O_2} and, 59–60
  point of inversion, 67
  species differences, 69, 71
  and stay at moderate altitude, 73–77
  systemic displacement of, 62
Oxygen efficiency, gross, 104
Oxygen extraction
  factors that alter, 62
  hematocrit and, 124–125
  oxygen dissociation curve and, 62
  P_{50} effect on, 56–59
Oxygen metabolism, abnormal cell, see Dysoxia
Oxygen pressure field, 46
Oxygen return, submaximal exercise and, 156
Oxygen supply
  anoxic zone and, 48–49
  arterial, compensatory mechanisms, 46–48
  flow regulation, 49–50
  to heart, 150–154
  hypoxic zone and, 48–49
  Krogh model, 45–51
  minimal, 45–51
  oxygen dissociation curve, 65–71
  oxygen pressure field, 46
  respiratory enzymes and, 49
Oxygen transport capacity, 117–121
  enhancement of, 120–121
  hemoconcentration in, 117–118
  to myocardium, 119
  peak, 119–120
  systemic, 118–120
  tissue, 154
Oxygen uptake, lung, see also Oxygen consumption, maximal
  diffusion, 16–20
  and lowland acclimatized highlanders, 21–26
  and maximal exercise, 84
Oxygen utilization, biochemical pathways involving, 34–36

P

P_{50}, 75–76
  definition of, 54–55
  maximal exercise and, 84, 86
  mixed-venous oxygen pressure and, 69–70
  oxygen extraction and, 56–59
  peripheral P_{O_2} and, 59–60
  starting, 69–71
Pentobarbital anesthesia, 160–161
Performance, exercise, see Exercise
Performance, maximal, see Oxygen consumption, maximal

Periodic breathing, 1–2, 6–7
pH blood
   alkalosis, 74
   drop in, 98–99
Phentolamine (Regitine), 158–161
Phosphates, high energy, splitting of, 90
Plasma viscosity, blood fluidity and, 113
Platelet survival, and lung sequestration, 179–180
Polycythemia
   hypoxia with, 119
   overt, 121
   treatment of, 121
Polymorphonuclear leukocytes, circulating, pulmonary edema and, 258–259
Potassium, plasma, 184
Power, maximal, *see also* Oxygen consumption, maximal
   assessment usefulness, 88
   cycloergometric exercise measurement of, 92
Pressure, systemic, 13
Promethazine, 285, 286
Propranolol, acute hypoxic pulmonary hypertension and, 159, 166
Pulmonary arterioles, hypoxic vasoconstriction of, 209
Pulmonary artery obstruction, pulmonary edema and, 256–259
Pulmonary artery pressure (PAP), 137, 246
   blood flow and, 143–149
   in chronic mountain sickness, 272–275
   in natives vs. newcomers, 142–149
Pulmonary blood flow, pulmonary artery pressure and, 143–149
Pulmonary circulation, 142–149
   cardiac output and, 145
   in right lung, 142–149
Pulmonary diffusing capacity, 19
Pulmonary edema, high altitude (HAPE), 196
   after lowland visit, 222
   age and, 220–221, 226, 233
   altitude and, 221, 227
   alveolar capillary membrane permeability and, 246
   blood viscosity and, 295–296
   cardiocirculatory data, 233–236, 239
   cases of, 219–230
   causes of, 240
   circulatory data, 235, 238
   cold pressor test and, 243
   constitutional susceptibility to, 242
   embolization and, 257–259
   furosemide in prevention of, 268
   geographical distribution of, 220, 225–226
   hematocrit and, 295–296
   hemodynamic and clinical findings, 240
   hemodynamics of, 231–240, 243
   histamine levels in, 246
   hypoxic stress and, 243
   incidence of, 229
   lung blood volume and, 250, 251
   lung density and, 249–251
   lung histology and, 242
   lung injury in, 257–259
   lung mass and, 251–253
   lung microvascular permeability in, 258
   lung volumes in, 248–252
   lymph protein clearance in, 258
   mechanisms of, 255–259
   mortality rate, 224
   muscular arterioles in, 244–246
   neutrophils in, circulating, 258–259
   pathogenesis of, 242–246
   pathophysiology of, 295–296
      hypothesis on, 266–267
   prevention of, 228–229, 268
   previous history of, 221, 223, 236, 239
   prognosis of, 229–230
   pulmonary artery obstruction and, 256–259
   pulmonary artery pressure, 246
   pure oxygen breathing and, 233, 236
   rate of ascent and, 221
   respiratory data on, 234
   sexual gender and, 220, 226
   single-breath nitrogen washout and, 248–251
   subclinical, with hypobaric hypoxia, 248–253
   symptoms of, 222–223, 228
   therapy for, 223–224, 228
   time lag of, 222, 227–228
   vasoconstriction of pulmonary arteries in, 255
   vasopressin and, 261–262
   ventilation and, distribution of, 253
Pulmonary hypertension
   chronic mountain sickness and, 272–275
   hypoxia and, 275
   pentobarbital anesthesia and, 160–161
   Phentolamine adrenergic blockade and, 158–161
   propranolol adrenergic blockade, 159, 166
Pulmonary resistance, 82
Pulmonary vasculature, anatomical structure of, 272–275
Pulmonary vasoconstriction, temperature and, 263–264
Pulmonary volumes, *see* Lung volumes
Pyruvate, blood, 95

R

Rapid eye movement sleep, 4
Red cell fluidity
   blood fluidity and, 110
   concept of, 110–112
   result of, 111–112
Red cells, nonnucleated, 109
Regional blood flow, and cardiac output, 129–137
Regitine (Phentolamine), acute hypoxic pulmonary hypertension and, 158–161
Renal circulations, 131
Respiratory diseases, 195

Respiratory enzymes, oxygen supply and, 49
Retinal hemorrhage, high altitude (HARH), 196, 197, 213–214, 294–296, 299
Retinopathy, 215–216; *see also* Retinal hemorrhage, high altitude

S

Sexual gender
   pulmonary edema and, 220, 226
   tolerance of altitude and, 284–285
Skeletal muscle, blood flow in, 134–135
Skin blood flow, thermoregulation and, 135–137
Skin circulation, 131–132
Sleep hypoxemia, at altitude, 1–7
   acetazolamide and, 4–7
   medroxyprogesterone and, 7
   staging of sleep in, 4
   theophylline derivatives and, 7
   types of sleep and, 4
Snowblindness, 212, 213
Sodium, plasma, 184
Splanchnic circulations, 131
Standardization of documentation, altitude, 301–305
Starling equation, 256
Stroke, 299–300
Stroke volume, and lowland acclimatized highlanders, 24
Substrate supply, 36–37, 39
Sunlight
   and cataracts, 213
   UV rays in, 212
Sweating rate, 136–137
Systemic pressure, 13, 82

T

Temperature, ambient, *see* Thermal conditions
Terminology, standardization of, 301–305
Theophylline derivatives, sleep hypoxemia and, 7
Thermal conditions, 105–106 135–137
   altitude tolerance and, 284
   thermoregulation for, 135–137
   vasoconstriction and, 263–264
Thermoregulation, 135–137
Thromboembolisms, high altitude (HATE), 196

Tissue oxygen transport, 154
Tissue oxygenation, 58–59
Tortuositas vasorum, 213
Trekking, 187–188
   definition of, 193
   hematologic parameters of, 187–188
   long-time, 188
   short-time, 187

U

Ultraviolet rays, 212
Urinary electrolytes, 183–184
Urinary elimination, 17-ketosteroids, 24
Urine osmolality, 183
Urine output, 24-hour, 183, 185
Urine production, trekking and, 187–188

V

Vascular disease, chronic mountain sickness as, 271–276
Vasoconstriction, pulmonary
   of arterioles, hypoxic, 209
   in pulmonary edema, 255
   thermal environment and, 263–264
Vasopressin, 261–262
Ventilation
   acute mountain sickness susceptibility and, 28–32
   hyperoxia and, 22
   in lowland acclimatized highlanders, 22
   pulmonary edema and, 253
Viscosity, blood, *see* Blood viscosity
Vitreous humor hemorrhages, 214

W

Water intake, 291
Waterbinding, total body, 296
Work index, 278–279

Z

Zinc, plasma, 185